Can a cup of tea help you have insights concerning the vector operator curl? Do kangaroos make use of the square root of the imaginary number i, to avoid burning themselves when sitting down in the Australian desert? Will gazing out of a window at night remind you about boundary conditions for the solutions of partial differential equations? Is there a connection between matrix multiplication, and washing and drying your hands? Is it possible to estimate Fourier coefficients simply by looking at a sketch of the function?

These are some of the questions that are addressed in the second volume of Dr Lyons' popular mathematics text for undergraduates in the physical sciences. Volumes 1 and 2 together cover the basic mathematics needed for a course in physics or engineering. The approach taken is to teach mathematics through real examples, so that it is used as a tool for analysing physical systems. The aim of these books is to instil a real understanding and appreciation of the topics discussed.

Dr Lyons is a lecturer in the Department of Physics, University of Oxford, and Fellow and Tutor in physics at Jesus College, Oxford.

ALL YOU WANTED TO KNOW
ABOUT MATHEMATICS
BUT WERE AFRAID TO ASK

ALL YOU WANTED TO KNOW
ABOUT MATHEMATICS
BUT WERE AFRAID TO ASK

Mathematics for science students
Volume 2

LOUIS LYONS

Jesus College, Oxford

CAMBRIDGE
UNIVERSITY PRESS

510
L991
v. 2

PUBLISHED BY THE PRESS SYNDICATE OF THE UNIVERSITY OF CAMBRIDGE
The Pitt Building, Trumpington Street, Cambridge CB2 1RP

CAMBRIDGE UNIVERSITY PRESS
The Edinburgh Building, Cambridge CB2 2RU, United Kingdom
40 West 20th Street, New York, NY 10011-4211, USA
10 Stamford Road, Oakleigh, Melbourne 3166, Australia

First published 1998

Printed in the United Kingdom at the University Press, Cambridge

Typeset in 10pt Times [TAG]

A catalogue record for this book is available from the British Library

ISBN 0 521 43466 1 hardback
ISBN 0 521 43601 X paperback

Contents

Contents

Chapter titles for Volume 1

ויכתבם על שני לחות אבנים

And he wrote them upon two tablets of stone

Deuteronomy iv 13

Preface

The ideas that motivated me to write this text, and my philosophy about what it should contain and how it should be used, were expounded at length in the Preface to Volume 1. Here I would like only to reemphasise that, in order to derive optimum benefit from this book, it is essential for the reader to work through the problems at the end of each chapter. This is the only way of making sure that the material covered has been absorbed. The problems are relatively few in number, so that it should be reasonable to attempt them all.

I wish to express my thanks to the various people who have offered me advice concerning the topics covered in this volume. They include David Acheson, Ian Aitchison, David Andrews, Peter Clifford, Gideon Engler, Raymond Hide, Moshe Kugler, Elaine Lyons, John Roe, Lee Segal, Robert Thorne and Andrew Tolley, as well as the Jesus College first-year physics students of 1996 who read and commented on the text.

I am most grateful to Brenda Willoughby for her sterling work in typing this document, and to Irmgard Smith for producing the beautiful diagrams.

Oxford   Louis Lyons
1997

9

Integrals

Simple one-dimensional integrals were discussed in Appendix A4 of Volume 1 of this book. Here we consider other types of integrals, starting with line integrals in two- or three-dimensional space, and then continuing with integrals involving more than one variable.

9.1 Line integrals

9.1.1 Introduction

A line integral in two dimensions is of the form

$$\int_C f(x, y)\,dl \tag{9.1}$$

where $f(x, y)$ is the function to be integrated. C is the curve $C(x, y) = 0$ in the x–y plane along which the integration is performed; it also incorporates points A and B, which are the limits of the integration region (see fig. 9.1(a)). The parameter l labels the length along the curve; a sensible convention is that l starts at zero at the starting point A, and increases towards B. To understand the significance of the integral, we imagine the path to be divided into lots of infinitesimal elements each of length δl, we multiply each by the value of the function $f(x, y)$ at that particular element, and then add up all these contributions for the whole of the relevant part of the curve.

Thus if the path is a track on the side of a mountain, and $f(x, y)$ is the gradient dz/dl of the track, then the integral gives just the change in height between the start and end of the journey. Alternatively, with $f = 1$, the integral is simply the length of the journey, since in this case we merely add the lengths of all the infinitesimal portions of the path.

1

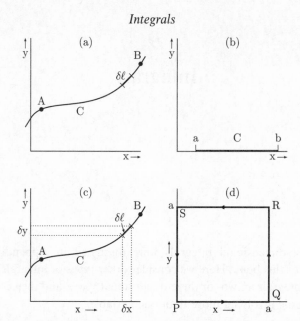

Fig. 9.1. Line integrals: (a) C is a path in the x–y plane. To evaluate the integral of $f(x, y)$ along C between points A and B, the path is divided into small segments of length δl. For each of these, we multiply δl by $f(x, y)$ evaluated at that position, and then add all these contributions; in the limit where the lengths of the segments become infinitesimal, this sum gives the integral. (b) The path C lies along the x axis, between the points with x coordinates a and b. The line integral now reduces to a standard one-dimensional integral. (c) A method of evaluating the line integral along the curve C is to replace it by one along the x axis. Then δl becomes $\delta x \sqrt{1 + (dy/dx)^2}$. Alternatively if dy/dx becomes infinite, the replacement of $\delta y \sqrt{1 + (dx/dy)^2}$ can be used. (d) The square path in the x–y plane along which the integrals of Table 9.1 are evaluated. The path is considered in an anticlockwise sense; this is the standard convention for a closed curve in two dimensions, since it corresponds to increasing polar angle θ (with respect to a point inside the curve).

Line integrals like (9.1) are not all that different from standard one-dimensional integrals. Thus in the special case where the path is along the x axis (see fig. 9.1(b)), the integral would reduce to $\int_a^b g(x)dx$, where $g(x) = f(x, 0)$; this is a standard one-dimensional integral.

Another aspect of this relationship emerges when we consider how to evaluate such an integral. Even when the path is a curve (rather than a line along one of the axes), we can reduce (9.1) to the standard case by some transformation of variables. For simple cases, the new choice is straightforward. Thus if the integral is around a complete circle of radius R, the variable would be chosen as the polar angle ϕ with respect to the centre of the circle; then dl is replaced by $Rd\phi$, and the range of ϕ is from zero to 2π. This approach works for most simply-specified curves.

The alternative is to write

$$\delta l = \left(\delta x^2 + \delta y^2\right)^{\frac{1}{2}} \approx \left[1 + \left(\frac{dy}{dx}\right)^2\right]^{\frac{1}{2}} \delta x \tag{9.2}$$

$$\approx \left[1 + \left(\frac{dx}{dy}\right)^2\right]^{\frac{1}{2}} \delta y \tag{9.2'}$$

Thus the integral along the path is replaced by one along either the x axis or the y axis (see fig. 9.1(c)). Then

$$\int f(x, y)dl \rightarrow \int f(x, y)\sqrt{1 + \left(\frac{dy}{dx}\right)^2} \, dx \tag{9.3}$$

It is important to remember that, although the integral is now in the variable x, the function $f(x, y)$ and the derivative dy/dx are to be evaluated at the relevant point in (x, y) space along the curve, and not along the x axis. However, despite first appearances, the integral (9.3) is not a function of y, since for the specified path, y is not an independent variable but is an explicit function of x.

Eqn (9.3) is likely to break down when dy/dx becomes infinite. In that case, eqn (9.2') can be used to transform the line integral to one in the variable y rather than x.

Of course there is no particular reason to confine the curve C to being in two-dimensional space. Thus it could, for example, be a circle or some fairly arbitrary wiggly non-planar curve in three-dimensions, i.e.

$$\int_C f(x, y, z)dl$$

where the curve C needs two equations to define it:

$$\left.\begin{array}{l} C_1(x, y, z) = 0 \\ \text{and} \quad C_2(x, y, z) = 0 \end{array}\right\}$$

9.1.2 *Extension to vectors*

The example of eqn (9.1) is not the only possible sort of line integral. Extensions arise when f or dl or both are vectors. Various possibilities are (see also fig. 9.2):

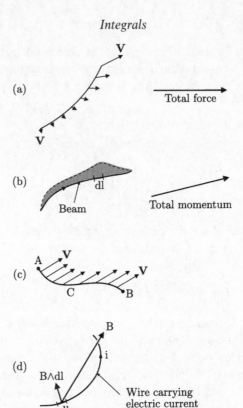

Fig. 9.2. Various line integrals: (a) $\int \mathbf{v}dl$: If $\mathbf{v}$ is the force per unit length of a wire, the integral gives the total force. (b) $\int f d\mathbf{l}$: Particles of mass m are travelling with speed s along a narrow beam. The shaded area represents not the transverse extent of the beam (which is taken to be negligible), but the number ρ of particles per unit length. The momentum of the particles in a short region is $s\rho m d\mathbf{l}$. With the path of integration chosen to coincide with the beam, and $f = s\rho m$, the integral is the total momentum of the beam. (c) $\int \mathbf{v} \cdot d\mathbf{l}$: $\mathbf{v}$ is the force pulling an object along a path C; it is shown as constant but need not be so. The integral is the work done by the force in moving the object from A to B. (d) $\int \mathbf{v} \wedge d\mathbf{l}$: If $\mathbf{v}$ is a magnetic field $\mathbf{B}$ and i is the constant current in a wire, the force on a small element $d\mathbf{l}$ is proportional to $i\mathbf{v} \wedge d\mathbf{l}$. Then the integral gives the total force on the wire.

(a) $\int \mathbf{v}dl$, with $\mathbf{v}$ being a vector which can depend on position, and dl a scalar as before. If $\mathbf{v}$ is the force per unit length of a wire, which lies along the curve C, then the total force on it is given by this integral, which like the force is a vector.

(b) $\int f(x, y)d\mathbf{l}$, where f is a scalar and $d\mathbf{l}$ a vector. The integral is a vector, and could represent, for example, the total momentum in a thin beam of particles, spread out along a curve. In that case $f(x, y) = s\rho m$, where s is the speed of the particles of mass m, and ρ gives the number per unit length of the beam.

(c) $\int \mathbf{v} \cdot d\mathbf{l}$ involves the dot product of two vectors and is hence a scalar. Here $\mathbf{v}$ could be a force $\mathbf{T}$ pulling an object along the path C. Then the integral is the total work done, a scalar.

(d) $\int \mathbf{v} \wedge d\mathbf{l}$ is a vector, because of the cross product. A wire carrying a constant current i in a magnetic field $\mathbf{v}$ experiences a force proportional to this integral. The cross product correctly allows for the fact that the vector force on each small element of wire is perpendicular to the directions of both the current and the magnetic field.

In all the above cases, we consider the integral as being equivalent to adding up lots of little contributions, one for each of the small elements of length δl for the curve C. This is performed whether the contributions are scalars or vectors. Naturally the integral has the same nature as the individual contributions.

Since we identify the distance element $d\mathbf{l}$ as being given simply by the difference in the radial vector $\mathbf{r}$ at the two ends of the infinitesimal element (i.e. $\mathbf{r} \equiv \mathbf{l}$), there is no ambiguity in the sign of $d\mathbf{l}$, and hence such integrals are absolutely defined. In contrast, as explained in Section 9.1.1, for integrals with a scalar dl, we adopt the convention of increasing l along the direction of integration.

To illustrate some of the different possibilities, we can consider the above examples for simple choices of $f(x, y)$ or $\mathbf{v}$, when the path of integration is the square of side a shown in fig. 9.1(d). The function f is taken as either unity or as the distance r from the origin, while $\mathbf{v}$ is chosen as either a constant vector $\mathbf{b}$ or as the position vector

$$\mathbf{r} = r\hat{\mathbf{r}}$$

where $\hat{\mathbf{r}}$ is a unit vector in the radial direction, i.e. $\mathbf{r}$ is simply a vector along r. In terms of Cartesian coordinates

$$\left. \begin{array}{r} r^2 = x^2 + y^2 \\ \text{and} \quad \mathbf{r} = x\mathbf{i} + y\mathbf{j} \end{array} \right\} \tag{9.4}$$

The various integrals are listed in Table 9.1.

Table 9.1. *Various line integrals.*

The curve along which the integrals are evaluated is the square shown in fig. 9.1(d). Because it is a closed curve, there is no need to specify where it starts and stops. The scalar variable of integration is the distance along the curve and is assumed to increase along the direction of integration; for the vector case, $\mathbf{l} \equiv \mathbf{r}$. In the integrals below, $\mathbf{b}$ is a constant vector, while $\mathbf{r}$ is the position vector and hence varies in length and direction.

$$\int dl = 4a$$
$$\int d\mathbf{l} = 0$$
$$\int \mathbf{b}dl = 4ab$$
$$\int \mathbf{b}.d\mathbf{l} = 0$$
$$\int \mathbf{b} \wedge d\mathbf{l} = 0$$
$$\int \mathbf{r}dl = 2a^2(\mathbf{i} + \mathbf{j})$$
$$\int rd\mathbf{l} = \frac{a^2}{2}(\beta + \sqrt{2} - 1)(\mathbf{i} - \mathbf{j}), \text{ where } \beta = \sinh^{-1}(1)$$
$$\int \mathbf{r} \cdot d\mathbf{l} = 0$$
$$\int \mathbf{r} \wedge d\mathbf{l} = 2a^2\mathbf{k}$$

9.1.3 Worked example

To illustrate how such integrals are performed, we demonstrate the evaluation of $\int \mathbf{r}dl$. The rest are left as an exercise for the reader.

Since the path consists of four distinct parts, we evaluate them separately. For the horizontal bit along the x axis (PQ of fig. 9.1(d))

$$\int (x\mathbf{i} + y\mathbf{j})dl = \int_0^a x\mathbf{i}dx$$
$$= \mathbf{i} \int_0^a xdx$$
$$= \mathbf{i}a^2/2 \qquad (9.5)$$

In the first equality, use has been made of the fact that $y = 0$ along the path of the integral. Since $\mathbf{i}$ is a constant vector it has been taken outside the integral in the second line, leaving a simple one-dimensional scalar integral to be performed.

For the vertical part starting at ($x = a$, $y = 0$),

$$\int (x\mathbf{i} + y\mathbf{j})dl = \int_0^a (a\mathbf{i} + y\mathbf{j})dy$$
$$= a^2\mathbf{i} + \frac{1}{2}a^2\mathbf{j} \qquad (9.6)$$

where the $x\mathbf{i}$ term is a constant along QR and is thus very easily integrated; and the integration of the $y\mathbf{j}dy$ term is performed analogously to that of $x\mathbf{i}dx$ in (9.5).

Now comes the RS section of the path. We need to be careful about signs here because x decreases, but l increases. Thus

$$\int_R^S \mathbf{v}dl = \int_a^0 \mathbf{v}(-dx)$$

$$= \int_0^a \mathbf{v}dx \qquad (9.7)$$

Alternatively we can imagine $\int_R^S \mathbf{v}dl$ as being replaced by the integral in the opposite direction, but with the l variable changed to increase from S to R; each of these modifications on its own would introduce a change in sign.

Hence

$$\int_R^S (x\mathbf{i} + y\mathbf{j})dl = \int_0^a (x\mathbf{i} + y\mathbf{j})dx$$

$$= \int_0^a (x\mathbf{i} + a\mathbf{j})dx$$

$$= \frac{1}{2}a^2\mathbf{i} + a^2\mathbf{j} \qquad (9.8)$$

where in the first line we have used eqn (9.7), and in the second the fact that $y = a$ along RS.

Finally, for SP

$$\int_S^P (x\mathbf{i} + y\mathbf{j})dl = \int_0^a y\mathbf{j}dy$$

$$= \frac{1}{2}a^2\mathbf{j} \qquad (9.9)$$

where again we have reversed the direction of the integration and remembered that $x = 0$ here.

We can check that the sign of the result (9.9) is correct. Our integrand $\mathbf{v} = \mathbf{r}$ is a vector which along SP always points in the positive y direction. As we are integrating along the direction of increasing l, we end up with an amount $\int \mathbf{r}dl$ which is also along the positive y axis. (If this is still not clear, imagine that $\mathbf{v}$ had been reversed so that it pointed along the direction of increasing l. Then the integral would also have been in this direction, i.e. along negative y. Our case is the reverse of this.)

What is at first sight disconcerting is that, had we integrated in the opposite direction, the result along PS would also have been in the positive y direction (and similarly all other contributions would have been unchanged). This seems to contradict the fact that the sign of an integral depends on the direction of integration. However, we have adopted the convention that, unless otherwise specified, our distance parameter increases along the direction of integration. Thus in the reversed direction, not only do the limits swap over but our infinitesimal element of path dl also changes sign. Thus with our convention about the choice of dl, the integral remains unchanged. Had we instead integrated along the direction PS, but with l increasing along SP, then the integral would indeed have changed sign.

The complete integral round the curve C is now given by adding the four contributions (9.5), (9.6), (9.8) and (9.9). Thus

$$\int_C \mathbf{r} dl = 2a^2(\mathbf{i} + \mathbf{j}) \qquad (9.10)$$

Comments on a couple of the other results in Table 9.1 are in order. Although it can be shown by direct integration that $\int \mathbf{r} \cdot d\mathbf{l} = 0$, this also follows (and rather more easily) from Stokes' Theorem (See Section 10.4.2). Thus

$$\int \mathbf{r} \cdot d\mathbf{l} = \int \operatorname{curl} \mathbf{r} \, da$$

where da is an element of area, and the region of integration for the right-hand side is the interior of the square of fig. 9.1(d). But curl $\mathbf{r} = 0$ (see the end of Section 10.2.3), so the integral is zero too. Using Stokes' Theorem thus shows that $\int \mathbf{r} \cdot d\mathbf{l} = 0$ for any closed curve, and not only for our square C.

It is perhaps a little surprising that $\int r d\mathbf{l}$ is in the direction $\mathbf{i} - \mathbf{j}$ rather than $\mathbf{i} + \mathbf{j}$, since the integrand and the region of integration appear to be symmetric in x and y. This is because the fact that the integration is performed anticlockwise around the square breaks the symmetry between x and y. (Thus in the first and the last steps, we travel along the x axis in the direction of increasing x, but along the y axis in the negative sense.)

9.1.4 Other line integrals

It is possible to have line integrals where the variable of integration is not the length along the curve (or an infinitesimal vector along it). Thus if we integrate around a circle centred on the origin, we could use the azimuthal angle θ in plane polar coordinates. Similarly a line integral along the x axis could have x^2 as the variable. In both cases, the variable is (very) simply related to the corresponding distance. In general the

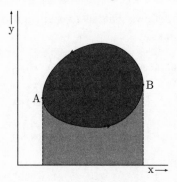

Fig. 9.3. The integral $\oint y\,dx$ around the closed curve can be thought of as consisting of two parts, the lower path from A to B, and the upper one back from B to A. The first part is equal to the lightly shaded area, while the second is equal to $-$(both shaded areas). The total integral is thus $-$(area enclosed by the curve). This result remains true even when part or all of the curve lies below the x axis, or indeed anywhere in the x–y plane.

infinitesimal dl or $d\mathbf{l}$ must be capable of being expressed in terms of where we are along the curve, in order for the integral to make sense. Then we can in principle convert the integral into one with the variable being the distance, although this may sometimes be algebraically difficult.

In these cases where a parameter like θ is defined along the curve, there is no sign ambiguity in its definition as we move along the curve, and hence we do not need to adopt a convention, as was required for the case where the variable of integration was the scalar length.

One specific integral with a simple interpretation is $\oint y\,dx$, where the circle on the integral sign indicates that the integration is to be performed around a closed curve, by convention in an anticlockwise direction. Fig. 9.3 shows such a curve, whose extremities in the x direction are A and B. Then $\int_A^B y\,dx$ is simply the area under the curve between A and B; if we consider the lower path from A to B, this corresponds to the lightly shaded area. Similarly the integral along the upper path from A to B would be the lightly shaded area plus the heavily shaded one. However, our path integral requires this second step to be performed in the opposite direction. Thus

$$\oint y\,dx = \left[\int_A^B y\,dx\right]_{\text{lower path}} + \left[\int_B^A y\,dx\right]_{\text{upper path}}$$
$$= \text{lightly shaded area} - \text{both shaded areas}$$
$$= - \text{ area enclosed by curve.}$$

A similar argument shows that $\oint x\,dy$ is equal to the enclosed area, but without the minus sign, i.e.

$$\text{area enclosed by curve} = \oint x\,dy$$

$$= -\oint y\,dx$$

$$= \frac{1}{2}\oint(x\,dy - y\,dx) \tag{9.11}$$

9.2 Multiple integrals

Now we have finished with line integrals. The rest of this chapter is devoted to multiple integrals.

9.2.1 What do they mean?

Let us assume that someone has performed a census of Britain, and has determined the population density over every rectangular area of 50×100 miles. These data could be presented as a list ρ_k, where k is simply a label denoting to which rectangle the density ρ_k refers. It is, however, more convenient to consider the density information to be arranged so that the values are specified as ρ_{ij}, where i and j give respectively the relevant row and column of the rectangles for which ρ_{ij} is the density (see fig. 9.4).

Now the number of people in the box labelled by i and j is simply $\rho_{ij}\delta x\delta y$, where δx and δy are the lengths of sides of the rectangles. Then the total population of Britain is obtained by adding up $\rho_{ij}\delta x\delta y$ for all the individual boxes.

We could imagine several different strategies for combining all these contributions. For example, we could first order them in magnitude, and then add them up starting with the largest and finishing with the smallest. Alternatively we could go through the ρ_{ij} starting in the middle of England, and then spiralling outwards. The following technique, however, is more appropriate for the consideration of double integrals, which will be discussed shortly.

We imagine performing the additions separately for each row one by one, and at the end of every row temporarily recording its sum. Thus we write

$$\sum_j \rho_{ij}\delta x\delta y = Y_i\delta y \tag{9.12}$$

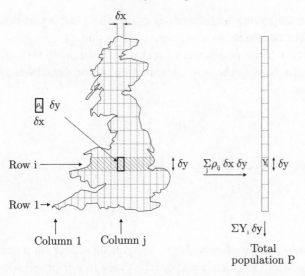

Fig. 9.4. Britain is divided by a rectangular web into boxes of area $\delta x \times \delta y$. The location of a box is determined by the labels i and j, which specify the relevant row and column respectively. The density of population in a given rectangle is ρ_{ij}, and the corresponding number of people is $\rho_{ij}\delta x \delta y$.

A procedure for determining the total population P is first to add the population in each row, and write the sub-totals in a new column; finally P is given by the sum of these sub-totals.

where the summation is over j at fixed i, i.e. it is a summation over the different boxes of one particular row.† Then $Y_i \delta y$ gives the number of people to be found in the ith row of rectangles on our map. (See fig. 9.4.) We have written this as $Y_i \delta y$, so that Y_i is the density of people per unit length of y. To find the total population P, we simply have to add up the numbers in each of these sub-totals, i.e.

$$P = \sum_i Y_i \delta y$$

$$= \sum_i \left(\sum_j \rho_{ij}\delta x \right) \delta y$$

$$= \sum_i \sum_j \rho_{ij}\delta x \delta y \tag{9.13}$$

† A slightly unfortunate feature of standard conventions becomes apparent at this point. We can refer to a box by its coordinates (x, y), or by labels (i, j) giving its row and column. Since i is a row label, it corresponds to the y coordinate, and j to x. The opposite way round for one of them would have been more convenient.

Instead of having a discrete set of numbers ρ_{ij}, some bright friend might have been able to write down a function $\rho(x, y)$ which describes the density in terms of distances x and y, east and north respectively from some origin. Now in the way a summation in one dimension approaches an integral:

$$\sum_i f_i \delta x \rightarrow \int f(x) dx$$

as the steps δx become progressively smaller, so here eqn (9.13) will tend to

$$P = \int \int \rho(x, y) dx dy \qquad (9.14)$$

This expression is written with two integral signs because it contains two quantities x and y which are allowed to vary in our problem.

Just as with definite integrals in one dimension, here too it is necessary to specify the region over which the integral is to be performed. In this case it is the whole of Britain. We shall return in Section 9.4 to a more explicit way of writing the integration limits in simple cases.

Another way of thinking about the relationship of the summation and the integral is that the integral (9.14) can be approximated numerically by the summation (9.13), where ρ_{ij} is now the value of the integrand $\rho(x, y)$ at the centre of the relevant box. Provided the function does not vary too rapidly and the boxes are reasonably small, the approximation should be a good one.†

If we want to perform the integral analytically for a specific function $\rho(x, y)$, we can use a procedure in complete analogy with that described for the summation in the paragraph above eqn (9.13). Thus we first integrate along a horizontal strip of infinitesimal width δy to obtain the population of this strip as

$$Y(y)\delta y = \left(\int \rho(x, y) dx \right) \delta y \qquad (9.15)$$

The population $Y(y)$ per unit length in y is of course a function of y. That is, it depends on which particular strip we are considering. Then the total population is obtained by integrating the density per unit y over

---

† This is not to suggest that the above is a particularly good recipe for the numerical evaluation of an integral, but it best illustrates the connection between the summation and the integral, and hence brings out the meaning of the double integral.

the relevant range in y. Hence

$$P = \int Y(y)dy$$

$$= \int \left(\int \rho(x,y)dx \right) dy$$

$$= \int \int \rho(x,y)dxdy \qquad (9.16)$$

Thus the double integral can be evaluated by simply performing two single integrals. However, care is needed in specifying the limits of integration (see Section 9.4).

As an example, we can consider the integral (9.16), with the population ρ being constant, but the integral being evaluated for the region as shown in fig. 9.6(a). Thus the integral is as given later in eqn (9.27), with $f(x,y)$ being the constant ρ. Then

$$\int_0^2 \left(\int_0^1 \rho dx \right) dy = \int_0^2 \rho dy = 2\rho$$

This gives the total population of the rectangle as its area (i.e. 2) times the population density ρ.

In both the summation (9.13) and in the integral (9.16), the described procedure involved adding or integrating first in a horizontal direction, and then combining the results for different y values. It is, of course, equally valid to perform the first summation or integration vertically, and then combine the results for different xs. Thus for the summation, we could determine the population of a column of rectangles as

$$X_j \delta x = \sum_i \rho_{ij} \delta y \delta x \qquad (9.17)$$

and then

$$P = \sum_j X_j \delta x$$

$$= \sum_j \left(\sum_i \rho_{ij} \delta y \right) \delta x$$

$$= \sum_i \sum_j \rho_{ij} \delta x \delta y \qquad (9.18)$$

where the order of the summations is unimportant. Similarly, for the continuous function $\rho(x, y)$, we define the horizontal density $X(x)$ by

$$X(x) = \int \rho(x, y)dy$$

and the total population

$$P = \int X(x)dx$$

$$= \int \left(\int \rho(x, y)dy \right) dx$$

$$= \int \int \rho(x, y)dydx \qquad (9.19)$$

The final result is, of course, the same as before, but the intermediate steps are different.

Once again, we will find we have to be very careful about the integration limits when we change the order of the separate one-dimensional integrals.

In the next section, some examples of multiple integrals will be given where the number of dimensions varies from two to six. Section 9.4 deals with the limits of integration, which are much more interesting than in the corresponding one-dimensional situation. Since the evaluation of multiple integrals is often facilitated by changing the variables, this topic is discussed in Section 9.5. Finally, worked examples of multiple integrals are provided in Section 9.6, in order to illustrate many of the points developed in this chapter.

9.3 Some more examples

9.3.1 From two to six dimensions

9.3.1.1 $\int \int f(x, y)dxdy$

In the previous section an example of this was discussed in which $f(x, y)$ was the population density and the integral was the population of Britain.

The simplest case is when $f(x, y)$ is unity. Then since $dxdy$ is a small element of area, $\int \int dxdy$ is just the total area of Britain (assuming we keep the same region of integration). This is an alternative to the earlier formula (9.11) for the area enclosed by a curve.

If instead $f(x, y)$ is simply the height of the ground at a particular x and y, the integral gives the total volume of Britain above sea level.

9.3.1.2 $\int \int \int g(x,y,z)dxdydz$

This is an integral over three dimensions. If the region of integration is that of the sun, and $g(x,y,z)$ is its density, the integral is its total mass. Alternatively, with $g(x,y,z)$ as its power output per unit volume, the integral is the total power generated from the whole sun.

Had we chosen $g(x,y,z)$ as unity, the triple integral would give just the volume of the sun.

9.3.1.3 $\int \int \int \int p(x,y,z,t)dxdydzdt$

We can imagine the variables x,y and z as being space coordinates and t as time, and again consider $p(x,y,z,t)$ as the power output of the sun. This varies with time over a scale of billions of years; in the past the sun initially warmed up enough for its nuclear fusion processes to take place, and then in the distant future it will have used up all its fuel and eventually cool down. Then as before

$$\int \int \int p(x,y,z,t)dxdydz$$

evaluated over the volume of the sun, gives its power output at a time t, and the complete four-dimensional integral is the total energy output of the sun over the whole time scale for which the integral is performed.

9.3.1.4 *Another four-dimensional integral*

Two friends are playing darts, and aiming for the bull's eye. If the first throws a lot of darts, the probability that they hit the board at a point with coordinates x_1 and y_1 in the plane of the dartboard is $p(x_1,y_1)dx_1dy_1$; if he is a good player, $p(x_1,y_1)$ will be centred on the bull's eye, and have a fairly peaked distribution. The second player produces a distribution $q(x_2,y_2)dx_2dy_2$, where the subscript 2 is used for the same coordinates to denote that they refer to the second player's darts. If he is not quite so expert, the maximum of q could be offset from the centre of the board, and wider than p especially in the vertical direction where he has not compensated for gravity so well.

Now the players each throw darts in pairs, one each. We may be interested in the average distance $\bar{d}$ between where they land. If the dart board is divided into finite size regions, $\bar{d}$ is given by the sum of every possible separation d of the darts, each multiplied by the probability $R(x_1,y_1,x_2,y_2)$ of that configuration, i.e.

$$\bar{d} = \sum \sum R(x_1,y_1,x_2,y_2)d \qquad (9.20)$$

where

$$d^2 = (x_1 - x_2)^2 + (y_1 - y_2)^2$$

The first and second summations are over every possible area of the board for the first and the second dart respectively. Assuming that the arrival points of the pair of darts from the two players are uncorrelated (and neglecting their possible improvements with time),

$$R(x_1, y_2, x_2, y_2) = p(x_1, y_1)\delta x_1 \delta y_1 q(x_2, y_2)\delta x_2 \delta y_2$$

where $\delta x \delta y$ is the area of the individual regions of the board.

As we go to the limit where the areas become infinitesimal, the summations of (9.20) become integrals, so

$$\bar{d} = \int \int \int \int \sqrt{(x_1 - x_2)^2 + (y_1 - y_2)^2} p(x_1, y_1)q(x_2, y_2)dx_1 dy_1 dx_2 dy_2$$

$$(9.21)$$

(The reason there are four integrals in (9.21), as opposed to two summations in (9.20) is that each of the latter represents a sum over areas, which as we have seen earlier is really equivalent to a double summation or integral over the two separate variables.)

Thus eqn (9.21) is a four-fold integral, where each of the variables is a spatial one. If we want to, we can think of each of the pair of locations for the two darts as being represented by a single point in four-dimensional space (with the axes labelled x_1, y_1, x_2 and y_2). Then eqn (9.21) involves an integral over this four-dimensional space.

It is, of course, a little difficult to picture a space with four mutually perpendicular axes, and it is recommended not to try to strain your imagination too much in this respect. It is best to think of it as a mathematical game.

Although we live in a world with three spatial dimensions, we often deal with mathematical problems involving only one or two variables. In a similar mathematical way, we can consider situations with more than three variables. Indeed the comparison of three-variable situations with the three spatial dimensions of the real world is not always relevant, since in many problems the variables are not spatial ones, but could involve, for example, time.

9.3.1.5 A six-dimensional integral

The potential energy V between two point electric charges e at a distance d apart is $e^2/4\pi\varepsilon_0 d$, where ε_0 is for our purposes simply a constant. In an atom, electrons are not localised at given positions, but are rather described by a quantum mechanical wave function which gives their probability distribution in space. We assume that these are $p(x_1, y_1, z_1)dx_1dy_1dz_1$ for the first electron and $q(x_2, y_2, z_2)dx_2dy_2dz_2$ for the second; and that these probabilities are independent† so that the probability of finding the electrons at (x_1, y_1, z_1) and (x_2, y_2, z_2) is $pq\,dx_1dy_1dz_1dx_2dy_2dz_2$. Then the mean potential energy $\bar{V}$, averaged over all possible positions of the two electrons, is

$$\bar{V} = \frac{e^2}{4\pi\varepsilon_0} \int \int \int \int \int \int \frac{p(x_1, y_1, z_1)q(x_2, y_2, z_2)dx_1dy_1dz_1dx_2dy_2dz_2}{\sqrt{(x_1 - x_2)^2 + (y_1 - y_2)^2 + (z_1 - z_2)^2}}$$
(9.22)

To perform the necessary averaging, we thus have to integrate over the six coordinates, three for each electron. We can think of this integral as being performed in the six-dimensional space which is needed to define the positions of the pair of electrons in the atom.

Another situation in which an expression of the form (9.22) would apply is for the gravitational attraction between two galaxies. Then p and q would represent the mass densities of the two galaxies; and $e^2/4\pi\varepsilon_0$ would be replaced by $-G$, the gravitational constant. $\bar{V}$ is now the gravitational potential energy of the two galaxies.

In an analogous manner, integrals can have any number of integral signs. Once we have become accustomed to more than one integration sign, triple integrals are no more of an intellectual challenge than double ones. And once you have grasped that quadruple integrals do not require us to believe that our world has four space dimensions, you will have no conceptual problem in accepting integrals with any large number of variables.

9.3.2 General surface integrals

Just as a line integral can be performed along a general curve in x, y, z space, so a double integral can be required over an arbitrary surface, rather than simply involving variables corresponding to two of the coordinate axes. For example, a house could have an interestingly shaped roof. If the mass of tiles per unit area of the roof is ρ (which can vary

---

† This is only approximately true, since an anticorrelation is produced by the electrons' repulsion. If we want to take this into account, we replace pq by a new function $f(x_1, y_1, z_1, x_2, y_2, z_2)$, which allows for this effect.

over the roof), the total mass M of tiles is given by†

$$M = \int \rho \, da \qquad (9.23)$$

where da is the area of an infinitesimally small region of the roof. This integral is to be performed over the whole area of the roof. We might have thought of writing instead $\int \rho \, dx dy$, where x and y are coordinates in the horizontal plane; this is incorrect, as it does not allow for the fact that the roof slopes (see next paragraph).

Corresponding to the ways discussed in Section 9.1.1 for attacking the analogous one-dimensional problem, such an integral can be performed either by finding a suitable pair of variables to describe the surface, or by converting it to an integral in x and y. The latter is achieved by writing

$$\int \rho \, da = \int \int \rho \, \frac{dx dy}{\cos \theta} \qquad (9.24)$$

where θ is the angle between the normal to the infinitesimal surface and the z axis (see fig. 9.5(a)). The normal to the surface $z = z(x, y)$ has direction cosine ratios‡ $(-\partial z/\partial x, -\partial z/\partial y, +1)$ (see the last paragraph of Section 7.5), and the value of $\cos \theta$ is obtained as the scalar product of unit vectors in this direction and in that of the z axis $(0, 0, 1)$. This gives

$$\cos \theta = 1 \left/ \sqrt{1 + \left(\frac{\partial z}{\partial x}\right)^2 + \left(\frac{\partial z}{\partial y}\right)^2} \right.$$

whence

$$\int \rho \, da = \int \int \rho \sqrt{1 + \left(\frac{\partial z}{\partial x}\right)^2 + \left(\frac{\partial z}{\partial y}\right)^2} \, dx dy \qquad (9.25)$$

For $\rho = 1$, this is a formula for the area of the surface $z = z(x, y)$. An alternative derivation of this formula is given in Problem 9.5(i). As always with multidimensional integrals, it is essential to be careful about the limits of integration (see Section 9.4).

Eqn (9.25) is equivalent to eqn (9.3) for the corresponding line integrals.

---

† By convention, the integral of (9.23) is written with just one integration sign, since the infinitesimal element is written as the single symbol da, even though the integration is a two-dimensional one over an area. Thus the identity (9.24) has a slightly curious look because, although the left- and right-hand sides involve equivalent integrals, one has one integration sign and the other has two.

‡ For example, if the surface were $z = x + 2y + 5$, the normal would be in the direction $(-1, -2, 1)$. As with all normals, there is an overall $\pm$ ambiguity in its direction (see also the next footnote). For such area problems as ours, it is usual to make the choice that results in the area coming out positive.

(a)

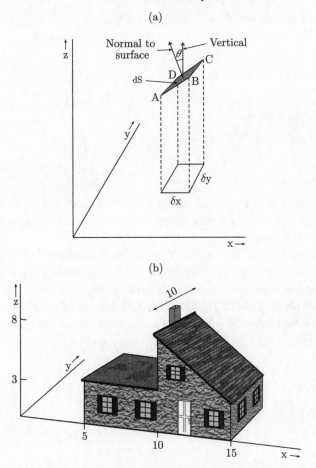

(b)

Fig. 9.5. Surface integration: (a) dS is an element of area in the surface $f(x, y, z) = 0$. Since $dxdy$ is the projection of dS in the x–y plane, $dxdy = dS \cos\theta$ (where θ is the angle between the z direction and the normal to the infinitesimal area). (b) A house, part of whose roof slopes. Dimensions are shown in metres.

Again we would require the derivatives in (9.25) to remain finite over the whole integration region.

As an example, we can calculate the area of the slanted part of the roof shown in fig. 9.5(b) by evaluating the integral on the right-hand side of (9.25), with $\rho = 1$. This part of the roof is defined by

$$z = 18 - x \text{ for } \begin{cases} 10 \le x \le 15 \\ \text{and} \quad 0 \le y \le 10 \end{cases}$$

(where all measurements are in metres). Then

$$\frac{\partial z}{\partial x} = -1$$

$$\text{and} \quad \frac{\partial z}{\partial y} = 0$$

so the integral becomes

$$\int_0^{10} \int_{10}^{15} \sqrt{1+1+0}\,dx\,dy = \sqrt{2} \int_0^{10} 5\,dy$$

$$= 50\sqrt{2} \text{ m}^2$$

As compared with the 50 m^2 of floor that it covers, the extra factor of $\sqrt{2}$ in the slanted roof's area arises because its normal makes an angle of 45° with the vertical.

An alternative for determining a surface area is to find a transformation to new variables, such that when these vary over suitable ranges the whole of the surface is mapped out. Thus for the surface of a sphere, these could be the polar and azimuthal angles. A small element of the surface is given by $r^2 \sin\theta\,d\theta\,d\phi$ (see fig. 9.9(c)), and so the complete surface is

$$\int_0^{2\pi} \int_0^{\pi} r^2 \sin\theta\,d\theta\,d\phi = r^2 \int_0^{2\pi} \left(\int_0^{\pi} \sin\theta\,d\theta \right) d\phi$$

$$= r^2 \int_0^{2\pi} 2\,d\phi$$

$$= 4\pi r^2 \qquad\qquad (9.26)$$

as expected.

9.3.3 *Vector integrals*

As with line integrals, the quantity to be integrated in multidimensional integrals can also be a vector. Thus for surface integrals, the infinitesimal element da can be considered as a vector in the direction of the normal to the surface.† Small volume elements dV are always taken as scalars.

Thus there are possibilites such as:

$\int da$: This is the total area of one side of a surface.
$\int d\mathbf{a}$: By adding the infinitesimal vector areas, the integral for a closed surface is zero.

---

† For closed surfaces, the usual convention is that this is taken as the direction of the outward-going normal. For open surfaces, there is an ambiguity as to which of the two possible antiparallel directions is chosen.

∫ *ρda*: If *ρ* is the amount of paint per unit area on a building, the integral
    is the total amount of paint required.

∫ **F***da*: This could be the total force exerted on a wall by a heavy wind
    blowing against it (**F** being the force per unit area).

∫ **F** · *d***a**: If the region of integration is part of the surface of the earth,
    and **F** is the flux of meteorites, the integral is the number of
    meteorites falling on that part of the earth.†

∫ *dV*: This is just the volume of the whole region over which the inte-
    gration is performed.

∫ *ρdV*: When *ρ* is the density, this is the mass.

∫ **F***dV*: For **F** being the force per unit volume on the element *δV* (e.g.
    due to the gravitation attraction on the mass within *δV* caused
    by a nearby black hole), the integral gives the total force vector
    on the whole volume of integration.

∫ div**F***dV*: The divergence div is a vector operator, which converts a
    vector into a scalar (see Chapter 10). Thus the integral is a
    scalar. It is in fact equal to ∫ **F** · *d***a**, evaluated over the closed
    surface surrounding the volume for the integral of div**F**.

9.4 Limits of integration

9.4.1 Limits in two dimensions

In one-dimensional situations, the limits of integration are never a real
problem. For integrals over several variables, more care is necessary.

We will start with a very simple example. We want to evaluate the integral of the function $f(x, y)$ over the rectangular region of fig. 9.6(a). This
area is defined by x covering the range from 0 to 1, and correspondingly
y from 0 to 2; not only is this obvious, it also happens to be correct.
Thus we write the integral as

$$\int_0^2 \int_0^1 f(x, y)dxdy \qquad (9.27)$$

The integral signs and their corresponding dx and dy are to be thought
of rather as pairs of brackets, so that the right-hand integral sign is

---

† For a uniform flux, a constant vector **F** in this integral would appear to give a result of
   zero for the total number of meteorites hitting the earth, since **F** · ∫ *d***a** = 0. The fallacy
   is that, although **F** is correct for the side on which meteorites hit the earth, on the other
   side the use of the same **F** would imply that meteorites are leaving the surface. This is
   clearly incorrect; the value for the flux there is zero, and so the integral with constant **F**
   is not over the complete surface.

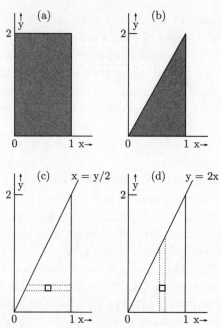

Fig. 9.6. Limits for the integration region: (a) A rectangular region, for which the limits of integration are written as in eqn (9.27). (b) A triangular region. Although the complete range of x is from 0 to 1, and for y from 0 to 2, the integral cannot be as given by eqn (9.27). (c) To perform the integration over the triangle, we could first combine infinitesimal areas at constant y. The relevant range of x is then from $y/2$ to 1, and so the double integral is as in (9.31). (d) Alternatively, areas at constant x could first be combined. The range of y for a given x is now from 0 to $2x$, and so (9.31') specifies the integral.

associated with the dx in (9.27). Initially it may even be worth writing some brackets explicitly, *viz*

$$\int_0^2 \left[\int_0^1 f(x, y) dx \right] dy \qquad (9.27')$$

This implies further that we are first integrating over x at constant y, and then integrating this result over the relevant range of y. It is worth noting that the *left-hand* integral sign in an expression like (9.27) refers to the integration to be performed *last*.

Although we have written these limits correctly without too much thought, we shall see immediately that perhaps we ought to have given them a little more attention.

Fig. 9.6(b) shows a different region of integration, which is triangular.

Again we want to integrate $f(x, y)$ over this area. Most people's first (but incorrect) reaction is that the region is defined by x going from 0 to 1, and y from 0 to 2. Thus it is tempting to write the integral as

$$\int_0^2 \int_0^1 f(x, y) dx dy$$

However, this must be wrong, in that it is exactly the same as (9.27) for the integral over the rectangular region of fig. 9.6(a). Since the limits are supposed to define the region of integration, we need to have something different here.

All we have to do is to think a little more carefully about what double integrals mean. Section 9.2.1 explained them by analogy with a summation process, in which it is necessary to decide on some specific procedure for ensuring that every little area within the region of integration is taken into account, but areas outside it are not. One of the strategies used there, and which can also be employed here, was first to combine all the infinitesimal areas at fixed y (i.e. integrate over x), and then to combine all the horizontal strips of different y values (i.e. integrate over y).

Thus if we start from a little area $\delta x \delta y$ somewhere in the middle of the region, we need to combine it with the other areas at fixed y (see fig. 9.6(c)). Now we realise the earlier mistake since, for a given (non-zero) y, x does *not* cover the range 0–1, but only from the diagonal line up to unity. Since the equation for this line is

$$y = 2x \tag{9.28}$$

the range of x is

$$y/2 \le x \le 1 \tag{9.29}$$

This defines the horizontal strip of fig. 9.6(c). Thus the inner part of the double integral should be

$$\int \left[\int_{y/2}^1 f(x, y) dx \right] dy \tag{9.30}$$

Once we have performed the integral over x to find the contribution from a given strip, we then need to combine all the strips at different y. Thus we have to integrate over the full range of y from 0 to 2. This means that the complete double integral is written as

$$\int_0^2 \int_{y/2}^1 f(x, y) dx dy \tag{9.31}$$

In the above procedure, we decided arbitrarily to integrate first over x and then over y. We could alternatively have decided to do the reverse; this involves first considering all the infinitesimal areas on a fixed vertical line at constant x (see fig. 9.6(d)), and then combining all the different vertical strips. Thus in the new first stage, we integrate over y which, from fig. 9.6(d) and from eqn (9.28), has a range

$$0 \le y \le 2x \qquad\qquad (9.32)$$

for a given x. The subsequent combination of the vertical strips is performed for values of x in the full range

$$0 \le x \le 1 \qquad\qquad (9.33)$$

Thus we can also write the double integral as

$$\int_0^1 \int_0^{2x} f(x,y)dydx \qquad\qquad (9.31')$$

where we have written the "dy" before the "dx" (and correspondingly the y limits on the right-hand integral sign) to show that the integral over y is to be performed first.

It is worth stressing that, despite their different appearances, eqns (9.31) and (9.31') both describe the integral of $f(x,y)$ over the same region of integration. The limits now differ from those of (9.27), which applies for the rectangular region of integration.

We now return to the earlier integral over the rectangular region of fig. 9.6(a). Rather than simply writing down the limits of integration as we did earlier, we really ought to have thought about the integration region in much the same way as we did for the triangular area of fig. 9.6(b). In fact it turns out that the rectangular region is such that the limits as written in eqn (9.27) are correct.

We could, of course, have decided to perform the integral for the rectangular region by first integrating over y (instead of x), in which case the integral would have been

$$\int_0^1 \int_0^2 f(x,y)dydx \qquad\qquad (9.27'')$$

In this case the limits on the integral signs are simply interchanged. This is usually not so (see eqns (9.31) and (9.31')). Here it is because each of the boundary lines of the integration region is such that one of the variables is constant along it.

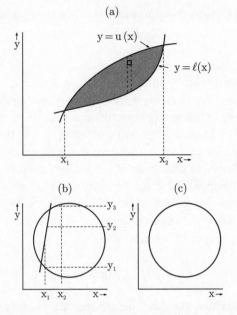

Fig. 9.7. Regions enclosed by curves: (a) The shaded region is bounded by the curve $y = l(x)$ below and $y = u(x)$ above. An integration over this region is written as in (9.35), when the integration is performed first over y. If instead x is integrated over first, then (9.35′) is appropriate. (b) The region is defined as that between the circumference of the circle and the line. For $x = x_1$, the y upper limit is on the line, but for x_2 it is on the circle. Similarly the x lower limit is on the circle for $y = y_1$ or y_3, but on the line for y_2. In cases where the functional form of a limit changes over the region of integration, care has to be taken to use the correct one. (c) The region is the interior of the circle, i.e. the same eqn (9.36) is used to define the upper and lower limits of either variable. In specifying these limits, it is necessary to select the correct solution in each case — see eqns (9.37) and (9.37′).

At this stage, the reader is strongly recommended to attempt Problem 9.3; the main point there is to insert the correct limits of integration.

9.4.2 Regions specified by curves

Fig. 9.7(a) shows a fairly arbitrary shaped region of integration. For a given x, the region is bounded by the curves

$$\left.\begin{array}{c} y = l(x) \\ \text{and } y = u(x) \end{array}\right\} \tag{9.34}$$

on the lower and upper sides respectively. The integral is then given by†

$$\int_{x_1}^{x_2} \int_{l(x)}^{u(x)} f(x, y) dy dx \qquad (9.35)$$

Even if it is fairly easy to integrate $f(x, y)$ with respect to y, if $u(x)$ and $l(x)$ are complicated functions, it can turn out that when they are used as the limits, the resulting function of x is difficult to integrate in the second stage. This feature of the integral being simple and the difficulties coming from the limits does not arise in single integrals.

For certain shapes of the region of integration, the problem can be simplified by an appropriate change of variables, for example from Cartesian to polar coordinates (see Section 9.5 for further details).

As usual we can rewrite the integral for the region of fig. 9.7(a), with the stages in the opposite order, as

$$\int_{y_1}^{y_2} \int_{u'(y)}^{l'(y)} f(x, y) dx dy \qquad (9.35')$$

where in this case

$$\left. \begin{array}{l} x = l'(y) \\ \text{and } x = u'(y) \end{array} \right\} \qquad (9.34')$$

are the inverses of eqns (9.34). Again this would need a slight modification in cases where the limit(s) of integration in x changed from one curve to the other, depending on the value of y (compare the previous footnote).

Sometimes the same equation can define the complete boundary of the integration region. This could, for example, be the circle

$$(x - a)^2 + (y - b)^2 = R^2 \qquad (9.36)$$

as shown in fig. 9.7(c). Then $l(x)$ and $u(x)$ of (9.34) are the same function, and we have to be sure that we use the appropriate solution of the equation in each place. Thus our integral is either

$$\int_{a-R}^{a+R} \int_{b-\sqrt{R^2-(x-a)^2}}^{b+\sqrt{R^2-(x-a)^2}} f(x, y) dy dx \qquad (9.37)$$

† This is not always quite the case. For some regions bounded by two curves (9.34), the limits on y could flip from one curve to the other, depending on the value of x (see, for example, fig. 9.7(b)). In that case, after the y integral has been performed, the limits would be inserted separately for the various x ranges; and these different results are taken into account when doing the x integral. This more or less corresponds to the way a one-dimensional integral is performed when the integrand is defined in different ways for different regions of the integration variables, e.g. $f(x) = x$ for $x \le 1$, and $f(x) = 1$ for $x > 1$ (see also Problem 9.8).

or

$$\int_{b-R}^{b+R} \int_{a-\sqrt{R^2-(y-b)^2}}^{a+\sqrt{R^2-(y-b)^2}} f(x,y)dxdy \qquad (9.37')$$

9.4.3 Limits for n-fold integrals

If we have an n-fold integral, which we perform first over x_1 and then over the other variables with x_n as the last, the limits of integration at any stage can be functions of all the variables that still remain to be integrated over, but not of the ones for which the integration has already been performed. Thus, for example, for a four-fold integral, we could have in general

$$I = \int_{l_4}^{u_4} \int_{l_3(x_4)}^{u_3(x_4)} \int_{l_2(x_3,x_4)}^{u_2(x_3,x_4)} \int_{l_1(x_2,x_3,x_4)}^{u_1(x_2,x_3,x_4)} f(x_1,x_2,x_3,x_4)dx_1 dx_2 dx_3 dx_4 \quad (9.38)$$

where, for example, l_2 and u_2 are functions giving the limits on the variable x_2 for the given values of x_3 and x_4, for which we still have to perform the integration. The last integral has limits l_4 and u_4 which do not depend on any of the other variables, and simply give the complete range of the variable x_4; these limits are thus the simplest ones.

As a specific example, consider the three-dimensional integral for the region bounded by the four planes

$$\left. \begin{array}{c} x = 0 \\ y = 0 \\ z = 0 \\ \text{and } x + y + z = 1 \end{array} \right\} \qquad (9.39)$$

This defines a pyramid with the three faces in the x–y, y–z and z–x planes being right-angled triangles with sides of length 1, 1 and $\sqrt{2}$; the fourth face is an equilateral triangle with sides of length $\sqrt{2}$. The integral over this region is given by

$$\int_0^1 \int_0^{1-z} \int_0^{1-y-z} f(x,y,z)dxdydz \qquad (9.40)$$

The first and last sets of limits are fairly obvious. The y upper limit corresponds to what the largest value of y is, for a given z, within the region, irrespective of x for which we have already integrated. It is readily seen that the limits in (9.40) satisfy the general requirements as displayed in (9.38). The calculation of the volume of this pyramid is one of the worked examples in Section 9.6.

Just as in one-dimensional integrals, some or all of the limits can be $\pm\infty$. For integrals to be meaningful over an infinite range, $f(x_1, x_2, ...)$ must approach zero suitably quickly for large values of the appropriate variables.

There is one case where multiple integral problems become particularly simple. This is when: (a) the function to be integrated factorises as the product of separate functions of one variable at a time; and (b) the limits of each integral are independent of each of the other variables. When both of these conditions are satisfied, the multiple integral becomes the product of a set of single integrals. Thus, if

$$\rho(x, y) = X(x)Y(y) \tag{9.41}$$

then

$$\int_a^b \int_c^d \rho(x, y)dxdy = \left[\int_c^d X(x)dx\right]\left[\int_a^b Y(y)dy\right] \tag{9.42}$$

For example, in the subject of kinetic theory, there are many multiple integrals involving such variables as a gas molecule's speed, azimuthal and polar angles of motion, and its distance from the last collision. In an elementary treatment of the subject, these integrals are assumed to satisfy the above conditions, with the result that they merely require the evaluation of a set of single-variable integrals. (See, for example, Problem 9.4.)

Finally we mention the inverse problem where we have been given all the limits in an n-dimensional integral, and we have to deduce what is the region of integration. This is achieved by finding the set of points in our n-dimensional space which are bounded by the specified limits, starting with the innermost integral and ending with the outer one. In all but the simplest cases, it may well be difficult to visualise the result.

9.5 Changing variables

9.5.1 *Why we need the Jacobian*

In Section 6.4 it has already been mentioned that if the variables of integration are changed from (x, y) to (u, v), then the integral transforms as

$$\int \int f(x, y)dxdy = \int \int f(u, v)|J|dudv \tag{6.68}$$

where the Jacobian J is given by

$$J = \frac{\partial(x,y)}{\partial(u,v)} = \begin{vmatrix} \dfrac{\partial x}{\partial u} & \dfrac{\partial x}{\partial v} \\ \dfrac{\partial y}{\partial u} & \dfrac{\partial y}{\partial v} \end{vmatrix} \qquad (6.67)$$

We now show why the general infinitesimal area $\delta x \delta y$ is to be replaced by $J \delta u \delta v$.

The proof we present is for two-dimensional transformations, simply because it is easier to draw the relevant diagrams. Formulae corresponding to eqns (6.67) and (6.68) carry over into larger numbers of dimensions (see Problem 9.5(ii) for the three-dimensional case).

Fig. 9.8(a) shows the region in the x–y plane that is bounded by the lines $u(x,y) = u_1$, and $u(x,y) = u_1 + \delta u$ on the one hand, and $v(x,y) = v_1$ and $v(x,y) = v_1 + \delta v$ on the other. It is the area of this small region that we need to calculate. The two lines at constant u are supposed to be very close together. To a good approximation they will then be parallel, and for short enough segments, they will be nearly straight; and similarly for the pair of constant v lines. In the figure, they are drawn at a reasonable separation for clarity.

Since the shaded region is approximately a parallelogram, we need the coordinates of the corners A, B and C in the x–y plane in order to calculate its area. In terms of u and v, they are:

$$\left. \begin{array}{l} \text{Point } A: \ (u_1, v_1) \\ \text{Point } B: \ (u_1, v_1 + \delta v) \\ \text{Point } C: \ (u_1 + \delta u, v_1) \end{array} \right\} \qquad (9.43)$$

The x and y coordinates of B and C relative to A are obtained via the two-dimensional Taylor series. Thus

$$\left. \begin{array}{l} \text{Point } A: \ (x_1, y_1) \\ \text{Point } B: \ \left(x_1 + \dfrac{\partial x}{\partial v}\delta v, \ y_1 + \dfrac{\partial y}{\partial v}\delta v \right) \\ \text{Point } C: \ \left(x_1 + \dfrac{\partial x}{\partial u}\delta u, \ y_1 + \dfrac{\partial y}{\partial u}\delta u \right) \end{array} \right\} \qquad (9.44)$$

The extra terms in B (as compared with A) contain only δv but not δu, since B has the same u coordinate as A (see eqn (9.43)).

Now that we have the coordinates of these points, the area of the shaded parallelogram can be found. This can be achieved by drawing a rectangle, with sides parallel to the x and y axes, around the shaded

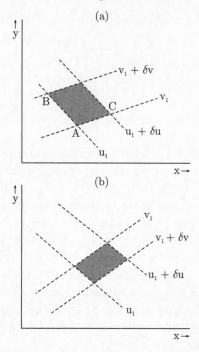

Fig. 9.8. (a) Demonstration of the need for the Jacobian J in transforming variables in a double integral. The shaded parallelogram is bounded by the lines $u = u_1$ and $u = u_1 + \delta u$, and $v = v_1$ and $v = v_1 + \delta v$. Its area is $J\delta u \delta v$, where $J = \partial(x, y)/\partial(u, v)$. (b) The area is now $-J\delta u \delta v$. If the area is deformed so that the lines of constant u and of constant v are parallel to those of constant x and y respectively, increasing u can be made to coincide with increasing x, but increasing v then corresponds to decreasing y.

area, and then subtracting triangular areas from that of the rectangle. Alternatively, the area can be obtained as the modulus of the vector product of the two vectors **AC** and **AB** (see Section 3.3.6.2). In an obvious notation, either approach gives the answer for the area S as

$$S = (y_B - y_A)(x_C - x_A) + (y_C - y_A)(x_A - x_B) \qquad (9.45)$$

Using (9.44) for the coordinates, we then obtain

$$
\begin{aligned}
S &= \left(\frac{\partial y}{\partial v}\delta v\right)\left(\frac{\partial x}{\partial u}\delta u\right) + \left(\frac{\partial y}{\partial u}\delta u\right)\left(-\frac{\partial x}{\partial v}\delta v\right) \\
&= \left(\frac{\partial x}{\partial u}\frac{\partial y}{\partial v} - \frac{\partial x}{\partial v}\frac{\partial y}{\partial u}\right)\delta u \delta v \\
&= J\delta u \delta v \qquad\qquad (9.46)
\end{aligned}
$$

where J is the Jacobian of the transformation, as defined by eqn (6.67) at the beginning of this section. This completes the proof that we need the factor J to convert the product $\delta u \delta v$ into the required area.

Thus J is the ratio between corresponding regions in the x–y and the u–v planes, defined by the four lines where u has two infinitesimally close values, and similarly for v. When we change variables in an integral, it is thus necessary to include this factor of J. Of course, we must also ensure that the limits of integration are changed so that the overall region in the u–v plane corresponds precisely to the original one in x–y (see examples in Section 9.6).

A simple example of this is provided by two-dimensional polar co-ordinates. An infinitesimal area enclosed between neighbouring lines at constant r and constant θ is given by $r dr d\theta$, and so this is what we need to replace $dxdy$ by when we change from rectangular to polar coordinates (see Sections 6.4 and 9.6).

Had we performed the inverse transformation, we would of necessity have obtained the result that

$$dudv \rightarrow |J'|dxdy \qquad (9.46')$$

where

$$J' = \frac{\partial(u,v)}{\partial(x,y)} \qquad (9.47)$$

It is a property of Jacobians that those of eqns (6.67) and (9.47) are related by

$$JJ' = 1 \qquad (6.80)$$

This is very satisfactory, as it ensures consistency of the relations (9.46) and (9.46'); if we perform a transformation from x–y to u–v and then back to x–y again, the overall factor we must multiply by is $JJ' = 1$.

9.5.2 Why we need the modulus sign

Eqn (6.68) at the beginning of this section actually has a modulus sign around J. This is because the determinant defining J in terms of the partial derivatives can turn out to be positive or negative, whereas we want the area of the shaded region of fig. 9.8(a) to be positive. If the contour lines of u and v are such that they look like fig. 9.8(b) rather than (a), the area is in fact given by *minus* the right-hand side of (9.45), and hence similarly for (9.46). The correct formula for the area in all cases is thus

$$S = |J| du dv \qquad (9.48)$$

The difference between figs 9.8(a) and (b) is as follows. We can imagine rotating and stretching the infinitesimal area of fig. 9.8(a) until the constant u lines are parallel to those of constant x, and the v lines to those of y. Then u increases in the same direction as x, and v increases with y. On the other hand, in (b) v decreases as y increases; we thus require a parity (or mirror) inversion before we can map one area onto the other. It is this that is the source of the negative sign in the area.

Because of the modulus sign in eqn (9.48), we have to be careful about integrating over regions where J changes sign; each sub-region in which J is of fixed sign should be integrated separately. Of course, taking the modulus of J does not imply that the integral itself will be positive.

Finally, in order to show that the sign of J is largely arbitrary, we can see that if we regard the transformation as being from x and y to v and u (rather than to u and v), the Jacobian would be

$$J_{vu} = \frac{\partial(x, y)}{\partial(v, u)} = \begin{vmatrix} \dfrac{\partial x}{\partial v} & \dfrac{\partial x}{\partial u} \\[2mm] \dfrac{\partial y}{\partial v} & \dfrac{\partial y}{\partial u} \end{vmatrix} \qquad (9.49)$$

which has just the opposite sign to our original Jacobian (6.67). This is not too surprising in terms of the argument about rotating the infinitesimal area. We now aim to make the constant v lines coincide with those of x. This results in the signs of the changes agreeing in fig. 9.8(b), but not in (a).

The conclusion is that we need to use eqn (9.48) for the area, in order to take care of the arbitrariness of the sign of the Jacobian.

9.5.3 *Some simple changes of variables*

In this section, we investigate some simple transformations and verify that the Jacobian is indeed necessary for giving the correct infinitesimal areas or volumes.

The Jacobian for the two-dimensional transformation from x, y to polar coordinates r, θ was given in Section 6.4 as

$$J = r \qquad (6.69)$$

That this is reasonable can be seen from fig. 9.9(a); the shaded area is $r \, dr \, d\theta$.

Extending this to three-dimensional cylindrical polars involves the relationships

$$\left. \begin{array}{l} x = r \cos \phi \\ y = r \sin \phi \\ \text{and } z = z \end{array} \right\} \qquad (A1.2)$$

of Appendix A1. Then the Jacobian is

$$\begin{vmatrix} \dfrac{\partial x}{\partial r} & \dfrac{\partial x}{\partial \phi} & \dfrac{\partial x}{\partial z} \\[2mm] \dfrac{\partial y}{\partial r} & \dfrac{\partial y}{\partial \phi} & \dfrac{\partial y}{\partial z} \\[2mm] \dfrac{\partial z}{\partial r} & \dfrac{\partial z}{\partial \phi} & \dfrac{\partial z}{\partial z} \end{vmatrix} = \begin{vmatrix} \cos \phi & -r \sin \phi & 0 \\ \sin \phi & r \cos \phi & 0 \\ 0 & 0 & 1 \end{vmatrix} = r \qquad (9.50)$$

Thus the volume enclosed by planes of constant r, ϕ and z is $r \, dr \, d\phi \, dz$. Fig. 9.9(b) shows that this is correct.

Finally we consider spherical polars, defined by

$$\left. \begin{array}{l} x = r \sin \theta \cos \phi \\ y = r \sin \theta \sin \phi \\ \text{and } z = r \cos \theta \end{array} \right\} \qquad (A1.3)$$

Now the Jacobian is

$$\begin{vmatrix} \dfrac{\partial x}{\partial r} & \dfrac{\partial x}{\partial \theta} & \dfrac{\partial x}{\partial \phi} \\[2mm] \dfrac{\partial y}{\partial r} & \dfrac{\partial y}{\partial \theta} & \dfrac{\partial y}{\partial \phi} \\[2mm] \dfrac{\partial z}{\partial r} & \dfrac{\partial z}{\partial \theta} & \dfrac{\partial z}{\partial \phi} \end{vmatrix} = \begin{vmatrix} \sin \theta \cos \phi & r \cos \theta \cos \phi & -r \sin \theta \sin \phi \\ \sin \theta \sin \phi & r \cos \theta \sin \phi & r \sin \theta \cos \phi \\ \cos \theta & -r \sin \theta & 0 \end{vmatrix} \qquad (9.51)$$

After expansion and simplification, this yields $r^2 \sin \theta$, implying that the relevant volume is $r^2 \sin \theta \, dr \, d\theta \, d\phi$. The sides of the cuboid in fig. 9.9(c)†

---

† Care is required if you have to reproduce this diagram (e.g. in an examination). Practice is recommended.

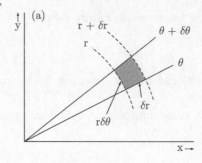

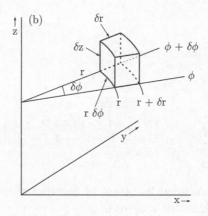

Fig. 9.9 (a) and (b) *For caption see facing page.*

are of lengths dr, $rd\theta$ and $r\sin\theta d\phi$. The volume thus again turns out to be correct.

9.6 Worked examples of multiple integrals

9.6.1 Area of circle

Here we evaluate the area enclosed by the circle

$$(x - a)^2 + (y - b)^2 = R^2$$

from

(i) $\frac{1}{2}\oint(xdy - ydx)$ (see eqn (9.11));
(ii) $\int\int dxdy$; and
(iii) transforming the double integral in (ii) to polar coordinates.

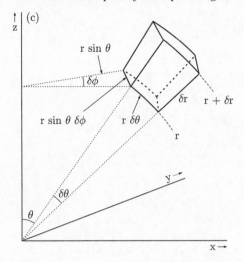

Fig. 9.9. Verifications of the Jacobian factor, for transformations of variables: (a) Two-dimensional polar coordinates. The shaded area lies between circles of constant r and lines of constant θ. (b) Three-dimensional cylindrical polars. The small volume is approximately cuboid; it is bounded by almost planar surfaces corresponding to each of the variables r, ϕ and z in turn being constant. The lengths of the edges of the cuboid are shown. (c) Three-dimensional spherical polars. The nearly planar surfaces are defined by r, θ and ϕ being constant. Again the lengths of the edges of the cuboid are shown. The area of the approximate rectangle at constant r is thus $r^2 \sin\theta\delta\theta\delta\phi$, and the volume is $r^2 \sin\theta\delta\theta\delta\phi\delta r$.

In (i), we must perform the line integral anticlockwise round the circle. This is most easily achieved by using the parametrisation

$$\left. \begin{array}{l} x = a + R\cos\theta \\ \text{and} \;\; y = b + R\sin\theta \end{array} \right\} \qquad (9.52)$$

with θ going from 0 to 2π. Then

$$\frac{1}{2} \oint (x\,dy - y\,dx)$$

$$= \frac{1}{2} \int_0^{2\pi} [(a + R\cos\theta)(R\cos\theta) - (b + R\sin\theta)(-R\sin\theta)]d\theta$$

$$= \frac{1}{2} \int_0^{2\pi} \left[R^2\cos^2\theta + R^2\sin^2\theta\right] d\theta$$

$$= \frac{1}{2}R^2 \cdot 2\pi \qquad (9.53)$$

(In the second line, the terms $aR\cos\theta$ and $bR\sin\theta$ in the integrand give

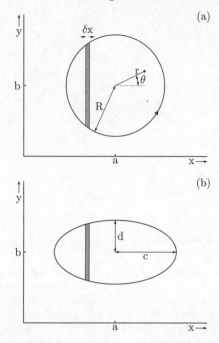

Fig. 9.10. (a) The circle whose area is required in the worked example of Section 9.6.1. In method (i), the line integral (9.11) is evaluated anticlockwise around the circle. For method (ii), the double integral is performed in x and y, while in (iii) the variables are the polar coordinates r and θ with respect to the centre of the circle. The shaded area is equal to δx multiplied by the height within the circle at that value of x. (b) The ellipse of Section 9.6.2. Its centre is at (a, b), and its axes have lengths $2c$ and $2d$.

zero contribution after integration over θ.) Thus this method successfully gives the area of the circle as πR^2.

In the form of the integral in (ii), it is essential to use the correct limits (compare eqn (9.37)). Then

$$\int_{a-R}^{a+R} \int_{b-\sqrt{R^2-(x-a)^2}}^{b+\sqrt{R^2-(x-a)^2}} dy\,dx = 2 \int_{a-R}^{a+R} \sqrt{R^2 - (x-a)^2}dx \qquad (9.54)$$

since the first integration over y is trivial. (Alternatively, we could have written the area as (9.54) directly, since $2\sqrt{R^2 - (x-a)^2}dx$ is the area within the circle of a strip of width δx at a position x — see fig. 9.10(a)).

Then changing the variable of integration to θ as defined by the first of eqns (9.52), we obtain

$$2 \int_{\pi}^{0} \sqrt{R^2 - R^2 \cos^2 \theta} (-R \sin \theta) d\theta = 2R^2 \int_{0}^{\pi} \sin^2 \theta d\theta$$
$$= \pi R^2 \qquad (9.55)$$

where, to obtain the last line, we have used the fact that the average value of $\sin^2 \theta$ over a range π in θ is $\frac{1}{2}$, and hence the integral is $\frac{1}{2} \cdot \pi$. Once again, the area of the circle comes out correctly.

If we change to polar coordinates† r and θ, we must incorporate the Jacobian of the transformation, and so the integral to be evaluated is

$$\int \int \frac{\partial(x, y)}{\partial(r, \theta)} dr d\theta$$

The Jacobian in this case is simply r (see Section 6.4), and so we require

$$\int_{0}^{2\pi} \int_{0}^{R} r dr d\theta = \frac{1}{2} R^2 \int_{0}^{2\pi} d\theta$$
$$= \pi R^2 \qquad (9.56)$$

The integrals we have to perform are simpler than in (ii) since here the limits are constants, and the double integral factorises into two separate single integrals. This illustrates the value of choosing the variables sensibly in multiple integral problems.

9.6.2 Area of ellipse

To find the area enclosed by the ellipse

$$\frac{(x - a)^2}{c^2} + \frac{(y - b)^2}{d^2} = 1 \qquad (9.57)$$

of fig. 9.10(b), we can proceed in analogy with the three methods used in Section 9.6.1 for the circle. Thus for the line integral of method (i), we use the parametrisation

$$\left. \begin{array}{c} x = a + c \cos \theta \\ \text{and} \quad y = b + d \sin \theta \end{array} \right\} \qquad (9.58)$$

Then the corresponding steps to those used to derive (9.53) here yield the area as πcd. This is very satisfactory since c and d are the semi-axes of the ellipse. If these are equal, the ellipse becomes a circle, and the area is πc^2 as expected.

† The polar coordinates are with respect to the centre of the circle (see fig. 9.10(a)), rather than the more usual choice of the origin.

For the integral (ii) over x and y, the limits for y (at fixed x) are $b \pm d\sqrt{1 - (x - a)^2/c^2}$, and for x are $a \pm c$. The y integral is readily performed, as before; for x, the substitution of (9.58) in terms of θ is again useful. Not surprisingly, the answer is πcd.

Rather than changing to polar coordinates directly, it is better first to use the transformation

$$\left.\begin{array}{c} \dfrac{x - a}{c} = x' \\[2mm] \dfrac{y - b}{d} = y' \end{array}\right\} \tag{9.59}$$

Then

$$\int\int dx dy = cd \int\int dx' dy'$$

Now going over to polar coordinates r and θ in (x', y') variables gives the area as

$$cd \int_0^{2\pi} \int_0^1 r\, dr\, d\theta = \pi cd$$

once again.

9.6.3 $\int_0^\infty e^{-x^2} dx$

The Gaussian or normal distribution is used extensively for the resolution or accuracy of experimental measurements: in many situations, it gives the probability of a particular measurement differing from its true value by x times the measurement accuracy. The distribution of speeds of gas molecules also involves this type of function.

The integration of e^{-x^2} over a finite range cannot be performed analytically, but when the limits are zero and infinity, a trick is available. Since x is a dummy variable (see Appendix A4.1), we can equally write

$$I = \int_0^\infty e^{-x^2} dx = \int_0^\infty e^{-y^2} dy \tag{9.60}$$

Then

$$I^2 = \int_0^\infty \int_0^\infty e^{-(x^2 + y^2)} dx\, dy$$

where the double integral extends over the whole of the first quadrant in the x–y plane. We now change to polar coordinates, and obtain

$$I^2 = \int_0^{\pi/2} \int_0^\infty e^{-r^2} r\, dr\, d\theta$$

What makes the integration possible is the extra factor of r in the integrand, coming from the Jacobian of the transformation.

Thus

$$I^2 = \int_0^{\pi/2} d\theta \int_1^0 -\frac{1}{2} d(e^{-r^2})$$

$$= \frac{\pi}{4}$$

so

$$I = \sqrt{\pi}/2 \tag{9.61}$$

Other useful extensions of this result are

$$\int_{-\infty}^{+\infty} e^{-x^2} dx = \sqrt{\pi} \tag{9.62}$$

$$\int_{-\infty}^{+\infty} e^{-\alpha x^2} dx = \sqrt{\pi/\alpha} \tag{9.63}$$

and

$$I_n = \int_{-\infty}^{+\infty} x^{2n} e^{-\alpha x^2} dx = \frac{2n-1}{2\alpha} I_{n-1} \tag{9.64}$$

9.6.4 Volume of pyramid

We now find the volume V of the pyramid, defined by the planes of eqns (9.39). This requires the evaluation of the integral (9.40), with f set equal to unity, i.e.

$$V = \int_0^1 \int_0^{1-z} \int_0^{1-y-z} dx dy dz \tag{9.40'}$$

The integration over x is trivial, giving $(1 - y - z)$. We thus now have to calculate

$$V = \int_0^1 \left[\int_0^{1-z} (1 - y - z) dy \right] dz$$

$$= \int_0^1 \left[y - \frac{y^2}{2} - yz \right]_0^{1-z} dz$$

$$= \int_0^1 \left(\frac{1}{2} - z + \frac{z^2}{2} \right) dz$$

$$= 1/6 \tag{9.65}$$

This agrees with the well-known formula for the volume of a pyramid being $\frac{1}{3}$ times the area of the base times the height. In our case the height in z is unity, while the base (a right-angled triangle with sides of $1, 1$ and $\sqrt{2}$) has area $\frac{1}{2}$.

The crucial step in finding the enclosed volume is to write down the limits of integration in equation (9.40′) correctly.

Further worked examples of multiple integrals (including vector cases) can be found in Sections 10.4.4.1 and 10.4.4.2.

Problems

9.1 A trajectory is defined in parametric form by

$$x = at$$
$$y = bt \cos t$$
$$z = bt \sin t$$

Find the distance along this curve, from the origin to the point $(2\pi a, 2\pi b, 0)$.

9.2 Verify that the entries in Table 9.1 are correct.

What would the entries in the table have been if the path of integration had been anticlockwise around the circle

$$x^2 + y^2 = R^2$$

rather than the square of fig. 9.1(d)?

9.3 Determine the area of the shaded region of fig. 9.6(b) by evaluating $\int \int dxdy$ and $\int \int r dr d\theta$, where the limits are suitably chosen for the required region. Evaluate each of the integrals twice, by using both of the possible choices of which variable to integrate over first.

9.4 According to kinetic theory, thermal conduction in a gas is due to the flow of energy caused by molecules moving from a hotter region to a cooler one transporting on average more energy than those travelling in the opposite direction. In an elementary treatment, it is possible to derive the contribution δQ to the heat flow per unit area per unit time coming from molecules which:

(i) have speed c (or more precisely, between c and $c + \delta c$);
(ii) are travelling with polar and azimuthal angles θ and ϕ with respect to an axis normal to the surface across which δQ is being determined; and
(iii) have travelled a distance r since their last collision.

When this contribution is integrated over all possible values of the relevant variables, we obtain

$$Q = \int_0^\infty \int_0^{2\pi} \int_0^{\pi} \int_0^{\infty} \left(E_0 + \frac{\partial E}{\partial y} r \cos \theta \right) \frac{e^{-r/\lambda}}{\lambda}$$
$$\times \frac{1}{4\pi} n p(c) \sin \theta \cos \theta \, dr d\theta d\phi dc$$

where n is the number of gas molecules per unit volume; $\partial E / \partial y$ is the gradient of the mean thermal energy of the molecules with distance y (and is caused by the temperature gradient); $p(c)\delta c$ is the probability of the speed being in the range c to $c + \delta c$; and λ is the mean free path of the molecules between collisions. Show that when the four integrations are performed

$$Q = \frac{1}{3} \frac{\partial E}{\partial y} n \bar{c} \lambda$$

where $\bar{c}$ is the mean speed of the molecules.

[By comparison with the definition of thermal conductivity

$$Q = \kappa \frac{\partial \theta}{\partial y}$$

this readily yields the gas's thermal conductivity coefficient κ in terms of its molecular properties as

$$\kappa = \frac{1}{3} \bar{c} \rho \lambda c_v$$

where ρ and c_v are respectively the density and the specific heat at constant volume for unit mass of the gas.]

9.5 (i) The points A, B, C and D at the corners of the shaded region δS of fig. 9.5(a) have the coordinates:

$$A = (x_0, y_0, z_0)$$
$$B = \left(x_0 + \delta x, y_0, z_0 + \frac{\partial z}{\partial x} \delta x \right)$$
$$C = \left(x_0 + \delta x, y_0 + \delta y, z_0 + \frac{\partial z}{\partial x} \delta x + \frac{\partial z}{\partial y} \delta y \right)$$
$$\text{and} \quad D = \left(x_0, y_0 + \delta y, z_0 + \frac{\partial z}{\partial y} \delta y \right)$$

For small $\delta x, \delta y$ and δz, the region approximates to a parallelogram. By taking a suitable vector product, show that its area is

$$\delta S = \sqrt{1 + \left(\frac{\partial z}{\partial x}\right)^2 + \left(\frac{\partial z}{\partial y}\right)^2}\, \delta x \delta y$$

(compare eqn (9.25)).

(ii) The variables u, v and w are defined in terms of the Cartesian coordinates x, y and z. The figure shows an approximate parallelepiped, whose edges each correspond to two of the three variables u, v and w being kept constant. The point A has $(x_0, y_0, z_0) = (u_0, v_0, w_0)$. The Cartesian coordinates of the other points are

$$B : \left(x_0 + \frac{\partial x}{\partial u}\delta u, y_0 + \frac{\partial y}{\partial u}\delta u, z_0 + \frac{\partial z}{\partial u}\delta u \right)$$

$$C : \left(x_0 + \frac{\partial x}{\partial v}\delta v, y_0 + \frac{\partial y}{\partial v}\delta v, z_0 + \frac{\partial z}{\partial v}\delta v \right)$$

$$\text{and} \quad D : \left(x_0 + \frac{\partial x}{\partial w}\delta w, y_0 + \frac{\partial y}{\partial w}\delta w, z_0 + \frac{\partial z}{\partial w}\delta w \right)$$

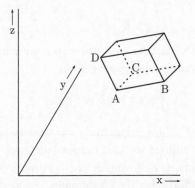

By considering a suitable triple scalar product, show that the volume of the parallelepiped is $J\delta u \delta v \delta w$, where

$$J = \begin{vmatrix} \dfrac{\partial x}{\partial u} & \dfrac{\partial y}{\partial u} & \dfrac{\partial z}{\partial u} \\[2mm] \dfrac{\partial x}{\partial v} & \dfrac{\partial y}{\partial v} & \dfrac{\partial z}{\partial v} \\[2mm] \dfrac{\partial x}{\partial w} & \dfrac{\partial y}{\partial w} & \dfrac{\partial z}{\partial w} \end{vmatrix}$$

and is the Jacobian of the transformation for x, y and z in terms of u, v and w.

9.6 The variables x and y are related to u and v by

$$x = u + v$$
$$y = uv$$

What region in the x–y plane corresponds to the interior of the triangle with vertices at $(0,0), (0,1)$ and $(1,1)$ in the u–v plane?

Verify that $\int \int dxdy$ for this region of the x–y plane agrees with $\int \int |J| dudv$ for the triangular region of the u–v plane, where

$$J = \frac{\partial(x,y)}{\partial(u,v)}$$

is the Jacobian of the transformation.

Repeat the above procedure for the triangle whose vertices have u–v coordinates $(0,0), (1,0)$ and $(1,1)$.

9.7 Two sets of variables are related by the equations

$$x = r \cosh \theta$$
$$y = r \sinh \theta$$

Evaluate independently the Jacobians

$$\frac{\partial(x,y)}{\partial(r,\theta)} \quad \text{and} \quad \frac{\partial(r,\theta)}{\partial(x,y)}$$

and verify that their product is 1.

Find the area with $y > 0$ and enclosed by the curves

$$y = 0, \ y = \frac{1}{3}x, \ x^2 - y^2 = a^2 \text{ and } x^2 - y^2 = 3a^2$$

9.8 Draw a diagram to show the region below the line

$$x + y = 7$$

and inside the circle

$$x^2 + y^2 = 25$$

Determine the area of this region by:
(i) integrating first over y, and then over x, being careful about the upper limit of integration in y; and
(ii) by using trigonometry to find the area excluded from the circle of radius 5.

9.9 The integral I is defined by

$$I = \int_0^1 \int_y^1 e^{(y/x)} dx dy$$

Draw a diagram showing the region of integration in the x–y plane.

By changing the order in which the integrations are performed, evaluate I.

9.10 Determine the value of

$$\int \int \left[x^3 \cos^2 y + x^2 \sin^3 y - 3(x - y) \tan x \tan y \, e^{-(x+y)} \right] dx dy$$

where the region of integration is the interior of the unit circle

$$x^2 + y^2 = 1$$

For other questions involving multiple integrals, see Problems 10.10, 10.12 and 10.13.

10

Vector operators

10.1 Introduction

In this chapter, we discuss three differential operators connected with vectors; they are called (and denoted by) grad, div and curl. The first two are short for gradient and divergence respectively. The curl is sometimes alternatively known as rot, short for rotation.

Since these operators all involve spatial derivatives, the interesting cases are when the things they operate on vary with position. Thus we need to deal with vector or with scalar fields. These require the relevant quantity to be defined at each point in the region of interest.

An example of a vector field, already discussed in Section 3.2, is the case of the electric field produced by a charge q situated at the origin (see eqn (3.12) and fig. 3.2). Here, the electric field is always directed radially, and hence varies in orientation. It also changes in magnitude, decreasing in strength at points further away from the charge.

An example of a scalar field is provided by the density of air in the earth's atmosphere or its pressure or temperature. All of these vary with position, in general decreasing with height; they also change with latitude and longitude (and with time too). However, density, pressure and temperature do not have a direction associated with them, and hence are scalars. In contrast, the wind velocity in the atmosphere is a vector, and hence is an example of a vector field.

Thus

$$\mathbf{v} = xy^2\mathbf{i} + x^2\mathbf{j} + 3e^x yz^2\mathbf{k} \qquad (10.1)$$

would be a vector field, while

$$\phi = y^2 + 6e^x yz \qquad (10.2)$$

is an example of a scalar field.

The definitions of grad, div and curl that we are about to introduce may at first sight seem a little arbitrary. However, it turns out that they are extremely useful in discussing physical phenomena, and provide a very neat and practical shorthand way of writing and manipulating many equations. This is perhaps most evident with the derivation of the wave equation for electromagnetic waves, as deduced from Maxwell's Equations. Maxwell himself achieved this without using this formalism, with the result that it required many pages of calculation. It needed a physicist of Maxwell's genius and persistence to see his way through the maze of algebra. By using vector operators, however, the derivation becomes almost trivial (see Section 10.5.3).

10.2 The vector operators

There are several different ways of defining the vector operators. Once one is adopted, other results then follow. However, an alternative could be chosen from one of the derived results, in which case the previous definition now becomes something to be derived. We will adopt a very physical way of introducing the operators.

10.2.1 Grad

The gradient operator acts on a scalar function ϕ and produces a vector field

$$\mathbf{v} = \mathrm{grad}\phi \qquad (10.3)$$

The simplest example is to consider that ϕ is a function of the two spatial coordinates x and y, describing the way the height of a hill depends on latitude and longitude. Then $\mathrm{grad}\phi$ is defined as being the vector which, at any point, is in the direction of the maximum gradient at that point, and whose magnitude gives the size of the gradient, i.e.

$$|\mathrm{grad}\phi| = \frac{\delta\phi}{\delta s} = \frac{\partial\phi}{\partial x}\cos\theta + \frac{\partial\phi}{\partial y}\sin\theta \qquad (10.4)$$

where δs is a short distance in the direction of the steepest gradient, which makes an angle θ with the x axis, and $\delta\phi = \phi_2 - \phi_1$ (see fig. 10.1).

Thus if you were standing on the side of this hill, as you looked around, each possible direction would be characterised by its own steepness, with some directions leading upwards and others downwards (assuming you were not at a maximum or a minimum). Then the vector $\mathrm{grad}\phi$ points in

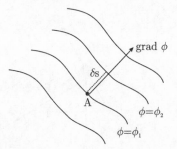

Fig. 10.1. The gradient vector: gradϕ points in the direction of the maximum upward gradient at the point A, and is perpendicular to the contour of constant ϕ passing through A; δs is the distance between nearby contours $\phi = \phi_1$ and $\phi = \phi_2$. The magnitude of gradϕ is the limit of $(\phi_2 - \phi_1)/\delta s$, as δs tends to zero.

the direction of the steepest climb, and is perpendicular to the direction in which the height stays constant. Its magnitude tells us the gradient in that direction. A ball placed on the hill will thus roll in the direction of $-$gradϕ, and its acceleration is proportional to $|$grad$\phi|$ (for small gradients). At a stationary point, gradϕ is zero, and its direction is undefined.

It is also possible to have gradients of three-dimensional functions. Thus if $\phi(x, y, z)$ represents the temperature in the atmosphere, gradϕ at a given point is a vector in the direction of the steepest increase in temperature at that point, and with a magnitude equal to its maximum temperature gradient there. This vector, defined at every point in the atmosphere, is a vector field. It determines the rate of heat flow at each position, and also its direction. In analogy with the two-dimensional case, gradϕ is perpendicular to the surfaces of constant temperature.

Returning to the case where ϕ represents the height of a hill, another useful way of thinking about ϕ is to regard it as proportional to the gravitational potential energy $mg\phi$ of a body on the hill, where g is the acceleration due to gravity and m is the mass of the body. Then $-mg$ gradϕ is a vector which for small slopes gives the net force acting on the body as a result of its being on the hillside.

This relationship generalises to other situations involving potentials. Thus, for example, if $\phi(x, y, z)$ is the electrostatic potential produced by some arrangement of electric charges and dipoles, the net force $\mathbf{F}$ on an electric charge q at (x_q, y_q, z_q) is $-q$gradϕ. This contrasts with the hillside situation of the previous paragraph, where the corresponding relationship was approximately true only for small gradients. That was

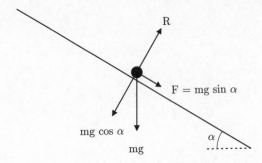

Fig. 10.2. The cross-section through a small portion of a hillside, such that the direction of greatest descent is in the plane of the page. Because only a small part of the hill is shown, the hillside is essentially straight, making a small angle α with the horizontal. There is a downward gravitational force mg on the body, and a reaction **R** due to the hillside. The component of mg normal to the hillside is balanced by **R**, while the component along the hillside provides the force **F** which causes the mass to move. The magnitude of **F** is $mg \sin \alpha$, which for small gradients is approximately equal to $mg \tan \alpha = mg|\mathrm{grad}\phi|$, where ϕ is the height h. Since **F** points down the hill while $\mathrm{grad}\phi$ is to the left, $\mathbf{F} \approx -mg\,\mathrm{grad}\phi$.

because on the hillside, the potential ϕ was a function of x and y only, but the motion was in all three dimensions, and of necessity not confined to an x–y plane only.

The vector function $\mathrm{grad}\phi$ can be written as

$$\mathrm{grad}\phi = \frac{\partial\phi}{\partial x}\mathbf{i} + \frac{\partial\phi}{\partial y}\mathbf{j} + \frac{\partial\phi}{\partial z}\mathbf{k} \tag{10.5}$$

That is, it is a vector whose x component is given by $\partial\phi/\partial x$, etc. In fact it is probably easiest to use this as the definition of $\mathrm{grad}\phi$, and to deduce the other properties from this. The reason we began the discussion in terms of gradients on hills or in temperature distributions was merely for ease of visualisation.

As a particular example, we can take the grad of the scalar function r^n, i.e. the contour lines are spheres centred on the origin. Then application of eqn (10.5) yields

$$\begin{aligned}
\mathrm{grad}(r^n) &= \frac{\partial}{\partial x}(r^n)\mathbf{i} + \frac{\partial}{\partial y}(r^n)\mathbf{j} + \frac{\partial}{\partial z}(r^n)\mathbf{k} \\
&= nr^{n-1}\frac{\partial r}{\partial x}\mathbf{i} + \cdots \\
&= nr^{n-2}x\mathbf{i} + \cdots \\
&= nr^{n-1}\hat{\mathbf{r}}
\end{aligned} \tag{10.6}$$

where as usual $\hat{\mathbf{r}}$ is a unit vector in the radial direction. As is required, $\text{grad}(r^n)$ is a vector. In this case it is in the radial direction, as it has to be perpendicular to the spherical contour surfaces (see Problem 10.1). Its magnitude (nr^{n-1}) looks very plausible. However, not all results of applying vector operators turn out to be so intuitively obvious.

The concept of $\text{grad}\phi$ is not tied to a particular coordinate system. Thus if we are using Cartesian coordinates, then $\text{grad}\phi$ is as defined in eqn (10.5). However, it is possible to use other systems; alternative expressions for $\text{grad}\phi$ and for the other vector operators are given later in Section 10.3.4.

For situations in which we are interested in the rate of change of ϕ in a direction not parallel to $\text{grad}\,\phi$, Taylor's series quickly gives the result that

$$\frac{d\phi}{dl} = \text{grad}\phi \cdot \hat{\mathbf{l}} \qquad (10.7)$$

where $\hat{\mathbf{l}}$ is a unit vector in the relevant direction, and l is the distance along this direction.

Another important property of $\text{grad}\phi$ is discussed in Section 10.5.1.

10.2.2 Div

In contrast to the grad operator, div acts on a vector function to produce a scalar result. We thus need to consider some vector field, such as $\mathbf{F}(x, y, z)$ giving the velocity flow vector at any point in a fluid in motion (e.g. water in a bath or air in the atmosphere). Then the divergence at any point P is defined as the limit (as the size of the region tends to zero) of the flux of $\mathbf{F}$ out of some small volume δV surrounding P, divided by δV. Thus

$$\text{div}\mathbf{F} = \lim_{\delta V \to 0} \left(\frac{1}{\delta V} \int \mathbf{F} \cdot d\mathbf{a} \right) \qquad (10.8)$$

Here the integration extends over the closed surface surrounding the small volume; $d\mathbf{a}$ is the area vector normal to a very small portion of this surface; and the dot product of the vector $\mathbf{F}$ with $d\mathbf{a}$ gives the contribution to the outward flux of the fluid.

Now liquids like water are almost incompressible, and so in most parts of a bath, as much water flows into any small region as flows out of it. The net flow across the closed surface is thus zero, and so

$$\text{div}\mathbf{F} = 0 \qquad (10.9)$$

The only exceptions to this are when the small region surrounds the plug hole, in which case there is a net inflow of water into the small region; or if it includes an inflow pipe, in which case there is net outflow from the region. (The water going down the plug hole or coming in from the inflow pipe is ignored.) In these cases, div**F** is negative and positive respectively (see Fig. 10.3(a)).

In a similar way, if we consider the electric field **E** produced by an arrangement of electric charges, the lines of force are such that $\int \mathbf{E} \cdot d\mathbf{a}$ is zero for integrals over closed surfaces in regions of space that do not include charges, because as many lines of force enter a small volume as leave it. Thus div**E** is zero, except for regions where the charges are located (see Fig. 10.3(b)). The relationship between div**E** and the charges is discussed further in Section 10.5.3.

We now return to the fluid flow situation, but instead consider air in the atmosphere. Air is compressible, and so more can flow into some region than flows out; this merely results in an accumulation of air in that region, and hence to a rise in the density there. Thus div**F** can be non-zero, and at any point is directly related to the rate of change of air density there (see also Section 10.5.2).

If an infinitesimal box is chosen to surround a point P, then from the way div**F** was introduced above, it is readily shown that

$$\mathrm{div}\mathbf{F} = \frac{\partial F_x}{\partial x} + \frac{\partial F_y}{\partial y} + \frac{\partial F_z}{\partial z} \qquad (10.10)$$

(See Problem 10.2(i).) Alternatively this can be used as the definition of div**F**, and the other properties deduced from it.

A simple example to illustrate the evaluation of the divergence is for the case $\mathbf{F} = \mathbf{r}$. Then, from the definition (10.10)

$$\mathrm{div}\mathbf{r} = \frac{\partial x}{\partial x} + \frac{\partial y}{\partial y} + \frac{\partial z}{\partial z} = 3 \qquad (10.11)$$

Similarly, for **v** and ϕ as defined in eqns (10.1) and (10.2), $\phi = \mathrm{div}\mathbf{v}$.

Expressions for the divergence in other coordinate systems are given in Table 10.4. Other properties are considered later in Section 10.4.1. Vectors with zero divergence are said to be "solenoidal".

10.2.3 Curl

The curl of a vector field **F** is another vector field. It is connected with the way the original vector field circulates around a given point, rather like the flow in a cup of tea that has been stirred with a spoon.

(a)

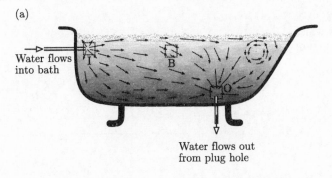

Water flows
into bath

Water flows out
from plug hole

(b)

(c)

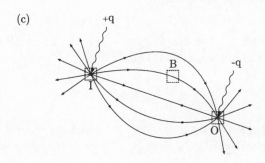

Fig. 10.3. (a) Water flowing around in a bath. Water enters from the pipe at the left, and drains out from the plug hole at the bottom. The arrows give the direction and the magnitude of the water flow at any point. At a general point B in the bath, the net ouflow of water from a small volume surrounding it is zero, because as much water flows in as flows out, since water is incompressible. Around the inflow and outflow pipes, however, there is a net flux (provided we ignore the flow in the pipes), and thus the divergence of $\mathbf{F}$ is positive at the inflow, and negative at the plug hole. (b) The regions around I, B and O shown seperately. (c) Electric lines of force, for two charges $+q$ and $-q$. The electric field $\mathbf{E}$ at any point is along the tangent to the line of force there, and is proportional to the density of lines of force in that region. The net flux of $\mathbf{E}$ out of a region is positive for the box I surrounding the positive charge, zero for B which encloses no charge, and is negative for O. Thus div$\mathbf{E}$ is non-zero only where there are charges.

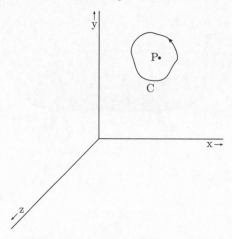

Fig. 10.4. To determine the z component of the curl of the vector field $\mathbf{F}$ at the point P, a small curve C enclosing P is drawn normal to the z axis, and the integral $\int \mathbf{F} \cdot d\mathbf{l}$ taken around this curve in an anticlockwise direction is evaluated. Then $(\text{curl}\mathbf{F})_z$ is the limit of the ratio of this integral to the area enclosed by C, as the curve becomes smaller and smaller.

We start by considering the line integral of $\mathbf{F}$ round a small planar path C (see fig. 10.4, and Section 9.1). Thus

$$s = \int_c \mathbf{F} \cdot d\mathbf{l} \tag{10.12}$$

The curl of $\mathbf{F}$ is defined to be such that its component normal to the plane of the path of integration is given by

$$(\text{curl}\mathbf{F})_\perp = \lim_{\delta a \to 0} (s/\delta a) \tag{10.13}$$

where δa, the area enclosed by the curve C, tends to zero. It is thus the limit of the ratio of the line integral to the enclosed area, as the curve enclosing the point of interest shrinks.

The reason why curl$\mathbf{F}$ is defined in this seemingly curious way will become clearer later. Basically it is because curl$\mathbf{F}$ as defined in this manner is very useful in a wide range of mathematical and physical applications (see Sections 10.4 and 10.5).

An example will help clarify how a curl is evaluated. We return to the case of the cup of tea, stirred so that the liquid is moving in a cylindrical manner (see fig. 10.5(a)). We assume that the velocity of each part of the liquid is in a tangential direction, and the speed is proportional to its distance from the central axis of the cup, i.e.

$$\mathbf{v} = \alpha\sqrt{x^2 + y^2}\hat{\tau} = \alpha r \hat{\tau} \tag{10.14}$$

where α is a constant, r is the radial distance in cylindrical polar coordinates, and the unit tangential vector is

$$\hat{\tau} = (-y\mathbf{i} + x\mathbf{j})/r \tag{10.15}$$

This means that the time taken for some small region of tea to complete one circuit of the cup is the same, independent of which small region of tea is considered.

To calculate the z component of curl**v** at the point P, we surround P with a closed path in an x–y plane, and which is chosen to be of such a shape as to simplify the evaluation of the line integral. In this case, we select a path made up of four separate parts, two at constant r ($r = R_1$ and $r = R_2$) and the others at constant ϕ ($= \phi_1$ and ϕ_2) — see fig. 10.5(b). Then along the straight sections at constant ϕ, the integrand vanishes because **v** is perpendicular to $d\mathbf{l}$ and hence $\mathbf{v} \cdot d\mathbf{l}$ is zero. For the part at $r = R_2$,

$$\int \mathbf{v} \cdot d\mathbf{l} = \int \alpha(R_2)(R_2 d\phi) = \alpha R_2^2 (\phi_2 - \phi_1)$$

and correspondingly at R_1,

$$\int \mathbf{v} \cdot d\mathbf{l} = -\alpha R_1^2 (\phi_2 - \phi_1)$$

where the minus sign arises because **v** and $d\mathbf{l}$ are antiparallel along this section. Thus for the whole curve

$$\int \mathbf{v} \cdot d\mathbf{l} = \alpha(R_2^2 - R_1^2)(\phi_2 - \phi_1) \tag{10.16}$$

To obtain (curl**v**)$_z$, we now have to divide by the area enclosed by the curve, which is

$$\delta a = \frac{1}{2}(R_2^2 - R_1^2)(\phi_2 - \phi_1) \tag{10.17}$$

and hence

$$(\text{curl}\mathbf{v})_z = \frac{\int \mathbf{v} \cdot d\mathbf{l}}{\delta a} = 2\alpha \tag{10.18}$$

In this case where the tangential speed depended linearly on r, it was not even necessary to go to the limit where δa vanishes. For other forms of the radial dependence of the tangential speed (e.g. $|\mathbf{v}| \propto r^n$, with $n \neq 1$),

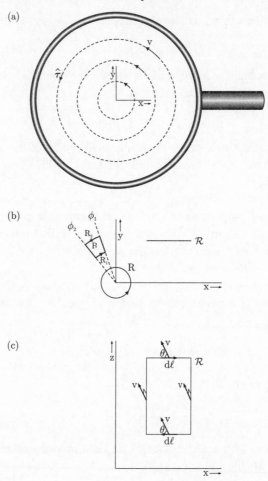

Fig. 10.5. Determining curl**v** for stirred tea in a cup: (a) Flow pattern of tea, as seen from above. The velocity **v** of the tea is independent of z (the height in the cup), and the speed depends only on its distance $\sqrt{x^2 + y^2}$ from the central axis. The velocity is in the tangential direction, defined by the unit vector $\hat{\tau}$ at any point. (b) The z component of curl**v** is calculated using a small path in an x–y plane. For calculating the curl at the origin, a circular path of radius R is used. For any other point P, the path is defined by the two neighbouring radii R_1 and $R_2 = R_1 + \delta R$, and by two radial lines at angles ϕ_1 and ϕ_2. (c) To determine the y component of curl**v**, we use a rectangle $\mathcal{R}$ in an x–z plane. (This is seen end-on in (b), and appears as just a horizontal line.) The velocity **v** is perpendicular to $d\mathbf{l}$ along the vertical parts of the path. For the two horizontal sections, **v** has the same value (since it is independent of z), but $d\mathbf{l}$ changes sign; hence these contributions cancel. The total integral is thus zero, and hence so is the y component of curl**v**.

it would be necessary to set $R_2 = R_1 + \delta R$, and then to find the limit as δR becomes vanishingly small — see Problem 10.4.

So far we have calculated only the z component of curl **v**. However, a similar analysis shows that the other components vanish. Thus for (curl **v**)$_y$, we choose a rectangular path in an x–z plane, with edges at two different constant z values, and two different constant x values (see fig. 10.5(c)). Then the integrals along the sides parallel to the z axis vanish, because **v** is perpendicular to d**l** there; while those along the sides parallel to the x axis (i.e. at the two different values of z) cancel, because **v** is independent of z, but the direction of the path of integration is opposite for the two sections.

Thus, in this example, curl**v** is a vector which has only a z component, i.e.

$$\text{curl} \mathbf{v} = 2\alpha \mathbf{k} \qquad (10.19)$$

That is, with the flow lines confined to the x–y plane and being independent of z, curl**v** is a vector in the z direction.

It is possible to extend the above analysis in two stages. The first is to replace the relationship (10.14) by the more general one

$$\mathbf{v} = \beta r^n \hat{\tau} \qquad (10.20)$$

where r is still the cylindrical polar coordinate $\sqrt{x^2 + y^2}$. Then a similar procedure gives the result

$$\text{curl} \mathbf{v} = \beta(n+1)r^{(n-1)}\mathbf{k} \qquad (10.21)$$

(see Problem 10.4). This is clearly consistent with the case $n = 1$ already discussed. Also the fact that applying a (vector) differential operator to a function whose dependence on r is r^n results in an answer with an $r^{(n-1)}$ form seems most reasonable. We might have guessed that the factor on the right-hand side of eqn (10.21) should have been n rather than $n + 1$. However, the vector differential operators are different from ordinary differentiation, and the $n + 1$ factor is indeed correct here. (We might also not have expected the result of curl **v** to have produced a vector in a different direction from **v**.)

A couple of warnings about eqn (10.21) are in order. First, the result for curl**v** is correct even for $n = 0$. In this case, **v** is constant in magnitude, but not in direction — it is always tangential — and so its derivative is not forced to vanish.

The second is that it is worth considering separately what happens at the origin. For this, the small circle of radius R (see fig. 10.5(b)) provides a suitable path for the integration. For most values of n, all is well, in that $(\text{curl}\mathbf{v})_{r=0}$ is as expected from the limit of curlv as r tends to zero. Thus it is zero when curlv varies as a positive power of r; and it diverges at the origin when curlv depends on a negative power of r. However, an exception occurs for the case $\mathbf{v} = r^{-1}\hat{\mathbf{t}}$. Here curlv is zero, except at the origin where it diverges (see Problem 10.4).

The other generalisation is to the situation in which instead of having $\mathbf{v}$ tangential and depending only on a power of r, as in eqn (10.21), we consider instead the curl of an arbitrary vector function $\mathbf{F}$ of x, y and z. Then again by considering a path C in a plane perpendicular to x, we can derive the result that

$$(\text{curl}\mathbf{F})_x = \frac{\partial F_z}{\partial y} - \frac{\partial F_y}{\partial z} \tag{10.22}$$

(see Problem 10.2). The other components of curlF are obtained from the above by cyclic permutation.

This result is not particularly easy to remember. It can, however, be rewritten in determinant form as

$$\text{curl}\mathbf{F} = \begin{vmatrix} \mathbf{i} & \mathbf{j} & \mathbf{k} \\ \dfrac{\partial}{\partial x} & \dfrac{\partial}{\partial y} & \dfrac{\partial}{\partial z} \\ F_x & F_y & F_z \end{vmatrix} \tag{10.22'}$$

An even more readily memorisable form is given in Section 10.3.

Functions for which curlF$= 0$ are said to be "irrotational". They have important properties that are discussed later (see Section 10.5.1). One simple example of this is curlr. This can be seen by considering integrals round suitably chosen contours, as was done earlier for the stirred tea in a cup. Alternatively eqn (10.22) can be used to derive

$$[\text{curl}\mathbf{r}]_x = \frac{\partial z}{\partial y} - \frac{\partial y}{\partial z}$$
$$= 0 - 0$$

and similarly for the y and z components. This result is not unreasonable, as the function $\mathbf{r}$ is a radial field which is spherically symmetric. In contrast, a non-zero curl requires some sort of circulation effect.

The curl of a constant vector $\mathbf{b}$ is also zero. This is because all the derivatives of $\mathbf{b}$ trivially vanish. Similarly divb is also zero.

Expressions for the curl in other coordinate systems are given in Table 10.4. Other properties are considered in Section 10.4.2

Table 10.1. *Scalar or vector nature of grad, div and curl.*

Nature of	$a = \text{grad}b$	$a = \text{div}\mathbf{b}$	$\mathbf{a} = \text{curl}\mathbf{b}$
b	scalar	vector	vector
a	vector	scalar	vector

10.2.4 Scalar or vector nature

It is absolutely essential to remember that grad, div and curl operate on fields which must have the correct scalar or vector nature; and they produce fields which have a specific scalar or vector type. These are summarised in Table 10.1.

While it is difficult to make a mistake in mixing scalars and vectors in an equation like $\mathbf{v} = 3$, the need for care is greater where vector operators are concerned. Thus the following, involving vectors $\mathbf{u}$, $\mathbf{v}$ and $\mathbf{w}$, and scalars s and t, are all forbidden:

$$\left.\begin{array}{r}
\mathbf{u} = \text{grad}v \\
\mathbf{u} = \text{div}\mathbf{v} \\
s = \text{div}(\mathbf{u} \cdot \mathbf{v}) \\
\text{div}\mathbf{u} + \text{curl}\mathbf{v} = \mathbf{w} \\
(\text{div}\mathbf{u}) \cdot (\text{grad}s) = 3 \\
\mathbf{u} = \text{grad}(\text{grad}s) \\
s = \text{div}(\text{div}\mathbf{v}) \\
\mathbf{u} = \text{curl}(\text{div}\mathbf{v}) \\
\mathbf{v} = \text{grad}(\text{curl}\mathbf{w})
\end{array}\right\} \quad \text{forbidden}$$

10.3 ∇, repeated operators and product functions

10.3.1 The ∇ operator

It is convenient to define the vector differential operator

$$\nabla = \mathbf{i}\frac{\partial}{\partial x} + \mathbf{j}\frac{\partial}{\partial y} + \mathbf{k}\frac{\partial}{\partial z} \tag{10.23}$$

It is called "del" and is useful in providing a neat notation for grad, div and curl. It also makes many of the equations relating the vector operators seem "obvious", by considering ∇ as more or less an ordinary vector (except that it implies differentiation of anything appearing immediately to its right). However, this is not so for all vector operator equations, and they should not be "derived" in this way. Rather the ∇ notation should be used as a mnemonic for most of these relations, and the exceptions should be remembered as such. Below we shall note explicitly those relations which do not follow directly from the mnemonic.

Eqns (10.5), (10.10) and (10.22′) can immediately be rewritten in terms of ∇ as

$$\text{grad}\phi = \nabla\phi \tag{10.24}$$

$$\text{div}\mathbf{v} = \nabla \cdot \mathbf{v} \tag{10.25}$$

and

$$\text{curl}\mathbf{v} = \nabla \wedge \mathbf{v} \tag{10.26}$$

Thus, for example,

$$[\text{curl}\mathbf{v}]_x = \left[\left(\mathbf{i}\frac{\partial}{\partial x} + \mathbf{j}\frac{\partial}{\partial y} + \mathbf{k}\frac{\partial}{\partial z} \right) \wedge (v_x\mathbf{i} + v_y\mathbf{j} + v_z\mathbf{k}) \right]_x$$

$$= \frac{\partial v_z}{\partial y} - \frac{\partial v_y}{\partial z}$$

in agreement with (10.22). Alternatively, eqn (10.26) appears to be a vector product, and so from eqn (3.22), it can be expressed as

$$\text{curl}\mathbf{v} = \begin{vmatrix} \mathbf{i} & \mathbf{j} & \mathbf{k} \\ \dfrac{\partial}{\partial x} & \dfrac{\partial}{\partial y} & \dfrac{\partial}{\partial z} \\ v_x & v_y & v_z \end{vmatrix}$$

which is identical to eqn (10.22′). The mnemonic works well in this case.

An example where it can be misleading is if, because curl**v** is written as ∇ ∧ **v**, we think that curl**v** must be perpendicular to **v**. However, this is readily seen not to be true. Thus even if, for some particular example, curl**v** is orthogonal to **v**, we can construct a new vector $\mathbf{v}' = \mathbf{v} + \mathbf{b}$, where **b** is a constant vector. This leaves curl**v**′ identically equal to curl**v**, but in general **v**′ is not in the same direction as **v**. Thus **v**′ and curl**v**′ need not be orthogonal.

More formally

$$\mathbf{v}' \cdot \text{curl}\mathbf{v}' = (\mathbf{v} + \mathbf{b}) \cdot [\text{curl}(\mathbf{v} + \mathbf{b})]$$

$$= (\mathbf{v} + \mathbf{b}) \cdot \text{curl}\mathbf{v}$$

$$= \mathbf{v} \cdot \text{curl}\mathbf{v} + \mathbf{b} \cdot \text{curl}\mathbf{v} \tag{10.27}$$

Thus, provided **b** has a non-vanishing component along curl**v**, $\mathbf{v}' \cdot \text{curl}\mathbf{v}'$ and **v** · curl**v** cannot both be zero.

10.3.2 Repeated operations

If we start out with a vector or a scalar field as appropriate, we can take the grad, div or curl of it, thus producing a new field. We can then take the grad, div or curl of this. Clearly there are nine combinations of these double operators. Of these, four are ruled out, as they involve an operation which is forbidden because of its vector or scalar character; they were listed in Section 10.2.4.

Of the remainder, two are identically zero:

$$\text{div}(\text{curl}\mathbf{v}) = \nabla \cdot (\nabla \wedge \mathbf{v}) = 0 \tag{10.28}$$

and

$$\text{curl}(\text{grad}\phi) = \nabla \wedge (\nabla\phi) = 0 \tag{10.29}$$

These are "obvious" from the ∇ mnemonic, but are derived by writing out the various expressions in components.

Two expressions that do not vanish are

$$\text{grad}(\text{div}\mathbf{v}) = \frac{\partial^2 v_x}{\partial x^2}\mathbf{i} + \frac{\partial^2 v_y}{\partial y^2}\mathbf{j} + \frac{\partial^2 v_z}{\partial z^2}\mathbf{k}$$
$$= \nabla(\nabla \cdot \mathbf{v}) \tag{10.30}$$

and

$$\text{div}(\text{grad}\phi) = \frac{\partial^2 \phi}{\partial x^2} + \frac{\partial^2 \phi}{\partial y^2} + \frac{\partial^2 \phi}{\partial z^2}$$
$$= \nabla \cdot (\nabla\phi)$$
$$= \nabla^2\phi \tag{10.31}$$

where

$$\nabla^2 = \frac{\partial^2}{\partial x^2} + \frac{\partial^2}{\partial y^2} + \frac{\partial^2}{\partial z^2} \tag{10.32}$$

In (10.30), $\nabla(\nabla\cdot\mathbf{v})$ is not written as $\nabla^2\mathbf{v}$ since, in analogy with the definition (10.32), that is reserved for

$$\nabla^2\mathbf{v} = (\nabla \cdot \nabla)\mathbf{v} = \frac{\partial^2\mathbf{v}}{\partial x^2} + \frac{\partial^2\mathbf{v}}{\partial y^2} + \frac{\partial^2\mathbf{v}}{\partial z^2} \tag{10.33}$$

which differs from (10.30). Again our mnemonic is useful here in that we should expect

$$\nabla(\nabla \cdot \mathbf{v}) \neq (\nabla \cdot \nabla)\mathbf{v}$$

since in general

$$\mathbf{a}(\mathbf{b} \cdot \mathbf{c}) \neq (\mathbf{a} \cdot \mathbf{b})\mathbf{c}$$

Table 10.2. *Double operators.*

List of the results of applying two vector operators in succession. The first operator is written to the right, and the second one on the left, e.g. the first entry on the bottom line is curl(gradϕ) = 0. A tenth combination is $\nabla^2 \mathbf{v}$ of eqn (10.33); it appears in the entry for curl(curl$\mathbf{v}$)

Second operator	First operator		
	grad ϕ	div$\mathbf{v}$	curl$\mathbf{v}$
grad	forbidden	$\nabla(\nabla \cdot \mathbf{v})$	forbidden
div	$\nabla^2 \phi$	forbidden	zero
curl	zero	forbidden	$\nabla(\nabla \cdot \mathbf{v}) - (\nabla \cdot \nabla)\mathbf{v}$

The final double operator is curl(curl$\mathbf{v}$). An identity relates this to (10.30) and (10.33), in that

$$\left.\begin{array}{r} \text{curl(curl}\mathbf{v}) = \text{grad(div}\mathbf{v}) - \nabla^2 \mathbf{v} \\ \text{or} \quad \nabla \wedge (\nabla \wedge \mathbf{v}) = \nabla(\nabla \cdot \mathbf{v}) - (\nabla \cdot \nabla)\mathbf{v} \end{array}\right\} \quad (10.34)$$

Again this is not unexpected from the mnemonic; again it must be proved by a more conventional method.

These double expressions are summarised in Table 10.2.

10.3.3 Vector operators for product functions

As well as applying the vector operators to a single vector or scalar field, they can instead operate on suitable products of them. These can be the product of two scalars, $\phi\psi$; of a scalar and a vector, $\phi\mathbf{v}$; a scalar product, $\mathbf{u} \cdot \mathbf{v}$; or a vector product $\mathbf{u} \wedge \mathbf{v}$. Given that it is obligatory to choose a field of the correct character for each vector operator, there are only six possibilities, *viz*: grad($\phi\psi$) or grad($\mathbf{u} \cdot \mathbf{v}$); div($\phi\mathbf{v}$) or div($\mathbf{u} \wedge \mathbf{v}$); and curl($\phi\mathbf{v}$) or curl($\mathbf{u} \wedge \mathbf{v}$).

Each of these can be expressed in terms of operations applied to single fields. These identities are presented in Table 10.3, both in grad, div and curl notation, and also in terms of ∇.

As usual, they are proved by writing them out fully in terms of components. Thus, as a simple example, the third relationship is verified by

$$\text{div}(\phi\mathbf{v}) = \frac{\partial}{\partial x}(\phi v_x) + \frac{\partial}{\partial y}(\phi v_y) + \frac{\partial}{\partial z}(\phi v_z)$$

Table 10.3. *Vector operators applied to product fields.*

(1)	$\begin{cases} \text{grad}(\phi\psi) = \phi\,\text{grad}\psi + \psi\,\text{grad}\phi \\ \nabla(\phi\psi) = \phi\nabla\psi + \psi\nabla\phi \end{cases}$
(2)	$\begin{cases} \text{grad}(\mathbf{u}\cdot\mathbf{v}) = (\mathbf{v}\cdot\nabla)\mathbf{u} + (\mathbf{u}\cdot\nabla)\mathbf{v} + \mathbf{v}\wedge\text{curl}\,\mathbf{u} + \mathbf{u}\wedge\text{curl}\,\mathbf{v} \\ \nabla(\mathbf{u}\cdot\mathbf{v}) = (\mathbf{v}\cdot\nabla)\mathbf{u} + (\mathbf{u}\cdot\nabla)\mathbf{v} + \mathbf{v}\wedge(\nabla\wedge\mathbf{u}) + \mathbf{u}\wedge(\nabla\wedge\mathbf{v}) \end{cases}$
(3)	$\begin{cases} \text{div}(\phi\mathbf{v}) = (\text{grad}\phi)\cdot\mathbf{v} + \phi\,\text{div}\,\mathbf{v} \\ \nabla\cdot(\phi\mathbf{v}) = (\nabla\phi)\cdot\mathbf{v} + \phi(\nabla\cdot\mathbf{v}) \end{cases}$
(4)	$\begin{cases} \text{div}(\mathbf{u}\wedge\mathbf{v}) = \mathbf{v}\cdot\text{curl}\,\mathbf{u} - \mathbf{u}\cdot\text{curl}\,\mathbf{v} \\ \nabla\cdot(\mathbf{u}\wedge\mathbf{v}) = \mathbf{v}\cdot(\nabla\wedge\mathbf{u}) - \mathbf{u}\cdot(\nabla\wedge\mathbf{v}) \end{cases}$
(5)	$\begin{cases} \text{curl}(\phi\mathbf{v}) = \phi\,\text{curl}\,\mathbf{v} + (\text{grad}\phi)\wedge\mathbf{v} \\ \nabla\wedge(\phi\mathbf{v}) = \phi(\nabla\wedge\mathbf{v}) + (\nabla\phi)\wedge\mathbf{v} \end{cases}$
(6)	$\begin{cases} \text{curl}(\mathbf{u}\wedge\mathbf{v}) = (\mathbf{v}\cdot\nabla)\mathbf{u} + (\text{div}\,\mathbf{v})\mathbf{u} - (\mathbf{u}\cdot\nabla)\mathbf{v} - (\text{div}\,\mathbf{u})\mathbf{v} \\ \nabla\wedge(\mathbf{u}\wedge\mathbf{v}) = (\mathbf{v}\cdot\nabla)\mathbf{u} + (\nabla\cdot\mathbf{v})\mathbf{u} - (\mathbf{u}\cdot\nabla)\mathbf{v} - (\nabla\cdot\mathbf{u})\mathbf{v} \end{cases}$

$$= \phi\frac{\partial v_x}{\partial x} + v_x\frac{\partial\phi}{\partial x} + \cdots$$
$$= \phi\,\text{div}\,\mathbf{v} + \mathbf{v}\cdot\text{grad}\phi \tag{10.35}$$

It is worth checking that, in each of the equalities of Table 10.3, the vector or scalar nature of each term is correct. Thus in the equation above, the divergence is applied to vector fields and the gradient to a scalar one; and the expression for div($\phi\mathbf{v}$) must have each of its terms as a scalar.

In applying the mnemonic involving ∇, it is important to remember that a differential operator acts on both factors of the product fields. Then the relationships (1) and (3) of Table 10.3 are derived without any difficulty. Items (4) and (5) need a little care to ensure that the vector product symbols appear in the right places. Thus in (5), when the ∇ operator is applied just to the scalar function ϕ, it must be as $\nabla\phi$ and then the vector product of this is taken with $\mathbf{v}$.

In the expression for div($\mathbf{u}\wedge\mathbf{v}$), the logic for the last term $\mathbf{u}\cdot\nabla\wedge\mathbf{v}$ appearing with a minus sign is that we have interchanged the $\mathbf{u}$ and ∇ in the original triple scalar product, and this causes a sign change. The first term has the ∇ applied only to $\mathbf{u}$. It cannot be $(\nabla\cdot\mathbf{u})\wedge\mathbf{v}$, which is forbidden; the alternative with the dot and cross in the triple scalar product interchanged is acceptable. Furthermore this then gives an answer which changes sign when $\mathbf{u}$ and $\mathbf{v}$ are interchanged, as is required for div($\mathbf{u}\wedge\mathbf{v}$).

Expression (6) involves four terms on the right-hand side. This is because a triple vector product can be written as the difference between two vectors (see eqn (3.34)); and the differential operator ∇ acting on the products converts each of these into two terms. Again the result is not unreasonable in terms of the ∇ mnemonic.

The remaining relationship is (2). There is no obvious way of "deriving" this using the ∇ mnemonic, which just does not seem to work in this case. Furthermore, it is not particularly obvious in advance that it should break down here. It is a good idea simply to remember that this is so.

In cases where the vector **u** is constant, the number of terms on the right-hand sides of expressions (2), (4) and (6) is halved in each case.

10.3.4 Vector operators in cylindrical and spherical polar coordinates

Because of the symmetries of a particular problem, it can be useful to rewrite vector operators in terms of coordinates other than the standard *xyz* ones. The results are listed in Table 10.4 for cylindrical and for spherical polar coordinates.

The derivation of these results is most readily achieved using tensor algebra methods, which are beyond the scope of this book. Alternative approaches are simpler in concept, but more tedious. Thus to obtain grad$s$ in spherical polars, we need to rewrite

$$\text{grad}s = \frac{\partial s}{\partial x}\mathbf{i} + \frac{\partial s}{\partial y}\mathbf{j} + \frac{\partial s}{\partial z}\mathbf{k} \tag{10.5}$$

with two types of substitutions. First we need to express $\partial s/\partial x$, etc. in terms of partial derivatives with respect to the new variables r, θ and ϕ, by

$$\frac{\partial s}{\partial x} = \frac{\partial s}{\partial r}\frac{\partial r}{\partial x} + \frac{\partial s}{\partial \theta}\frac{\partial \theta}{\partial x} + \frac{\partial s}{\partial \phi}\frac{\partial \phi}{\partial x} \tag{6.30}$$

(see Section 6.3.4). Then we must replace the Cartesian unit vectors **i**, **j** and **k** by those along the new orthogonal axes $\hat{\mathbf{r}}, \hat{\boldsymbol{\theta}}$ and $\hat{\boldsymbol{\phi}}$. We now digress to do this.

Digression: The new unit vectors

The new unit vectors $\hat{\mathbf{r}}$, $\hat{\boldsymbol{\theta}}$ and $\hat{\boldsymbol{\phi}}$ are such that, for small distances along each vector, the other two variables remain constant (to first order). A little thought reveals that $\hat{\mathbf{r}}$ can be written in terms of **i**, **j** and **k** as

$$\hat{\mathbf{r}} = (x\mathbf{i} + y\mathbf{j} + z\mathbf{k})/r \tag{10.36}$$

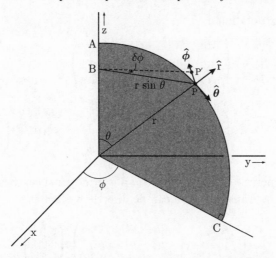

Fig. 10.6. The unit vectors $\hat{\mathbf{r}}, \hat{\boldsymbol{\theta}}$ and $\hat{\boldsymbol{\phi}}$, appropriate for spherical polar coordinates, for the point P. $\hat{\mathbf{r}}$ simply points radially away from the origin; along this direction, θ and ϕ are constant. $\hat{\boldsymbol{\theta}}$ is along the tangent to the circle APC of radius r, centred at the origin, and passing through P, and which has both r and ϕ constant. Finally $\hat{\boldsymbol{\phi}}$ is tangential to the circle centred at B (the foot of the perpendicular from P onto the polar z axis) and passing through P and P'; this circle is of radius $r \sin \theta$, and has constant r and θ.

where the factor $1/r$ is introduced to ensure that $\hat{\mathbf{r}}$ is a unit vector.

Now since $\hat{\boldsymbol{\phi}}$ is parallel to the x–y plane and is perpendicular to $\hat{\mathbf{r}}$ (see Fig. 10.6), it must be†

$$\hat{\boldsymbol{\phi}} = (-y\mathbf{i} + x\mathbf{j})/\sqrt{x^2 + y^2} \tag{10.37}$$

Finally

$$\hat{\boldsymbol{\theta}} = \hat{\boldsymbol{\phi}} \wedge \hat{\mathbf{r}}$$
$$= (xz\mathbf{i} + yz\mathbf{j} - (x^2 + y^2)\mathbf{k})/(r\sqrt{x^2 + y^2}) \tag{10.38}$$

We can rewrite the equations for the three new unit vectors, without any explicit reference to x, y and z, as

$$\left. \begin{array}{l} \hat{\mathbf{r}} = \sin\theta \cos\phi\,\mathbf{i} + \sin\theta \sin\phi\,\mathbf{j} + \cos\theta\,\mathbf{k} \\ \hat{\boldsymbol{\theta}} = \cos\theta \cos\phi\,\mathbf{i} + \cos\theta \sin\phi\,\mathbf{j} - \sin\theta\,\mathbf{k} \\ \hat{\boldsymbol{\phi}} = -\sin\phi\,\mathbf{i} + \cos\phi\,\mathbf{j} \end{array} \right\} \tag{10.39}$$

† $\hat{\boldsymbol{\phi}}$ is chosen in this direction (rather than the antiparallel one) so as to correspond to increasing ϕ.

or in matrix notation as

$$
\begin{pmatrix} \hat{\mathbf{r}} \\ \hat{\boldsymbol{\theta}} \\ \hat{\boldsymbol{\phi}} \end{pmatrix} = \begin{pmatrix} a_{11} & a_{12} & a_{13} \\ a_{21} & a_{22} & a_{23} \\ a_{31} & a_{32} & a_{33} \end{pmatrix} \begin{pmatrix} \mathbf{i} \\ \mathbf{j} \\ \mathbf{k} \end{pmatrix} \tag{10.40}
$$

$$
= \begin{pmatrix} \sin\theta\cos\phi & \sin\theta\sin\phi & \cos\theta \\ \cos\theta\cos\phi & \cos\theta\sin\phi & -\sin\theta \\ -\sin\phi & \cos\phi & 0 \end{pmatrix} \begin{pmatrix} \mathbf{i} \\ \mathbf{j} \\ \mathbf{k} \end{pmatrix}
$$

Now the matrix $\mathbf{A}$ (with elements a_{ij}) is orthogonal (see Section 15.9.4), so the inverse transformation can be directly written as

$$
\begin{pmatrix} \mathbf{i} \\ \mathbf{j} \\ \mathbf{k} \end{pmatrix} = \begin{pmatrix} a_{11} & a_{21} & a_{31} \\ a_{12} & a_{22} & a_{32} \\ a_{13} & a_{23} & a_{33} \end{pmatrix} \begin{pmatrix} \hat{\mathbf{r}} \\ \hat{\boldsymbol{\theta}} \\ \hat{\boldsymbol{\phi}} \end{pmatrix} \tag{10.41}
$$

Another property of the matrix $\mathbf{A}$ becomes apparent when we consider the partial derivatives like $\partial x/\partial r$ of the transformation equations relating x, y and z to r, θ and ϕ (see eqns (A1.3) of Appendix A). We find that the matrix $\mathbf{A}$ is given by

$$
\mathbf{A} = \begin{pmatrix} \dfrac{\partial x}{\partial r} & \dfrac{\partial y}{\partial r} & \dfrac{\partial z}{\partial r} \\[2ex] \dfrac{1}{r}\dfrac{\partial x}{\partial\theta} & \dfrac{1}{r}\dfrac{\partial y}{\partial\theta} & \dfrac{1}{r}\dfrac{\partial z}{\partial\theta} \\[2ex] \dfrac{1}{r\sin\theta}\dfrac{\partial x}{\partial\phi} & \dfrac{1}{r\sin\theta}\dfrac{\partial y}{\partial\phi} & \dfrac{1}{r\sin\theta}\dfrac{\partial z}{\partial\phi} \end{pmatrix} \tag{10.42}
$$

This is verified by explicitly evaluating the partial derivatives, and then comparing them with eqn (10.39′). However, this can be understood because $\mathbf{i}\partial x/\partial r + \mathbf{j}\partial y/\partial r + \mathbf{k}\partial z/\partial r$ is simply the change in the vector $\mathbf{r}$ when r increases but θ and ϕ are kept constant. This is just the unit vector $\hat{\mathbf{r}}$.

Similarly $(\partial x/\partial\theta, \partial y/\partial\theta, \partial z/\partial\theta)$ defines a direction with constant r and constant ϕ. Here we divide by r simply to convert $(\partial x/\partial\theta, \partial y/\partial\theta, \partial z/\partial\theta)$ to a unit vector. This factor of r, and the analogous ones of 1 and $r\sin\theta$ for the $\hat{\mathbf{r}}$ and the $\hat{\boldsymbol{\phi}}$ unit vectors, correspond to the fact that, for small changes in r, θ and ϕ, the distance δs is given by

$$
\delta s^2 = \delta r^2 + (r\delta\theta)^2 + (r\sin\theta\delta\phi)^2 \tag{10.43}
$$

(see fig. 9.9(c)).

Thus eqns (10.42) and (10.40) provide a method for expressing the new unit vectors in terms of the old ones, without detailed consideration of a diagram like fig. 10.6.

By symmetry, the inverse matrix

$$\mathbf{A}^{-1} = \tilde{\mathbf{A}} = \begin{pmatrix} \dfrac{\partial r}{\partial x} & r\dfrac{\partial \theta}{\partial x} & r\sin\theta\dfrac{\partial \phi}{\partial x} \\[2mm] \dfrac{\partial r}{\partial y} & r\dfrac{\partial \theta}{\partial y} & r\sin\theta\dfrac{\partial \phi}{\partial y} \\[2mm] \dfrac{\partial r}{\partial z} & r\dfrac{\partial \theta}{\partial z} & r\sin\theta\dfrac{\partial \phi}{\partial z} \end{pmatrix} \tag{10.44}$$

can be used to write $\mathbf{i}$, $\mathbf{j}$ and $\mathbf{k}$ in terms of $\hat{\mathbf{r}}, \hat{\boldsymbol{\theta}}$ and $\hat{\boldsymbol{\phi}}$. The elements of $\mathbf{A}^{-1}$ are such that, for example, $\partial r/\partial x$ is of course not equal to $(\partial x/\partial r)^{-1}$ (compare Section 6.4).

We now return to the derivation, from which we digressed at the beginning of Section 10.3.4. We rewrite eqn (10.5) using (6.30) and (10.42) as

$$\begin{aligned}
\text{grad}s = & \left(\frac{\partial s}{\partial r}\frac{\partial r}{\partial x} + \frac{\partial s}{\partial \theta}\frac{\partial \theta}{\partial x} + \frac{\partial s}{\partial \phi}\frac{\partial \phi}{\partial x} \right)(a_{11}\hat{\mathbf{r}} + a_{21}\hat{\boldsymbol{\theta}} + a_{31}\hat{\boldsymbol{\phi}}) \\
& + \left(\frac{\partial s}{\partial r}\frac{\partial r}{\partial y} + \frac{\partial s}{\partial \theta}\frac{\partial \theta}{\partial y} + \frac{\partial s}{\partial \phi}\frac{\partial \phi}{\partial y} \right)(a_{12}\hat{\mathbf{r}} + a_{22}\hat{\boldsymbol{\theta}} + a_{32}\hat{\boldsymbol{\phi}}) \\
& + \left(\frac{\partial s}{\partial r}\frac{\partial r}{\partial z} + \frac{\partial s}{\partial \theta}\frac{\partial \theta}{\partial z} + \frac{\partial s}{\partial \phi}\frac{\partial \phi}{\partial z} \right)(a_{13}\hat{\mathbf{r}} + a_{23}\hat{\boldsymbol{\theta}} + a_{33}\hat{\boldsymbol{\phi}}) \\
= & \frac{\partial s}{\partial r}\left[(a_{11}^2 + a_{12}^2 + a_{13}^2)\hat{\mathbf{r}} + (a_{11}a_{21} + a_{12}a_{22} + a_{13}a_{23})\hat{\boldsymbol{\theta}} \right. \\
& \left. + (a_{11}a_{31} + a_{12}a_{32} + a_{13}a_{33})\hat{\boldsymbol{\phi}} \right] \\
& + \frac{1}{r}\frac{\partial s}{\partial \theta}\left[(a_{21}a_{11} + a_{22}a_{12} + a_{23}a_{13})\hat{\mathbf{r}} + (a_{21}^2 + a_{22}^2 + a_{32}^2)\hat{\boldsymbol{\theta}} \right. \\
& \left. + (a_{21}a_{31} + a_{22}a_{32} + a_{23}a_{33})\hat{\boldsymbol{\phi}} \right] \\
& + \frac{1}{r\sin\theta}\frac{\partial s}{\partial \phi}\left[(a_{31}a_{11} + a_{32}a_{12} + a_{33}a_{13})\hat{\mathbf{r}} \right. \\
& + (a_{31}a_{21} + a_{32}a_{22} + a_{33}a_{23})\hat{\boldsymbol{\theta}} \\
& \left. + (a_{31}^2 + a_{32}^2 + a_{33}^2)\hat{\boldsymbol{\phi}} \right]
\end{aligned} \tag{10.45}$$

where (10.44) has been used in order to write

$$\left.\begin{array}{r} \dfrac{\partial r}{\partial x} = a_{11} \\[2mm] r\dfrac{\partial \theta}{\partial x} = a_{21} \\[2mm] r\sin\theta\dfrac{\partial \phi}{\partial x} = a_{31} \\[2mm] \text{etc.} \end{array}\right\}$$

Now the matrix $\mathbf{A}$ is orthogonal†, and so

$$\sum_i a_{ij}a_{ik} = \begin{cases} 0 \text{ if } j \neq k \\ 1 \text{ if } j = k \end{cases} \tag{10.46}$$

Thus, at long last,

$$\text{grad}s = \frac{\partial s}{\partial r}(\hat{\mathbf{r}} + 0 + 0) + \frac{1}{r}\frac{\partial s}{\partial \theta}(0 + \hat{\boldsymbol{\theta}} + 0) + \frac{1}{r\sin\theta}\frac{\partial s}{\partial \phi}(0 + 0 + \hat{\boldsymbol{\phi}}) \tag{10.47}$$

which agrees with the entry in Table 10.4.

In a similar way, we can convert $\text{div}\mathbf{v}$ from Cartesian coordinates to spherical polar ones. We start by writing

$$\text{div}\mathbf{v} = \frac{\partial v_x}{\partial x} + \frac{\partial v_y}{\partial y} + \frac{\partial v_z}{\partial z} \tag{10.10}$$

We again need to change the partial derivatives with respect to x, y and z, to those involving r, θ and ϕ. We also must reexpress v_x, v_y and v_z in terms of v_r, v_θ and v_ϕ, the components of $\mathbf{v}$ along the new axes. This is achieved by noting that

$$\left.\begin{array}{l} \mathbf{v} = v_x\mathbf{i} + v_y\mathbf{j} + v_z\mathbf{k} \\ \quad = v_r\hat{\mathbf{r}} + v_\theta\hat{\boldsymbol{\theta}} + v_\phi\hat{\boldsymbol{\phi}} \end{array}\right\} \tag{10.48}$$

If, for example, we now take the scalar product of these equations with $\mathbf{i}$, we obtain

$$\left.\begin{array}{l} v_x = v_r\hat{\mathbf{r}}\cdot\mathbf{i} + v_\theta\hat{\boldsymbol{\theta}}\cdot\mathbf{i} + v_\phi\hat{\boldsymbol{\phi}}\cdot\mathbf{i} \\ \quad = a_{11}v_r + a_{21}v_\theta + a_{31}v_\phi, \end{array}\right\} \tag{10.49}$$

where we have made use of eqns (10.40) to evaluate $\hat{\mathbf{r}}\cdot\mathbf{i}$, etc. Then (10.49), and equations like (6.30) for the transformation of the partial derivatives, are substituted into (10.10).

---

† If you are not yet happy with the concept of orthogonal matrices, you can simply check that for the explicit matrix $\mathbf{A}$ of eqn (10.39′), eqns (10.46) are indeed satisfied.

Table 10.4. *Expressions for vector operators in different coordinate systems.*

Vector operation	Cartesian coordinates	Cylindrical polar coordinates	Spherical polar coordinates
grad s	$\dfrac{\partial s}{\partial x}\mathbf{i} + \dfrac{\partial s}{\partial y}\mathbf{j} + \dfrac{\partial s}{\partial z}\mathbf{k}$	$\dfrac{\partial s}{\partial r}\hat{\mathbf{r}} + \dfrac{1}{r}\dfrac{\partial s}{\partial \theta}\hat{\boldsymbol{\theta}} + \dfrac{\partial s}{\partial z}\hat{\mathbf{z}}$	$\dfrac{\partial s}{\partial r}\hat{\mathbf{r}} + \dfrac{1}{r}\dfrac{\partial s}{\partial \theta}\hat{\boldsymbol{\theta}} + \dfrac{1}{r\sin\theta}\dfrac{\partial s}{\partial \phi}\hat{\boldsymbol{\phi}}$
div v	$\dfrac{\partial v_x}{\partial x} + \dfrac{\partial v_y}{\partial y} + \dfrac{\partial v_z}{\partial z}$	$\dfrac{1}{r}\dfrac{\partial}{\partial r}(rv_r) + \dfrac{1}{r}\dfrac{\partial v_\theta}{\partial \theta} + \dfrac{\partial v_z}{\partial z}$	$\dfrac{1}{r^2}\dfrac{\partial}{\partial r}(r^2 v_r) + \dfrac{1}{r\sin\theta}\dfrac{\partial}{\partial \theta}(\sin\theta\, v_\theta) + \dfrac{1}{r\sin\theta}\dfrac{\partial v_\phi}{\partial \phi}$
curl v	$\begin{vmatrix} \mathbf{i} & \mathbf{j} & \mathbf{k} \\[2pt] \dfrac{\partial}{\partial x} & \dfrac{\partial}{\partial y} & \dfrac{\partial}{\partial z} \\[4pt] v_x & v_y & v_z \end{vmatrix}$	$\left(\dfrac{1}{r}\dfrac{\partial v_z}{\partial \theta} - \dfrac{\partial v_\theta}{\partial z}\right)\hat{\mathbf{r}} + \left(\dfrac{\partial v_r}{\partial z} - \dfrac{\partial v_z}{\partial r}\right)\hat{\boldsymbol{\theta}}$ $\qquad + \dfrac{1}{r}\left[\dfrac{\partial}{\partial r}(rv_\theta) - \dfrac{\partial v_r}{\partial \theta}\right]\hat{\mathbf{z}}$	$\dfrac{1}{r\sin\theta}\left(\dfrac{\partial}{\partial \theta}(v_\phi\sin\theta) - \dfrac{\partial v_\theta}{\partial \phi}\right)\hat{\mathbf{r}} + \dfrac{1}{r}\left[\dfrac{1}{\sin\theta}\dfrac{\partial v_r}{\partial \phi} - \dfrac{\partial}{\partial r}(rv_\phi)\right]\hat{\boldsymbol{\theta}}$ $\qquad + \dfrac{1}{r}\left[\dfrac{\partial}{\partial r}(rv_\theta) - \dfrac{\partial v_r}{\partial \theta}\right]\hat{\boldsymbol{\phi}}$
$\nabla^2 s$	$\dfrac{\partial^2 s}{\partial x^2} + \dfrac{\partial^2 s}{\partial y^2} + \dfrac{\partial^2 s}{\partial z^2}$	$\dfrac{1}{r}\dfrac{\partial}{\partial r}\left(r\dfrac{\partial s}{\partial r}\right) + \dfrac{1}{r^2}\dfrac{\partial^2 s}{\partial \theta^2} + \dfrac{\partial^2 s}{\partial z^2}$	$\dfrac{1}{r^2}\dfrac{\partial}{\partial r}\left(r^2\dfrac{\partial s}{\partial r}\right) + \dfrac{1}{r^2\sin\theta}\dfrac{\partial}{\partial \theta}\left(\sin\theta\dfrac{\partial s}{\partial \theta}\right) + \dfrac{1}{r^2\sin^2\theta}\dfrac{\partial^2 s}{\partial \phi^2}$

An extra phase of the algebra occurs here because, for example, the $\partial v_x / \partial x$ term becomes

$$\left(\frac{\partial r}{\partial x} \frac{\partial}{\partial r} + \frac{\partial \theta}{\partial x} \frac{\partial}{\partial \theta} + \frac{\partial \phi}{\partial x} \frac{\partial}{\partial \phi} \right) (a_{11} v_r + a_{21} v_\theta + a_{31} v_\phi) \qquad (10.50)$$

Now v_r, v_θ and v_ϕ can, of course, be functions of r, θ and ϕ, but the coefficients a_{ij} also depend on θ and ϕ. Thus on performing the differentiation as required in (10.50), we obtain two sorts of terms such as

$$\frac{\partial \theta}{\partial x} \left(\frac{\partial a_{21}}{\partial \theta} v_\theta + v_{21} \frac{\partial v_\theta}{\partial \theta} \right)$$

where we have considered just the second term in each of the brackets in (10.50).

Eventually, on collecting all the terms of the transformed eqn (10.10), using the orthogonality relations (10.46) and also the explicit expressions (10.39′) for **A**, we obtain the result given in Table 10.4.

The other entries in the table are derived in an analogous manner.

Maybe after all there is something to be said in favour of learning tensor algebra.

10.4 Theorems involving integrals

In this section, the final one before we consider applications, we discuss theorems involving integrals of vector operators over various regions of space. They are known as the Divergence Theorem, Stokes' Theorem and Green's Theorems.

10.4.1 Divergence Theorem

The divergence theorem states that

$$\int \mathrm{div} v \, dV = \int \mathbf{v} \cdot d\mathbf{a} \qquad (10.51)$$

Here dV is an infinitesimal volume element, and the integral on the left-hand side is evaluated over the whole of some finite volume in three-dimensional space. The second integration is carried out over the surface enclosing the previously used volume, and $d\mathbf{a}$ is an infinitesimal surface area vector (pointing *out* of the volume).

The relationship (10.51) arises from the definition of the divergence (see eqn (10.8)). Just before going to the limit of infinitesimal size, we can write

$$\text{div}\mathbf{v}\delta V \approx \int \mathbf{v} \cdot d\mathbf{a} \tag{10.52}$$

for a very small region of volume δV, and where the integration is evaluated over its very small surface. Thus if a finite size volume is divided into very many small regions, eqn (10.52) will apply to each of them (see fig. 10.7). The summation over all small regions which make up the larger one then yields

$$\sum \text{div}\mathbf{v}\delta V \approx \sum \int_{\text{walls of small surface}} \mathbf{v} \cdot d\mathbf{a} \tag{10.53}$$

Now in the evaluation of the summation of the surface integrals, interior walls and exterior ones play different rôles. Each interior wall will contribute twice to the summation, once for each of the small volumes that share this common surface. For these two contributions, $\mathbf{v}$ is the same but $d\mathbf{a}$ reverses sign. These interior walls thus make a net contribution of zero to the sum. This is not so for the exterior walls, each of which appears only once in the summation for the right-hand side of eqn (10.53). Thus

$$\sum \text{div}\mathbf{v}\delta V \approx \int_{\text{exterior surfaces}} \mathbf{v} \cdot d\mathbf{a} \tag{10.54}$$

On proceeding to the limit where each small volume becomes infinitesimal, the sum on the left-hand side becomes an integral, and we obtain the Divergence Theorem (10.51).

In essence, the Divergence Theorem looks like applying the definition of the divergence to a large region, rather than just at a point. (This makes it easy to remember the theorem.) It is important to remember the part of the derivation that allows the contributions of interior surfaces to be discarded.

Mathematical examples of using the divergence theorem are found later in Section 10.4.4.1 and in the problems. Its application to physical situations is described in Section 10.5.

10.4.2 Stokes' Theorem

Stokes' Theorem relates a line integral around a finite size closed curve C, to a surface integral evaluated over a region S bounded by the same

Vector operators

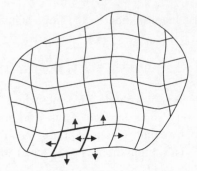

Fig. 10.7. Illustration of the Divergence Theorem. A finite size volume is divided into lots of small volumes δV (for simplicity of drawing, this is displayed in two dimensions only). From the definition of the divergence, for each small region $\mathrm{div}\mathbf{v}\delta V \approx \int \mathbf{v} \cdot d\mathbf{a}$. The heavy arrows on the boundary walls of the outlined box represent the area vectors $\delta \mathbf{a}$ involved in the surface integral for that box. The dashed arrows similarly show the vectors for a neighbouring box. On summation over all the small boxes, the two contributions of $\mathbf{v} \cdot d\mathbf{a}$ for any interior wall cancel, and we are left with just the integral over the external surface. This gives the Divergence Theorem.

curve (see fig. 10.8(a)). It states that

$$\int_S \mathrm{curl}\mathbf{v} \cdot d\mathbf{a} = \int_C \mathbf{v} \cdot d\mathbf{l} \tag{10.55}$$

where the line integral is performed anticlockwise around the curve.

This is derived in a manner very analogous to that used for the Divergence Theorem, except that here we start with the definition of $\mathrm{curl}\mathbf{v}$ for one of the small area elements. Thus

$$\mathrm{curl}\mathbf{v} \cdot \delta \mathbf{a} \approx \int_{\text{small line path}} \mathbf{v} \cdot d\mathbf{l} \tag{10.56}$$

On summing over all the small regions, and remembering that internal contributions cancel (see fig. 10.8(b)), we are left with

$$\sum \mathrm{curl}\mathbf{v} \cdot \delta \mathbf{a} \approx \int_C \mathbf{v} \cdot d\mathbf{l} \tag{10.57}$$

Finally on going to the limit where the regions are infinitesimal, we obtain Stokes' Theorem (10.55).

Stokes' Theorem can be used to obtain other relationships between line and surface integrals. Thus, for example, when the theorem is applied to the vector field $\psi\mathbf{b}$, where $\mathbf{b}$ is a constant vector and ψ is a scalar field,

(a)

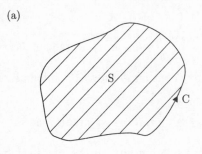

(b)

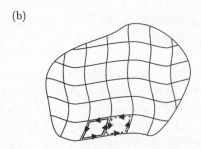

Fig. 10.8. Stokes' Theorem: (a) It equates the line integral $\int \mathbf{v} \cdot d\mathbf{l}$ for a vector $\mathbf{v}$ around the closed path C, to the surface integral $\int \text{curl}\mathbf{v} \cdot d\mathbf{a}$ evaluated over any surface S which is bounded by C. (b) For the derivation of Stokes' Theorem, the area S is divided into lots of very small areas $\delta\mathbf{a}$. From the definition of curl, for each small region $\text{curl}\mathbf{v} \cdot \delta\mathbf{a} \approx \int \mathbf{v} \cdot d\mathbf{l}$. The heavy arrows surrounding one of the areas show the path for the line integral around its boundary; the corresponding path for a neighbouring region is shown dashed. On summation over all the small regions, the two contributions on any interior line cancel, and just the integral around the outer curve C remains. This gives Stokes' Theorem.

it yields

$$\int \psi\mathbf{b} \cdot d\mathbf{l} = \int \text{curl}(\psi\mathbf{b}) \cdot d\mathbf{a}$$

$$= \int \psi\text{curl}\mathbf{b} \cdot d\mathbf{a} + \int (\text{grad}\psi \wedge \mathbf{b}) \cdot d\mathbf{a}$$

$$= -\int \mathbf{b} \cdot (\text{grad}\psi \wedge d\mathbf{S}) \qquad (10.58)$$

In the second line, we have made use of the identity (5) for $\text{curl}(\psi\mathbf{b})$ from Table 10.3; and in the third line we have used the fact that, since $\mathbf{b}$ is a constant vector, its curl vanishes.

Now since $\mathbf{b}$ is constant, it can be taken outside the integration sign to give

$$\mathbf{b} \cdot \int \psi d\mathbf{l} = \mathbf{b} \cdot \int -\mathrm{grad}\psi \wedge d\mathbf{a} \qquad (10.59)$$

Finally, since **b** is *any* constant vector,† the integrals themselves must be equal, i.e.

$$\int \psi d\mathbf{l} = -\int \mathrm{grad}\psi \wedge d\mathbf{a} \qquad (10.60)$$

Similarly, another relationship can be obtained by applying Stokes' Theorem to the vector field $\mathbf{b} \wedge \mathbf{c}$. The Divergence Theorem can also be applied to such product fields to obtain equations relating volume and surface integrals (see Green's Theorems below, and the problems at the end of the chapter).

Applications of Stokes' Theorem are found later in the chapter. It is said that the theorem first appeared as part of an examination set by Stokes, at which one of the students was James Clerk Maxwell. He later made extensive use of this and the Divergence Theorem in his study of electricity and magnetism (see Section 10.5).

10.4.3 Green's Theorems

These are a series of relations between surface and volume integrals, or between line and surface ones. They are closely related to the Divergence and Stokes' Theorems.

The equation

$$\int (P dx + Q dy) = \int \int \left(\frac{\partial Q}{\partial x} - \frac{\partial P}{\partial y} \right) dx dy \qquad (10.61)$$

is known as "Green's Theorem in the Plane". The left-hand integral is taken anticlockwise round a contour C in the x–y plane, which encloses an area that defines the region of integration on the right-hand side. Here P and Q are functions of x and y.

This can be deduced immediately as a special case of Stokes' Theorem (10.55), when the contour of fig. 10.8(a) lies in the x–y plane, and the vector **v** has components (P, Q, R), where R is irrelevant.

Alternatively, it can be derived by considering first

$$\int \frac{\partial Q}{\partial x} dx = Q_R - Q_L$$

---

† This is an essential part of the argument. Otherwise it would be invalid to deduce from $\mathbf{b} \cdot \mathbf{c} = \mathbf{b} \cdot \mathbf{d}$, that $\mathbf{c} = \mathbf{d}$ (see Section 3.3.5).

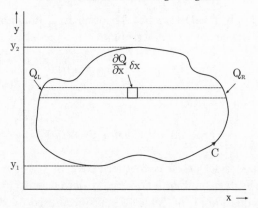

Fig. 10.9. Proof of Green's Theorem in the plane. Integration of $\partial Q/\partial x$ with respect to x yields $Q_R - Q_L$, where Q_R and Q_L are the values of Q on the right- and left-hand sides of the boundary curve C. A second integration of $Q_R - Q_L$ with respect to y (between the limits y_1 and y_2) then is equivalent to integrating Q anticlockwise all the way round C. The minus sign in front of Q_L is reproduced because the integral goes *down* the left-hand side of curve C.

(see fig. 10.9). Then

$$\int \left(\int \frac{\partial Q}{\partial x} dx \right) dy = \int_{y_1}^{y_2} (Q_R - Q_L) dy$$

$$= \int_C Q dy \qquad (10.62)$$

where the last integral is evaluated all the way round the curve C.

In an analogous manner

$$\int \left(\int \frac{\partial P}{\partial y} dy \right) dx = - \int_C P dx, \qquad (10.63)$$

where the minus sign arises because the anticlockwise sense of integration means that the upper part of the curve corresponds to decreasing x. Eqns (10.62) and (10.63) together give Green's Theorem (10.61).

Green's Theorem in Space is

$$\int \int (P\,dydz + Q\,dxdz + R\,dxdy) = \int \int \int \left(\frac{\partial P}{\partial x} + \frac{\partial Q}{\partial y} + \frac{\partial R}{\partial z} \right) dxdydz \qquad (10.64)$$

where the integrations are over a closed surface and the volume within it respectively. This is simply the Divergence Theorem (10.51) for the vector $\mathbf{v} = (P, Q, R)$. Eqn (10.64) can alternatively be proved by direct integration in a manner analogous to that used for eqn (10.61).

If we now apply the Divergence Theorem to a function $\mathbf{v} = s\,\mathrm{grad}t$ (where s and t are scalar fields), we obtain

$$\int s\,\mathrm{grad}t \cdot d\mathbf{a} = \int \mathrm{div}(s\,\mathrm{grad}t)dV$$

This is more usually written as

$$\int s\frac{\partial t}{\partial n}da = \int (s\nabla^2 t + \nabla s \cdot \nabla t)dV \qquad (10.65)$$

On the left-hand side, $\partial t/\partial n$ is the derivative of t along the outward normal to the surface, and da is now the scalar element of surface area. On the right, equation (3) of Table 10.3 has been used for $\mathrm{div}(s\,\mathrm{grad}t)$.

Finally, if s and t are interchanged and the result subtracted from the original form of (10.65), we obtain

$$\int \left(s\frac{\partial t}{\partial n} - t\frac{\partial s}{\partial n} \right) da = \int (s\nabla^2 t - t\nabla^2 s)dV \qquad (10.66)$$

This is known as Green's Two-Dimensional Formula.

As a very simple example of the use of eqn (10.61), we set $P = -y$ and $Q = x$. Then Green's formula reduces to

$$\int (-ydx + xdy) = \int\int 2dxdy \qquad (10.67)$$

Each side of the above equation is recognisable as twice the area enclosed by the curve C. Similarly in eqn (10.64), if $P = x$, $Q = y$ and $R = z$, each side of the equation is simply three times the volume within the closed surface.

10.4.4 *Mathematical examples*

We here give examples in which the Divergence and Stokes' Theorems are checked for specific cases.

10.4.4.1 *Divergence Theorem*

We first evaluate the integral

$$I_1 = \int \mathrm{div}\mathbf{v}dV$$

for

$$\mathbf{v} = x\mathbf{i} + (y+2)\mathbf{j} + z\mathbf{k} \qquad (10.68)$$

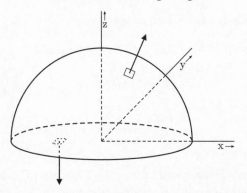

Fig. 10.10. The regions of integration in the example of Section 10.4.4.1 are the complete surface of the hemisphere for the surface integral, and the whole interior of the hemisphere for the volume one. Unit vectors normal to the surface are shown for the curved and for the flat surfaces of the hemisphere; they are in the radial and the $-z$ directions respectively.

calculated for the interior of the hemisphere sitting on the x–y plane

$$\left. \begin{array}{c} x^2 + y^2 + z^2 = R^2 \\ z \geq 0 \end{array} \right\}$$

This is then compared with $\int \mathbf{v} \cdot d\mathbf{a}$ for the closed surface surrounding the hemisphere. By the Divergence Theorem, these should be equal.

The calculation of I_1 is trivial. For $\mathbf{v}$ given by eqn (10.68),

$$\begin{aligned} \mathrm{div}\,\mathbf{v} &= \frac{\partial v_x}{\partial x} + \frac{\partial v_y}{\partial y} + \frac{\partial v_z}{\partial z} \\ &= 1 + 1 + 1 = 3 \end{aligned}$$

so I_1 is simply three times the volume of the hemisphere, i.e. $2\pi R^3$.

To calculate the surface integral, we first need an explicit expression for a unit vector normal to the small element of surface. For the curved surface (see fig. 10.10), this is

$$(x\mathbf{i} + y\mathbf{j} + z\mathbf{k})/R \tag{10.69}$$

Then the integral over the curved surface is

$$I_2 = \int [x\mathbf{i} + (y+2)\mathbf{j} + z\mathbf{k}] \cdot (x\mathbf{i} + y\mathbf{j} + z\mathbf{k})da/R, \tag{10.70}$$

where da is now the scalar area of an element. Then

$$I_2 = \int ((x^2 + y^2 + z^2) + 2y)da/R$$

$$= \int (R^2 + 2y)da/R$$

$$= R \int da$$

$$= 2\pi R^3 \qquad (10.71)$$

In the third line, we have made use of the facts: that R is a constant and so can be brought outside the integral; and that $2y/R$ is an odd function of y, and hence its integral over the symmetric area vanishes (compare Appendix A8). In the last line, we have simply written the total area of the curved surface of the hemisphere as $2\pi R^2$. This agrees with I_1, and so we may think we have verified the Divergence Theorem for this case. However, this involves a significant error of principle.

The Divergence Theorem relates integrals over a *closed* surface and over the volume within it. Our hemisphere has not only a curved surface, but also a flat bottom. We must thus also calculate

$$I_3 = \int [x\mathbf{i} + (y + 2)\mathbf{j} + z\mathbf{k}] \cdot (-\mathbf{k}da) \qquad (10.72)$$

where the integral extends over the circular area in the x–y plane, and the $-\mathbf{k}$ factor is the unit vector normal to this surface. Then

$$I_3 = - \int z\,da = 0$$

because $z = 0$ for this flat surface.

Thus

$$I_1 = I_2 + I_3 \qquad (10.73)$$

and this indeed confirms the Divergence Theorem. It was simply bad luck that in this case $I_3 = 0$, and hence we were in danger of being satisfied with $I_1 = I_2$. The important message is always to make sure to use a closed surface in Divergence Theorem questions.

An interesting point about this example is that, although it involved a surface and a volume integral, we did not actually perform any explicit integration. The actual integrals involved either vanished by symmetry, or corresponded to the surface area or volume of a hemisphere; the results could be written down by inspection. It is a common feature of simple problems of this type that any integrals will be either completely trivial, or at worst very simple. The essence of these problems is in the ideas involved, rather than in dexterity in evaluating difficult integrals.

Of course, in real-life situations the integrations could be more difficult. One of the beauties of the Divergence Theorem is that it provides two different routes to the answer; if one of them is complicated, there is the hope that the other may be more tractable.

This theorem can also be applied to a wide variety of physical situations, some of which are discussed in Section 10.5.

10.4.4.2 Stokes' Theorem

Here we return to eqn (10.60), which was derived by applying Stokes' Theorem to the vector field $\psi \mathbf{b}$. We will verify it explicitly for the case where $\psi = y$, and the region of integration is defined by the circle

$$\left.\begin{array}{c} x^2 + y^2 = R^2 \\ z = 0 \end{array}\right\} \tag{10.74}$$

We first determine $\int y d\mathbf{l}$, evaluated around the circle of radius R. Along this contour†

$$d\mathbf{l} = (-y\mathbf{i} + x\mathbf{j})d\theta \tag{10.75}$$

so

$$\int y d\mathbf{l} = \int (-y^2\mathbf{i} + xy\mathbf{j})d\theta$$

$$= R^2 \int (-\sin^2\theta\mathbf{i} + \cos\theta\sin\theta\mathbf{j})d\theta$$

Now the average value of $\sin^2\theta$ round the complete circle is $1/2$, while that for $\cos\theta\sin\theta$ is zero (as perhaps can be seen even more easily for xy itself). Thus

$$\int y d\mathbf{l} = R^2 \left(-\tfrac{1}{2}\right)(2\pi)\mathbf{i} = -\pi R^2\mathbf{i} \tag{10.76}$$

Next, we consider the right-hand side of (10.60). Since $\psi = y$, gradψ is simply $\mathbf{j}$. Now for the circle in the x–y plane, the infinitesimal area element is normal to the $\mathbf{k}$ direction, so

$$d\mathbf{a} = \mathbf{k} da \tag{10.77}$$

† This result for $d\mathbf{l}$ can be obtained simply by writing $\mathbf{r} = x\mathbf{i} + y\mathbf{j}$, with $x = R\cos\theta$ and $y = R\sin\theta$. Then $d\mathbf{l} = d\mathbf{r} = (-R\sin\theta\mathbf{i} + R\cos\theta\mathbf{j})d\theta$, which is the same as (10.75). Alternatively (10.75) is a vector of length $Rd\theta$ and is perpendicular to the radius vector, as required. A third approach, which is somewhat heavy-handed for this simple problem, is to use eqns (10.42) and (10.40) for constructing a unit vector in the required direction.

where *da* is the scalar area. Then

$$\int \text{grad}\psi \wedge d\mathbf{a} = \int \mathbf{j} \wedge \mathbf{k} da$$

$$= \int \mathbf{i} da = \pi R^2 \mathbf{i} \tag{10.78}$$

since the integral of *da* over the complete circle is just its area. Eqns (10.76) and (10.78) thus are consistent with (10.60).

Once again, we completed the problem without performing any explicit integration.

10.5 Physics applications

In this section, we discuss a few of the very many physics applications of the vector operators.

10.5.1 Potentials

The sun's gravitation field **G** produces a force **F** on nearby objects (where $\mathbf{F} = m\mathbf{G}$ and *m* is the mass of the object). This field is related to the gravitational potential ϕ by

$$\mathbf{G} = -\text{grad}\phi \tag{10.79}$$

Because curl(gradϕ) is zero (see Table 10.2), if any field **G** can be related to a scalar potential according to eqn (10.79), then curl**G** must be zero. Conversely, if curl**G** is zero, **G** can be expressed in terms of a scalar potential. This then provides a suitable way of testing whether any field is such that it is derivable from a scalar potential.

If it is, it follows from Stokes' Theorem that $\int_A^B \mathbf{F} \cdot d\mathbf{l}$ is independent of the path† between *A* and *B* (see fig. 10.11). This is not surprising since

$$\int_A^B \mathbf{F} \cdot d\mathbf{l} = -\int_A^B m\text{grad}\phi \cdot d\mathbf{l} = m(\phi_A - \phi_B) \tag{10.80}$$

i.e. the work done by the force in moving a mass *m* from *A* to *B* is equal to *m* times the difference in potential between these points.

Thus, for $\mathbf{G} = -\kappa\mathbf{r}/r^3$, curl**G** = 0 so that a potential ϕ exists. To find

---

† A particular example of this is that if $(\text{curl}\mathbf{F})_z = \partial F_y/\partial x - \partial F_x/\partial y = 0$, then $\int_A^B (F_x dx + F_y dy)$ is independent of the path of integration from *A* to *B* in the *x*–*y* plane, i.e. $F_x dx + F_y dy$ is an exact differential (compare Section 5.7).

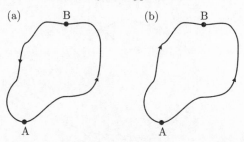

Fig. 10.11. A vector field **F** defines a force at every point in space. If this is derivable from a scalar potential ϕ according to $\mathbf{F} = -\text{grad}\phi$, then $\text{curl}\mathbf{F} = 0$. This implies that $\int \mathbf{F} \cdot d\mathbf{l}$ is zero round any closed curve, such as the one shown in (a) passing through the points A and B. If the direction of integration along one of the paths between A and B is reversed (see (b)), we deduce that $\int_A^B \mathbf{F} \cdot d\mathbf{l}$ is independent of the path between A and B, and is equal to the difference in the potential ϕ between these two points (which clearly makes no reference to the path between A and B).

ϕ, we first calculate

$$-\frac{\partial \phi}{\partial x} = G_x = -\kappa x/(x^2 + y^2 + z^2)^{3/2} \qquad (10.81)$$

so that

$$\phi = -\kappa/(x^2 + y^2 + z^2)^{1/2} + f(y, z) \qquad (10.82)$$

(Concerning the $f(y, z)$ term, compare the discussion below eqn (5.59).) The arbitrary function $f(y, z)$ is determined by comparing (10.82) with the corresponding results obtained from the other components of **G**. Then

$$\phi = -\kappa/r + c \qquad (10.83)$$

This then is the scalar potential, such that $\mathbf{G} = -\kappa\mathbf{r}/r^3$ is determined from it via eqn (10.79).

In contrast, for $\mathbf{G} = \lambda\mathbf{i}/r^2$, $\text{curl}\mathbf{G} = 2\lambda(-z\mathbf{j} + y\mathbf{k})/r^4 \neq 0$, so this **G** cannot be written in terms of a scalar potential, and the work done is not independent of the path between two points.

The above discussion of scalar potentials can also be applied to electrostatic or magnetostatic fields, provided $\text{curl}\mathbf{E}$ or $\text{curl}\mathbf{B}$ respectively is zero. Then a scalar electric or magnetic potential ϕ_E or ϕ_B exists, with the fields being related to them via eqn (10.79).

However, electromagnetism can be more complicated, because the presence of currents can make $\text{curl}\mathbf{B}$ non-zero (see eqn (M4) of Section 10.5.3), and similarly changing magnetic fields can result in $\text{curl}\mathbf{E} \neq 0$ (see

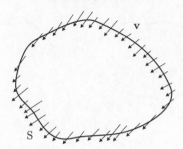

Fig. 10.12. Molecules moving in a gas. Their number per unit volume is given by the scalar field n, and their velocity by the vector field **v**. Within the region enclosed by the surface S, there will be an accumulation of molecules if there is a net flow inwards when the whole of the surface S is considered. The inward flow across a small area $d\mathbf{a}$ is $-n\mathbf{v} \cdot d\mathbf{a}$ (The scalar product is required because it is only the component of **v** normal to the surface area which is relevant. The minus sign is needed because we are considering the *inward* flow, while $d\mathbf{a}$ points *outwards*.) Then the total flow is $-\int n\mathbf{v} \cdot d\mathbf{a}$, and this is equal to $(\partial/\partial t) \int n \, dV$, the rate of increase of the number of molecules in the enclosed volume.

eqn (M3)). In these cases, it is necessary to introduce a more complicated vector potential.

It can readily be seen that a scalar potential ϕ does not work for **B** in the presence of currents because the lines of force for **B** then circulate around the currents. Thus work is required for taking a (mythical) magnetic monopole along a closed path round the current. However, a scalar potential would have returned to its original value, and hence the work done would be zero.

For further discussion of vector potentials, text books on electromagnetism should be consulted.

10.5.2 *Continuity Equation and other partial differential equations*

The Continuity Equation applies to many different situations. Its basic message is that the rate of accumulation of some quantity within a volume is equal to its flux through the surrounding surface (see fig. 10.12).

Thus the increase in density of gas molecules within a region is determined by the flow of molecules into it, i.e.

$$\frac{\partial}{\partial t} \int n \, dV = - \int n\mathbf{v} \cdot d\mathbf{a} \qquad (10.84)$$

where n is the number of molecules per unit volume, and **v** is their velocity. The regions of integration are over the volume being considered, and its surrounding closed surface. The minus sign on the right-hand side is because $d\mathbf{a}$ conventionally is in the direction out the surface, so positive $\mathbf{v} \cdot d\mathbf{a}$ corresponds to an outflow as gas, which contributes to a decrease in the gas density.

The right-hand side of (10.84) can be rewritten using the Divergence Theorem, to give

$$\frac{\partial}{\partial t} \int ndV = -\int \text{div}(n\mathbf{v})dV \tag{10.85}$$

The region of integration on the right-hand side is now over the enclosed volume, and hence is identical to that on the left. In addition, eqn (10.85) applies for any volume of integration, and so the two integrands can be equated to give the continuity equation as

$$\frac{\partial n}{\partial t} = -\text{div}(n\mathbf{v}) \tag{10.86}$$

Similarly if we consider the change in the electric charge density ρ in a region due to a flow of electric current with current density $\mathbf{j}$ per unit area, then

$$\frac{\partial \rho}{\partial t} = -\text{div}\mathbf{j} \tag{10.87}$$

In much the same way, heat conduction into a region within a body causes a rise in temperature T there, given by

$$\int \mathbf{Q} \cdot d\mathbf{a} = -\frac{\partial}{\partial t} \int \rho s T dV \tag{10.88}$$

where $\mathbf{Q}$ is the rate of heat flow per unit area, and ρ and s are the body's density and its specific heat capacity. Then

$$\text{div}\mathbf{Q} = -\frac{\partial}{\partial t}(s\rho T) \tag{10.89}$$

Now heat conduction is related to the temperature gradient by

$$\mathbf{Q} = -k\text{grad}T \tag{10.90}$$

where k is the thermal conductivity. Assuming k is constant, we finally obtain the heat conduction equation as

$$k\nabla^2 T = s\rho \frac{\partial T}{\partial t} \tag{10.91}$$

A particularly simple case of this arises when thermal equilibrium is reached. Since T does not vary with time, (10.91) reduces to Laplace's Equation

$$\nabla^2 T = 0 \tag{10.92}$$

The example of the wave equation for pressure waves in a gas is mentioned in Section 11.1.2.2.

10.5.3 Electromagnetism and Maxwell's Equations

One of the richest fields for physical applications of vector operators is electromagnetism. The major laws of this field are encapsulated in Maxwell's four equations, which can then be used to derive the wave equation for electromagnetic waves.

The first law is based on Gauss' Theorem, which in turn is derived from Coulomb's Inverse Square Law. This states that the electric displacement **D** at a position **r**, caused by an electric charge q at the origin, is

$$\mathbf{D} = \frac{1}{4\pi} \frac{q\mathbf{r}}{r^3} \tag{10.93}$$

From this Gauss deduced that

$$\int \mathbf{D} \cdot d\mathbf{a} = q \tag{10.94}$$

where the integration is carried out over any surface completely enclosing the charge q.

Now if instead of a discrete charge q there is a continuous distribution of charge density ρ, the right-hand side of (10.94) is replaced by $\int \rho dV$, where the integral extends over the volume enclosed by the surface. The left-hand side of (10.94) can be transformed by the Divergence Theorem to give

$$\int \mathrm{div}\mathbf{D} dV = \int \rho dV$$

The usual argument about the integrals being evaluated over the same volume, and their being equal for any arbitrary region of integration, then implies that the integrands must be equal too, i.e.

$$\mathrm{div}\mathbf{D} = \rho \tag{M1}$$

In passing we note that, in cases where **D** can be defined in terms of the electric potential ϕ by

$$\mathbf{D} = -\varepsilon\varepsilon_0 \mathrm{grad}\phi \tag{10.95}$$

(where ε and ε_0 are the relative dielectric constant of the medium and the permittivity of free space respectively), ϕ obeys Poisson's Equation

$$\nabla^2\phi = -\rho/\varepsilon\varepsilon_0 \tag{10.96}$$

In free space, the charge density is zero, and ϕ then satisfies Laplace's Equation (compare eqn (10.92)).

We now return to Maxwell's Equations. Since magnetism also obeys an inverse square law, a similar equation must apply for the magnetic flux density **B**, except that since free magnetic monopoles do not exist, the magnetic charge density must be zero. Thus

$$\mathrm{div}\mathbf{B} = 0 \qquad\qquad (\mathrm{M2})$$

Faraday's investigations of the electromotive force (emf) produced round a closed loop of wire through which there was a changing magnetic flux led him to the law that

$$\mathrm{emf} = \int \mathbf{E} \cdot d\mathbf{l} = -\frac{\partial}{\partial t} \int \mathbf{B} \cdot d\mathbf{a} \qquad\qquad (10.97)$$

The first integral is around the loop of wire, while the second is over any area bounded by it. The line integral of the electric field **E** can be transformed using Stokes' Theorem to $\int \mathrm{curl}\mathbf{E} \cdot d\mathbf{a}$. Again the two are equal when evaluated over identical but arbitrary areas, and hence the integrands are equal, whence

$$\mathrm{curl}\mathbf{E} = -\frac{\partial \mathbf{B}}{\partial t} \qquad\qquad (\mathrm{M3})$$

Maxwell's last equation is obtained from the Biot–Savart Law, giving the magnetic field **H** produced by a steady electric current. It is

$$\int \mathbf{H} \cdot d\mathbf{l} = i \qquad\qquad (10.98)$$

where the line integral is around any curve enclosing the current. This equation is then rewritten, using Stokes' Theorem on the left-hand side, and replacing the current i in a wire by the integral of the current density **j** over the same area as used for the Stokes' integral. Thus

$$\int \mathrm{curl}\mathbf{H} \cdot d\mathbf{a} = \int \mathbf{j} \cdot d\mathbf{a} \qquad\qquad (10.99)$$

Again the integrals can be equated to give

$$\mathrm{curl}\mathbf{H} = \mathbf{j} \qquad\qquad (10.100)$$

This is not yet Maxwell's Equation. What he realised was that, for varying currents, an extra term is needed in order to satisfy the Continuity Equation (see Problem 10.16). Thus his last equation is

$$\mathrm{curl}\mathbf{H} = \mathbf{j} + \frac{\partial \mathbf{D}}{\partial t} \qquad\qquad (\mathrm{M4})$$

This also has the effect of restoring a certain symmetry between the electric and magnetic fields.

Further justification for the extra term $\partial \mathbf{D}/\partial t$ in eqn (M4) can be found in textbooks on electromagnetism.

For the specific case of a vacuum, Maxwell's Equations simplify, because there are no free charges or electric currents. Also $\mathbf{D} = \varepsilon_0 \mathbf{E}$ and $\mathbf{B} = \mu_0 \mathbf{H}$, where μ_0 is the permeability of free space. Then

$$\text{div}\mathbf{E} = 0 \tag{10.101a}$$

$$\text{div}\mathbf{H} = 0 \tag{10.101b}$$

$$\text{curl}\mathbf{E} = -\mu_0 \frac{\partial \mathbf{H}}{\partial t} \tag{10.101c}$$

and

$$\text{curl}\mathbf{H} = \varepsilon_0 \frac{\partial \mathbf{E}}{\partial t} \tag{10.101d}$$

To see what sort of electric fields the vacuum is capable of supporting, we can eliminate $\mathbf{H}$ from eqns (10.101). This is most easily achieved by observing that $\mathbf{H}$ is operated on by $\partial/\partial t$ in eqn (10.101c), and by curl in (10.101d). Thus we need to take the curl of eqn (10.101c), and differentiate (10.101d) partially with respect to time. Then

$$\text{curl}(\text{curl}\mathbf{E}) = -\mu_0 \text{curl}\left(\frac{\partial \mathbf{H}}{\partial t}\right) = -\mu_0 \varepsilon_0 \frac{\partial^2 \mathbf{E}}{\partial t^2} \tag{10.102}$$

We finally use the relation in Table 10.2 for curl(curlE):

$$\text{curl}(\text{curl}\mathbf{E}) = \text{grad}(\text{div}\mathbf{E}) - \nabla^2 E = -\nabla^2 \mathbf{E} \tag{10.103}$$

since $\text{div}\mathbf{E} = 0$ (eqn (10.101a)).

Thus from eqns (10.102) and (10.103)

$$\nabla^2 \mathbf{E} = \mu_0 \varepsilon_0 \frac{\partial^2 \mathbf{E}}{\partial t^2} \tag{10.104}$$

which is the wave equation† for the vector field $\mathbf{E}$. Thus Maxwell's Equations predict that such waves can be transmitted through a vacuum, and that their speed should be $1/\sqrt{\mu_0 \varepsilon_0} \sim 3 \times 10^8$ m s^{-1}. Slightly different manipulations with eqns (10.101) show that the electric field $\mathbf{E}$ of such waves is accompanied by magnetic fields $\mathbf{H}$; and that both $\mathbf{E}$ and $\mathbf{H}$ are polarised perpendicular to the direction of propagation of the waves (i.e. they are transverse waves), and perpendicular to each other.

† The great simplification achieved by using vector operators in the above derivation of the wave equation becomes more vividly obvious when it is compared to the full horror of the algebra required for a derivation without the use of vector operators.

The existence of such electromagnetic waves was spectacularly demonstrated by Hertz a few years later. Other examples of such radiation include light and heat from the sun, radio and television transmissions, the cosmic microwave background radiation, etc. Whenever you enjoy any of these, you should remember and appreciate the rôle played by vector operators in the derivation of their wave equation.

Problems

10.1 $\phi(x, y, z)$ is a scalar function of the coordinates x, y and z. By considering grad$\phi \cdot d\mathbf{l}$, where $d\mathbf{l}$ is an infinitesimal vector along a contour surface for ϕ, show that grad ϕ is perpendicular to the contour surface.

10.2 (i) The divergence of a vector field $\mathbf{F}$ is defined by the limiting process of eqn (10.8). By considering the flux of $\mathbf{F}$ out of an infinitesimal cuboid box, with sides of length δx, δy and δz parallel to the coordinate axes, show that (10.8) results in eqn (10.10).

(ii) Similarly curl $\mathbf{F}$ is defined by (10.13). By considering the integral (10.12) round an infinitesimal rectangular path, with sides of length δy and δz parallel to the y and z axes, show that (10.13) yields eqn (10.22), and hence that curl$\mathbf{F}$ can be written as in (10.22′).

10.3 The curl of a vector $\mathbf{v}$ can be written as $\nabla \wedge \mathbf{v}$. This could be taken as implying that curl$\mathbf{v}$ is perpendicular to $\mathbf{v}$. Either prove that this must be so, or provide an explicit example that demonstrates that this is false.

10.4 Use the definition (10.13) for curl$\mathbf{v}$ to deduce that, for

$$\mathbf{v} = \beta r^n \hat{\tau} \qquad (10.20)$$

(where $\hat{\tau}$ is a unit vector, tangential to the cylindrical polar coordinates r and z), curl$\mathbf{v}$ is as given in eqn (10.21). Do this by explicit evaluation of $\int \mathbf{v} \cdot d\mathbf{l}$ round a small contour as shown in fig. 10.5(b), with $R_2 = R_1 + \delta R$.

Confirm this result by using formula (10.22′) for curl$\mathbf{v}$.

Evaluate curl$\mathbf{v}$ at the origin by performing the integration $\int \mathbf{v} \cdot d\mathbf{l}$ round a small circle of radius R, centred on the origin. Compare your result with that obtained above from the general formula for curl$\mathbf{v}$, when letting R_1 tend to zero.

10.5 Show that:

    (i) $\text{grad}[f(r)] = \dfrac{df}{dr}\hat{\mathbf{r}}$;

    (ii) $\text{div}[f(r)\mathbf{r}] = r\dfrac{df}{dr} + 3f$;

    (iii) $\text{curl}[f(r)\mathbf{r}] = 0$;

    (iv) $\text{div}(\text{grad}r^n) = n(n+1)r^{(n-2)}$;

    (v) $\text{grad}(\mathbf{b}\cdot\mathbf{r}) = \mathbf{b}$; and

    (vi) $\text{curl}(\mathbf{b}\wedge\mathbf{r}) = 2\mathbf{b}$,

where $\mathbf{b}$ is a constant vector.

10.6 For each of the following vector fields $\mathbf{F}$, evaluate curl$\mathbf{F}$ and div$\mathbf{F}$:

    (i) $\mathbf{F} = y^3\mathbf{i} - xy^2\mathbf{j}$;

    (ii) $\mathbf{F} = [2x\tan y - (\ln z)/x^2]\mathbf{i} + x^2\sec^2 y\mathbf{j} + (1/xz)\mathbf{k}$;

    (iii) $\mathbf{F} = e^{-x}\mathbf{i} + \cos y\cos z\mathbf{j} - \sin y\sin z\mathbf{k}$.

For which of the above examples does there exist a scalar field ϕ, such that

$$-\text{grad}\phi = \mathbf{F}?$$

Obtain the scalar potential ϕ for the case(s) where it exists.

    For each of the three vector fields above, what is the value of the integral $\int \mathbf{F}\cdot d\mathbf{l}$, where the path of integration is taken anticlockwise round the circle

$$x^2 + y^2 = 1$$

and

$$z = 0?$$

10.7 (i) Explain why

$$\text{curlgrad} \neq \text{gradcurl}$$

and

$$\text{divgrad} \neq \text{graddiv}$$

(ii) Show that

$$\text{curl}(\text{grad}\phi) = 0$$

and

$$\text{div}(\text{curl}\mathbf{v}) = 0$$

What is the relevance of one of these equations for determining whether a vector field can be represented by a scalar potential?

(iii) Derive the relationship

$$\mathrm{curl}(\mathrm{curl}\mathbf{v}) = \mathrm{grad}(\mathrm{div}\mathbf{v}) - \nabla^2\mathbf{v}$$

10.8 Verify the relationships:

(i) $\mathrm{grad}(\mathbf{u} \cdot \mathbf{v}) = (\mathbf{v} \cdot \nabla) \cdot \mathbf{u} + (\mathbf{u} \cdot \nabla)\mathbf{v} + \mathbf{v} \wedge \mathrm{curl}\mathbf{u} + \mathbf{u} \wedge \mathrm{curl}\mathbf{v}$

(ii) $\mathrm{div}(\mathbf{u} \wedge \mathbf{v}) = \mathbf{v} \cdot \mathrm{curl}\mathbf{u} - \mathbf{u} \cdot \mathrm{curl}\mathbf{v}$

(iii) $\mathrm{curl}(\mathbf{u} \wedge \mathbf{v}) = (\mathbf{v} \cdot \nabla)\mathbf{u} - (\mathbf{u} \cdot \nabla)\mathbf{v} + (\mathrm{div}\mathbf{v})\mathbf{u} - (\mathrm{div}\mathbf{u})\mathbf{v}$; and

(iv) $\mathrm{curl}(\phi\mathbf{v}) = \phi\mathrm{curl}\mathbf{v} + (\mathrm{grad}\phi) \wedge \mathbf{v}$.

Rewrite the above four relationships for the case where $\mathbf{v}$ is a constant vector.

10.9 The electric field $\mathbf{E}$ arising from an electric dipole at the origin is given by

$$\mathbf{E} = -\mathrm{grad}\phi$$

where

$$\phi = \frac{1}{4\pi\varepsilon_0}\frac{\boldsymbol{\mu} \cdot \mathbf{r}}{r^3}$$

and $\boldsymbol{\mu}$ is a constant vector. Find $\mathbf{E}$, and show that div $\mathbf{E} = 0$.

The vector potential $\mathbf{A}$ is defined by

$$\mathbf{A} = \frac{1}{4\pi\varepsilon_0}\frac{\boldsymbol{\mu} \wedge \mathbf{r}}{r^3}$$

Evaluate curl $\mathbf{A}$ and thus show that it is equal to $\mathbf{E}$.

10.10 Verify the Divergence Theorem for the vector

$$\mathbf{v} = y\mathbf{i} + x\mathbf{j} + z^2\mathbf{k}$$

for the cylindrical region bounded by

$$x^2 + y^2 = R^2$$

$$z = 0$$

$$\text{and} \quad z = h$$

10.11 Evaluate

$$\int y^2 dx + 2(xy + z)dy + 2y dz$$

along the path (i) from the origin along a straight line to $(1, 1, 1)$; and (ii) from the origin along a straight line to $(1, 2, 1)$, and then along a second straight line to $(1, 1, 1)$.

Investigate whether a vector $\mathbf{v}$ with components $(y^2, 2(xy+z), 2y)$ can be expressed as grad ϕ. What is the relevance of this to the above integrals?

10.12 Calculate curl **v** for the vector field

$$\mathbf{v} = y\mathbf{i} + 2x\mathbf{j} + 3xy\mathbf{k}$$

Verify Stokes' Theorem for **v**, where the path of integration is around the sides of a triangle, whose corners have the coordinates $(0, 0, 0), (1, 1, 0)$ and $(0, 1, 0)$.

10.13 Apply the Divergence Theorem to the vector field $\mathbf{v} = \phi \text{grad} \phi$ (where ϕ is a scalar field), to show that

$$\int \phi(\text{grad}\phi) \cdot d\mathbf{a} = \int \left[(\text{grad}\phi)^2 + \phi \nabla^2 \phi \right] dV$$

where as usual the regions of integration are a closed surface and the volume within it respectively.

Verify the above equation for the case where

$$\phi = x + y + 2z$$

and the region of integration is as in Problem 10.10.

10.14 Show that the Divergence Theorem applied to the function $\mathbf{b}\phi$ (where **b** is a constant vector and ϕ is a scalar field) yields the result

$$\int \text{grad}\phi \, dV = \int \phi \, d\mathbf{a}$$

Show also that, when applied to $\mathbf{b} \wedge \mathbf{c}$ (where as before **b** is constant, but **c** is a vector field), it gives

$$\int \text{curl} \mathbf{c} \, dV = - \int \mathbf{c} \wedge d\mathbf{a}$$

The above two results, together with the Divergence Theorem itself, can be summarised in the single formula

$$\int \nabla \circ f \, dV = \int d\mathbf{a} \circ f$$

where the $\circ$ symbol represents ordinary multiplication if f is a scalar field, or either a dot or a cross if f is a vector field.

Similarly, three results relating a line and an area integral, as derived from Stokes' Theorem, can be combined as

$$\int d\mathbf{l} \circ f = \int (d\mathbf{a} \wedge \nabla) \circ f$$

10.15 A vector field

$$\mathbf{F} = g(xyz)[h(x)\mathbf{i} + h(y)\mathbf{j} + h(z)\mathbf{k}]$$

is such that

$$\text{curl}\mathbf{F} = 0$$

and

$$\text{div}\mathbf{F} = 0$$

(Here g is a function of the product xyz, and h is a function of just one variable.) Given that it is not just a constant, determine **F**.

10.16 Show that eqn (10.100) (the first attempt to write down the fourth Maxwell Equation) violates the Continuity Equation (10.87) for electric charge; but that when the extra term $\partial \mathbf{D}/\partial t$ is added as in eqn (M4), the Continuity Equation is satisfied. (All that is required is to use Table 10.2 for the double vector operators, and another of Maxwell's Equations.)

10.17 Derive the wave equation for **H**, starting from Maxwell's Equations in a vacuum (10.101).

11

Partial differential equations

11.1 Introduction

11.1.1 What is a partial differential equation?

It is tea-time, and you have decided to make some smoked salmon sandwiches. A slight inconvenience is that you have only just removed the loaf of bread from the freezer and need to let it defrost. How long is this going to take?

The way in which the temperature rises for any small region in the interior of the loaf depends on the rate at which heat is conducted into that region. This in turn depends on the temperature gradients within the loaf. Thus there is a relationship between the spatial and the time derivatives of the temperature T of the bread. This relationship is a partial differential equation in that it involves partial derivatives of T (with respect to x and with respect to t).

This is typical of partial differential equations. In contrast to ordinary differential equations, which have only one independent variable (see Chapter 5), here we consider differential equations which involve at least two independent variables. Because the dependent variable of our partial differential equation (T in the above example) is a function of these independent variables, the derivatives are necessarily partial ones. The solution of the equation then involves finding a specific functional dependence for, say, T in terms of position and time, which satisfies the particular requirements of the given problem.

11.1.2 Specific examples

Here we describe and derive some of the more common examples of partial differential equations. Although their physical origins are outlined,

90

it is of course possible to consider partial differential equations from a purely mathematical point of view. That is, they can be regarded simply as equations that have to be solved, subject to some specific conditions, without any mention of a physical connection. However, relating the equation to a specific situation has the advantages that it may be possible to make use of physical intuition to help solve the problem; or to see whether or not a supposed solution is physically reasonable.

11.1.2.1 Heat conduction equation

We now derive the heat conduction equation, which was alluded to in the above example. Here we restrict ourselves to a long thin uniform rod of cross-sectional area A, made of material with thermal conductivity κ and heat capacity $s\rho$ per unit volume. The sides of the rod are insulated so that no heat can escape or enter through them but can only be conducted along the length of the rod. Its temperature T is a function of just the two variables x and t, the position along the rod and the time, respectively.

If we consider a short length of the rod between x and $x + \delta x$ (see fig 11.1(a)), the rate at which heat is conducted into the region through the end at x is

$$Q_x = -A\kappa \left(\frac{\partial T}{\partial x} \right)_x \tag{11.1}$$

where the minus sign reflects the fact that if $\partial T/\partial x$ is positive, heat is conducted to the left. Similarly the rate of heat conduction out at the far end of the small region is

$$Q_{x+\delta x} = -A\kappa \left(\frac{\partial T}{\partial x} \right)_{x+\delta x} \tag{11.1'}$$

As in most problems of setting up partial differential equations, we now use Taylor's Theorem (see Section 7.5 and eqn (7.37)) to relate the relevant quantities — in this case Q — at x and at $x + \delta x$:

$$Q_{x+\delta x} \approx Q_x + \left(\frac{\partial Q}{\partial x} \right)_x \delta x$$

$$= Q_x - A\kappa \frac{\partial^2 T}{\partial x^2} \delta x \tag{11.2}$$

Then the net rate of heat flow into the small region is

$$Q_x - Q_{x+\delta x} \approx A\kappa \frac{\partial^2 T}{\partial x^2} \delta x \tag{11.3}$$

This is responsible for producing a temperature rise in this region, such

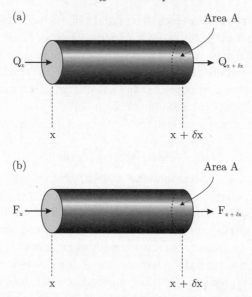

Fig. 11.1. (a) Heat conduction along an insulated rod. For the small region of length δx, the rate of heat flow is Q_x in at the left-hand end, and $Q_{x+\delta x}$ out at the right. The difference between these is responsible for causing the temperature of this part of the rod to rise or fall. (b) Diffusion of perfume in air within a cylindrical region. There are nV perfume molecules within the volume. This number is augmented by the flux F_x across the plane at x, as given by eqn (11.8); it is decreased by the flux $F_{x+\delta x}$ out of the plane at $x+\delta x$. The difference between these fluxes is responsible for the change in the number of perfume molecules within the small volume, as given by eqn (11.9).

that

$$s\rho A\delta x\frac{\partial T}{\partial t} = Q_x - Q_{x+\delta x} \tag{11.4}$$

Equating the two expressions (11.3) and (11.4) results in the heat conduction equation

$$\frac{\partial^2 T}{\partial x^2} = \frac{s\rho}{\kappa}\frac{\partial T}{\partial t} \tag{11.5}$$

If we consider a three-dimensional conductor rather than a one-dimensional rod, a similar analysis will yield the result

$$\nabla^2 T = \frac{s\rho}{\kappa}\frac{\partial T}{\partial t} \tag{11.6}$$

where

$$\nabla^2 = \frac{\partial^2}{\partial x^2} + \frac{\partial^2}{\partial y^2} + \frac{\partial^2}{\partial z^2} \tag{11.7}$$

This can alternatively be derived rather neatly by vector operator methods (see Chapter 10). Thus the net rate of heat flow into an infinitesimal volume δV is $-\mathrm{div}\mathbf{F}\delta V$, where $\mathbf{F}$ is the rate of heat flow per unit area. As before this is equal to $s\rho\delta V\,\partial T/\partial t$. But

$$\mathbf{F} = -\kappa\,\mathrm{grad}\,T$$

and so

$$s\rho\frac{\partial T}{\partial t} = -\mathrm{div}(-\kappa\,\mathrm{grad}\,T)$$
$$= \kappa\nabla^2 T \tag{11.6'}$$

An equation of exactly the same structure is obtained in diffusion processes. Again in the one-dimensional situation, we imagine a gas containing a non-uniform distribution of some particular vapour (e.g. air containing perfume). The rate at which the vapour molecules diffuse across a small plane area A is given by

$$F = -AD\frac{\partial n}{\partial x} \tag{11.8}$$

That is, the flux F is proportional to the gradient of the density of the perfume — if there is no density gradient, there is no net diffusion — and to the area through which the diffusion occurs; the constant of proportionality is the diffusion coefficient D. The number of perfume molecules per unit volume is n, and hence the total number within the small volume $V = A\delta x$ is nV.

The effect of the flux into a region is to increase the number of molecules within a volume V according to

$$\frac{\partial}{\partial t}(nV) = F$$

Thus the net rate at which molecules accumulate in a given region of length δx depends on the difference between the flux in at x, and that going out at $x + \delta x$, i.e.

$$\frac{\partial(nV)}{\partial t} = F_x - F_{x+\delta x}$$
$$= -\frac{\partial F}{\partial x}\delta x$$
$$= AD\frac{\partial^2 n}{\partial x^2}\delta x$$

whence

$$\frac{\partial n}{\partial t} = D\frac{\partial^2 n}{\partial x^2} \tag{11.9}$$

Apart from the meaning of the symbols, this has the identical structure to eqn (11.5), and hence can be solved in a completely analogous manner.

The three-dimensional version of (11.9) follows in an exactly similar way to (11.6).

11.1.2.2 Wave equation

The equation for a transverse wave on a string stretched along the x direction is given by

$$\frac{\partial^2 z}{\partial x^2} = \frac{\rho}{T}\frac{\partial^2 z}{\partial t^2} = \frac{1}{c^2}\frac{\partial^2 z}{\partial t^2} \qquad (11.10)$$

where T is the tension in the string, ρ is its linear density and c is the speed of the waves along the string.† This is obtained by considering a short portion of the string, and equating the upward force on it (due to the tension in the string) to its mass times its acceleration. The derivation is given in more detail in Section 14.3 (see also fig. 14.5).

In analogy with the heat conduction equation, if we instead consider the transverse oscillations of a stretched membrane (e.g. the skin of a drum), we obtain

$$\frac{\partial^2 z}{\partial x^2} + \frac{\partial^2 z}{\partial y^2} = \frac{\rho}{T}\frac{\partial^2 z}{\partial t^2} \qquad (11.10')$$

where ρ is now the mass per unit area. We cannot go to the complete three-dimensional analogy involving the ∇^2 operator of eqn (11.7) because there is no way of having displacements that are transverse to all of x, y and z.

Longitudinal waves give rise to a very similar equation

$$k\frac{\partial^2 w}{\partial x^2} = \rho\frac{\partial^2 w}{\partial t^2} \qquad (14.81)$$

where k is a constant relating to the driving force that is responsible for the displacement w (see Section 14.8). It is now possible to have a fully three-dimensional wave (e.g. sound waves in air), when the equation becomes

$$k\nabla^2 w = \rho\frac{\partial^2 w}{\partial t^2} \qquad (11.11)$$

† The fact that c is the speed will be demonstrated later. However, just from a dimensional point of view the units of c must be distance/time.

11.1.2.3 Poisson's and Laplace's Equations

The electric field $\mathbf{E}$ is related to the electrostatic potential V by

$$\mathbf{E} = -\mathrm{grad}\,V \qquad (11.12)$$

This is the vector operator extension of the one-dimensional equation

$$E = -\frac{\partial V}{\partial x} \qquad (11.12')$$

Now the electric field in a region that contains electric charges with charge density ρ is given by Gauss' Theorem as

$$\epsilon_0\epsilon\,\mathrm{div}\mathbf{E} = \rho \qquad (11.13)$$

This is the vector operator version of the statement that the surface integral of the normal component of electric flux out of a small volume containing a charge $\int \rho dV$ is given by

$$\epsilon_0\epsilon \int \mathbf{E} \cdot d\mathbf{a} = \int \rho dV \qquad (11.13')$$

In eqns (11.13) and (11.13'), ϵ is the relative dielectric constant of the medium and ϵ_0 is the permittivity of free space.†

By combining eqns (11.12) and (11.13), we obtain

$$\rho = \epsilon_0\epsilon\,\mathrm{div}(\mathrm{grad}V)$$
$$= \epsilon_0\epsilon\nabla^2 V \qquad (11.14)$$

with ∇^2 as given in eqn (11.7) for the three-dimensional case, or by a smaller number of terms for a two- or one-dimensional problem. This is Poisson's Equation.

A special case is obtained for a region that is free of electric charge. Then $\rho = 0$, and so eqn (11.14) reduces to Laplace's Equation

$$\nabla^2 V = 0 \qquad (11.15)$$

Another situation where Laplace's equation applies is for the equilibrium temperature distribution throughout a body whose temperature is defined along its boundary. Of course, the heat conduction equation applies, but because this is an equilibrium situation, the time derivative of the temperature is zero, so that eqn (11.6) reduces to the form of (11.15).

It is the solutions of the heat conduction equation, the wave equation

---

† Beware. The symbol V is used for both the potential and for the volume.

and Laplace's Equation that will concern us for the remainder of this chapter.

11.1.3 Boundary conditions

Each of the above equations as it stands has a large variety of possible solutions. Thus the solutions for the wave equation for an infinite string include stationary harmonic waves, travelling harmonic waves and non-harmonic stationary waves which change shape as time goes by. Similarly Laplace's Equation in three dimensions has solutions which grow with distance r and others which fall with increasing r; and solutions which have complicated angular dependences, corresponding to various electrostatic multipole configurations, or are independent of angle and are functions of r only (e.g. corresponding to an electric charge at the origin). What serve to define each particular problem are the particular boundary conditions. When enough of these are specified, the solution to the problem is unique (see Section 11.1.4).

Thus for the wave equation, we may be told the initial transverse displacement and velocity at each point on the string (with, for example, the initial velocity possibly being zero). A suitable set of boundary conditions for the heat conduction equation for an insulated rod could be its initial temperature distribution, and the way the temperatures at its ends are varied (or kept constant) as functions of time. An electrostatic problem involving Laplace's Equation could specify the value of the potential V on surfaces bounding the region of interest.

The solution of a particular problem must then incorporate the given boundary conditions. It is this that selects the required solution from among the myriad of ones otherwise possible. In some cases, the geometric arrangement of the boundary conditions may make it appropriate to change the independent variables from, say, x, y and z to r, θ and ϕ (see Section 11.2.2).

Doing all this requires a lot of practice. Some specific worked examples are given in Section 11.5; other problems for the reader are provided at the end of this chapter.

11.1.4 Uniqueness Theorem

The Uniqueness Theorem is most important theoretically, although it has only an indirect effect on the way partial differential equations are solved. It states that, given suitable boundary conditions, there is only one possible solution of the equation. This implies that if by any method we find a specific function that satisfies the differential equation and the boundary conditions, then this is in fact the unique solution to the problem, and there is no need to look for anything more general.

The notion of what constitute "suitable boundary conditions" is a rather subtle matter. In some cases, it is sufficient to specify boundary conditions around some closed region in the space of the independent variables; then the solution of the problem within the closed region is unique. An example of this is Laplace's Equation, with the potential V given for some closed surface.

The concept of the "closed surface in space" should not be taken too literally. First, if the partial differential equation has independent variables x and t (e.g. the heat conduction equation), then the boundary conditions need to be specified over a curve in '(x, t)' space. Alternatively, if the variables are x, y, z and t (see eqn (11.6)), then the closed region is in that four-dimensional space.

A second type of generalisation is that, in some circumstances, the theorem still applies, even if part of the "enclosed" region is at infinity. Thus in the heat conduction problem discussed in Section 11.3.2.1, the boundary conditions specify

(i)   the temperature T at $t = 0$ along the rod (from $x = 0$ to $x = L$);
(ii) the fact that $T = 0$ for all t at $x = 0$ and at $x = L$; and
(iii) the implication that T remains finite along the bar as t tends to infinity.

These conditions apply along the sides of an infinite rectangle in (x, t) space (see fig. 11.2), resulting in a unique solution to the heat conduction equation within this region, which is the physically interesting one.

For the wave equation, specifying the displacement and velocity at all relevant positions at some specific time is sufficient to define the solution uniquely.

In any problem in which you are required to find an explicit solution, it is probably reasonable to assume that this is unique.

11.1.5 Connections with other branches of mathematics

Partial differential equations are clearly related to ordinary differential equations. Indeed in many methods for solving partial differential equations in n independent variables, the aim is to reduce the problem to n separate ordinary differential equations, at which point the specific

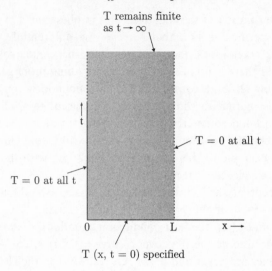

T remains finite
as t → ∞

t

T = 0 at all t

T = 0 at all t

0   L   x →

T (x, t = 0) specified

Fig. 11.2. Boundary conditions and the Uniqueness Theorem. The temperature $T(x, t)$ of the rod is specified along the boundary of the semi-infinite rectangle in (x, t) space. This is sufficient to ensure that the solution of the heat conduction equation is unique within the physically interesting shaded area.

difficulties associated with partial differential equations have been overcome. We shall come across several examples of this in Section 11.3. Nevertheless we shall in general go on to solve such ordinary differential equations, so that the explicit form of the solution is apparent. Indeed, this is a necessary step on the way to implementing the specified boundary conditions of any given problem. We also emphasise when various tricks are worth remembering for solving the ordinary differential equations.

Needless to say, the solutions of the wave equation are discussed at length in Chapter 14 on waves. This should be referred to as crossreferenced throughout this chapter.

The method of Fourier analysis can be used for solving the problem of the subsequent temperature distribution $T(x, t)$ on a thermally insulated rod whose initial temperature distribution $T(x, 0)$ is known. The method consists of expressing $T(x, 0)$ as a sum of harmonic spatial functions with various amplitudes. The reason this is useful is that it is particularly easy to derive the time dependence of each component with a harmonic spatial dependence for its temperature.

Furthermore, this property of simple behaviour for harmonic spatial components applies also to waves on a string, and hence the Fourier series approach is most useful there too. Thus Chapter 12 on Fourier series has a substantial overlap with the material discussed here.

The simplest behaviour for waves on a string consists in every part of it oscillating at the same frequency. This is analogous to a normal mode situation, for a system of N coupled objects (see Chapter 13). The definition of a normal mode requires all N components of the system to oscillate at the same frequency. The main difference is that in the normal mode problem there are N modes (assuming motion in just one dimension for each component), while for a string the number of natural frequencies is infinite† because there are an infinite number of infinitesimal elements on the string. Then just as any motion of the N-component system can be expressed as a sum of the various normal modes, each with its own coefficient, so too can any arbitrary motion of the string be defined in terms of the harmonic vibrations; this takes us back again to Fourier series (or Fourier analysis†).

We thus see that partial differential equations have connections with a wide range of mathematics and physics. The study of these topics from their various viewpoints and of their interrelationships provides an enhanced understanding of each of the subjects.

11.2 Changing the independent variables

There are two reasons why it may be desirable to change the independent variables in a partial differential equation. The first is to make the form of the equation easier to solve; an example of this is provided in Section 11.2.1. The second is that the new variables may be more appropriate for the boundary conditions of a given problem (see Section 11.2.2).

11.2.1 *Using* $x \pm ct$ *in the wave equation*

A trick due to d'Alembert uses the transformation to new variables

$$\left. \begin{aligned} u &= x + ct \\ \text{and } v &= x - ct \end{aligned} \right\} \tag{11.16}$$

in order to solve the one-dimensional wave equation (11.10).

To discover how the differential equation is modified, we use the rule of Section 6.3.4 for changing the variable in a partial derivative. Adapted

---

† For a finite string with its ends fixed these frequencies are discrete, while for an infinite string, they form a continuum — see Section 11.3.2.5.

to our variables, this gives

$$\frac{\partial z}{\partial x} = \frac{\partial z}{\partial u}\frac{\partial u}{\partial x} + \frac{\partial z}{\partial v}\frac{\partial v}{\partial x} \tag{6.30}$$

$$= \frac{\partial z}{\partial u} + \frac{\partial z}{\partial v} \tag{11.17}$$

where in the last line we have used the fact that $\partial u/\partial x$ and $\partial v/\partial x$ are both unity (see eqns (11.16)). Similarly

$$\frac{\partial z}{\partial t} = \frac{\partial z}{\partial u}\frac{\partial u}{\partial t} + \frac{\partial z}{\partial v}\frac{\partial v}{\partial t}$$

$$= c\left(\frac{\partial z}{\partial u} - \frac{\partial z}{\partial v}\right) \tag{11.18}$$

So far so good, but we now need to express $\partial^2 z/\partial x^2$ in terms of derivatives with respect to u and v. This is achieved by writing

$$\frac{\partial^2 z}{\partial x^2} = \frac{\partial}{\partial x}\left(\frac{\partial z}{\partial x}\right)$$

$$= \frac{\partial}{\partial u}\left(\frac{\partial z}{\partial u} + \frac{\partial z}{\partial v}\right)\frac{\partial u}{\partial x} + \frac{\partial}{\partial v}\left(\frac{\partial z}{\partial u} + \frac{\partial z}{\partial v}\right)\frac{\partial v}{\partial x}$$

$$= \frac{\partial^2 z}{\partial u^2} + 2\frac{\partial^2 z}{\partial u \partial v} + \frac{\partial^2 z}{\partial v^2} \tag{11.19}$$

In the second line, eqn (11.17) for $\partial z/\partial x$ has been used.

Similarly, after analogous manipulations

$$\frac{\partial^2 z}{\partial t^2} = c^2\left(\frac{\partial^2 z}{\partial u^2} - 2\frac{\partial^2 z}{\partial u \partial v} + \frac{\partial^2 z}{\partial v^2}\right) \tag{11.20}$$

Substitution of (11.19) and (11.20) into (11.10) yields

$$4\frac{\partial^2 z}{\partial u \partial v} = 0 \tag{11.21}$$

Although this may not appear to be so much simpler than (11.10), it in fact immediately yields solutions for travelling waves (see Section 11.4).

The above procedure is very straightforward but somewhat tedious. In fact it is rather easier than most transformations (compare Section 11.2.2), since here the partial derivatives ($\partial u/\partial x$, etc.) of the transformation are all constants. Thus when changing the variables, you should not despair if you have not completed the calculation in a couple of minutes. Furthermore, since it is easy to make an algebraic slip, it is worth memorising the results for a few simple transformations, in order to be able to check that your answer is correct.

In fact the derivation could have been shortened somewhat by writing (11.17) and (11.18) as operator equations:

$$\frac{\partial}{\partial x} = \frac{\partial}{\partial u} + \frac{\partial}{\partial v} \tag{11.17'}$$

and

$$\frac{\partial}{\partial t} = c\left(\frac{\partial}{\partial u} - \frac{\partial}{\partial v}\right) \tag{11.18'}$$

Then

$$\frac{\partial^2}{\partial x^2} = \left(\frac{\partial}{\partial u} + \frac{\partial}{\partial v}\right)^2$$

$$= \frac{\partial^2}{\partial u^2} + 2\frac{\partial^2}{\partial u \partial v} + \frac{\partial^2}{\partial v^2} \tag{11.19'}$$

and similarly for $\partial^2/\partial t^2$. Again this is somewhat deceptively easy because of the constancy of $\partial u/\partial x$, etc.

11.2.2 From Cartesian to polar coordinates

The wave equation, heat conduction equation and Laplace's Equation all involve the operator ∇^2 in its three-, two- or one-dimensional form, as appropriate to the particular problem. Up till now we have thought of this as being expressed in terms of the Cartesian coordinates x, y and z (see eqn (11.7) for the three-dimensional form). Solutions of these equations would then naturally turn out to be expressed in these variables. It may well be that, because of the specific boundary conditions of a given problem, a different choice of variables is more useful. These could be, for example, two-dimensional polar, spherical polar or cylindrical polar coordinates. Thus if the temperature distribution is required within a uniform medium when the temperature is specified along boundaries with constant r, constant θ and constant z, clearly it would be advantageous to express ∇^2 in terms of cylindrical polar coordinates.

We now change ∇^2 from two-dimensional x,y coordinates to two-dimensional polar coordinates, i.e. we want to re-express

$$\frac{\partial^2}{\partial x^2} + \frac{\partial^2}{\partial y^2}$$

in terms of partial derivatives with respect to r and θ. The transformation equations between the sets of variables are

$$\left. \begin{array}{r} r^2 = x^2 + y^2 \\ \text{and} \quad \tan\theta = y/x \end{array} \right\} \tag{A1.1}$$

The method used is in complete analogy with that for the transformation of Section 11.2.1, but it is a bit more involved algebraically because the partial derivatives of the transformation here (e.g. $\partial\theta/\partial x$) are not simply constants, but are functions of the variables r and θ.

As before, we start by writing

$$\frac{\partial}{\partial x} = \frac{\partial r}{\partial x}\frac{\partial}{\partial r} + \frac{\partial\theta}{\partial x}\frac{\partial}{\partial\theta}$$

$$= \cos\theta\frac{\partial}{\partial r} - \frac{\sin\theta}{r}\frac{\partial}{\partial\theta} \tag{11.22}$$

Here we have substituted the relevant expressions for $\partial r/\partial x$ and $\partial\theta/\partial x$, obtained by differentiating eqns (A1.1). Similarly

$$\frac{\partial}{\partial y} = \sin\theta\frac{\partial}{\partial r} + \frac{\cos\theta}{r}\frac{\partial}{\partial\theta} \tag{11.22'}$$

The second derivatives are obtained by using the above operators twice. Thus

$$\frac{\partial^2}{\partial x^2} = \left(\cos\theta\frac{\partial}{\partial r} - \frac{\sin\theta}{r}\frac{\partial}{\partial\theta}\right)\left(\cos\theta\frac{\partial}{\partial r} - \frac{\sin\theta}{r}\frac{\partial}{\partial\theta}\right) \tag{11.23}$$

In expanding terms like $(\cos\theta\,\partial/\partial r)[-(\sin\theta/r)\,\partial/\partial\theta]$, it is important to remember that $\partial/\partial r$ operates on the second bracket and yields

$$\cos\theta\left(\frac{\sin\theta}{r^2}\frac{\partial}{\partial\theta} - \frac{\sin\theta}{r}\frac{\partial^2}{\partial r\partial\theta}\right)$$

Finally the whole expression gives

$$\frac{\partial^2}{\partial x^2} = \cos^2\theta\frac{\partial^2}{\partial r^2} + \frac{\sin^2\theta}{r}\frac{\partial}{\partial r} + \frac{\sin^2\theta}{r^2}\frac{\partial^2}{\partial\theta^2} + \frac{2\sin\theta\cos\theta}{r^2}\frac{\partial}{\partial\theta} \tag{11.24}$$

There is also in principle a $\partial^2/\partial r\partial\theta$ term, but it in fact vanishes.

In a completely analogous manner, we also can derive

$$\frac{\partial^2}{\partial y^2} = \sin^2\theta\frac{\partial^2}{\partial r^2} + \frac{\cos^2\theta}{r}\frac{\partial}{\partial r} + \frac{\cos^2\theta}{r^2}\frac{\partial^2}{\partial\theta^2} - \frac{2\sin\theta\cos\theta}{r^2}\frac{\partial}{\partial\theta} \tag{11.24'}$$

Adding (11.24) and (11.24'), we at last obtain

$$\frac{\partial^2}{\partial x^2} + \frac{\partial^2}{\partial y^2} = \frac{\partial^2}{\partial r^2} + \frac{1}{r}\frac{\partial}{\partial r} + \frac{1}{r^2}\frac{\partial^2}{\partial\theta^2}$$

$$= \frac{1}{r}\frac{\partial}{\partial r}\left(r\frac{\partial}{\partial r}\right) + \frac{1}{r^2}\frac{\partial^2}{\partial\theta^2} \tag{11.25}$$

This derivation of a rather simple-looking result has involved a large amount of algebra.

The extension to cylindrical polar coordinates is straightforward, in that the extra term $\partial^2/\partial z^2$ on the left-hand side of eqn (11.25) simply becomes $\partial^2/\partial z^2$ on the right-hand side,† i.e.

$$\frac{\partial^2}{\partial x^2} + \frac{\partial^2}{\partial y^2} + \frac{\partial^2}{\partial z^2} = \frac{1}{r}\frac{\partial}{\partial r}\left(r\frac{\partial}{\partial r}\right) + \frac{1}{r^2}\frac{\partial^2}{\partial\theta^2} + \frac{\partial^2}{\partial z^2} \qquad (11.26)$$

Now we want the expression for ∇^2 in spherical polar coordinates. Given that the transformation equations are

$$\left.\begin{array}{l} r^2 = x^2 + y^2 + z^2 \\ \tan\theta = \sqrt{x^2 + y^2}/z \\ \text{and } \tan\phi = y/x \end{array}\right\} \qquad (A1.3)$$

we can derive

$$\frac{\partial}{\partial x} = \sin\theta\cos\phi\frac{\partial}{\partial r} + \frac{\cos\phi\cos\theta}{r}\frac{\partial}{\partial\theta} - \frac{\sin\phi}{r\sin\theta}\frac{\partial}{\partial\phi} \qquad (11.27)$$

With some considerable effort, we can then obtain $\partial^2/\partial x^2$ in terms of first and second derivatives with respect to r, θ and ϕ. After similar processes for $\partial^2/\partial y^2$ and $\partial^2/\partial z^2$, we finally express ∇^2 in spherical polar coordinates.

This process is very lengthy and highly prone to error. A better approach is to realise that $\nabla^2 V$ is simply div(grad V); in fact use was made of this in the derivation of Poisson's Equation in Section 11.1.2.3. Now in spherical polar coordinates

$$\text{grad} V = \left(\frac{\partial V}{\partial r}, \frac{1}{r}\frac{\partial V}{\partial\phi}, \frac{1}{r\sin\theta}\frac{\partial V}{\partial\phi}\right)$$

while

$$\text{div}\mathbf{a} = \frac{1}{r^2}\frac{\partial}{\partial r}(r^2 a_r) + \frac{1}{r\sin\theta}\frac{\partial}{\partial\theta}(\sin\theta a_\theta) + \frac{1}{r\sin\theta}\frac{\partial a_\phi}{\partial\phi}$$

Here the components of grad V are along vectors in the directions of increasing r, θ and ϕ respectively; while a_r, a_θ and a_ϕ refer to components of $\mathbf{a}$ along these directions (see Section 10.3.4 and particularly Table 10.4).

Combining these expressions for div and grad yields

$$\nabla^2 V = \frac{1}{r^2}\frac{\partial}{\partial r}\left(r^2\frac{\partial V}{\partial r}\right) + \frac{1}{r^2\sin\theta}\frac{\partial}{\partial\theta}\left(\sin\theta\frac{\partial V}{\partial\theta}\right) + \frac{1}{r^2\sin^2\theta}\frac{\partial^2 V}{\partial\phi^2} \qquad (11.28)$$

as our result for spherical polar coordinates.‡

† This is not completely trivial because the partial derivative on the left-hand side implies that x and y are kept constant, whereas on the right it is the cylindrical polar coordinates r and θ which are constant. However, constant x and y is equivalent to constant r and θ.

‡ This method is much shorter because all the tedious algebra is hidden away in the derivation of the expressions for grad V and div$\mathbf{a}$ in spherical polar coordinates.

11.3 Separation of variables

Now we will actually solve some partial differential equations. Section 11.4 deals with the special method of d'Alembert for finding solutions of the wave equation. Here we discuss 'separating the variables' which is the most common approach. This is not to be confused with the method of solving an ordinary differential equation when its variables are separable (see Section 5.3).

11.3.1 Basic idea

The basic idea for this method is that we try to find a solution that can be written as the product of separate functions, each of which depends on just one of the variables. For example, for the wave equation (11.10), the most general solution for the displacement is $z(x,t)$, a function of the two independent variables of the partial differential equation. In the "separation of variables" method, we look for solutions of the type

$$z(x,t) = F(x)G(t) \qquad (11.29)$$

where $F(x)$ is a function only of x, and $G(t)$ only of t.

Of course not all solutions of a partial differential equation will be of this form. However, since the partial differential equations we are considering are linear and homogeneous (see Section 5.1.8), any linear combination of solutions is also a solution. Thus if we have several separable solutions, distinguished by suffix i, then

$$z(x,t) = \sum_i \alpha_i F_i(x)G_i(t) \qquad (11.30)$$

also satisfies the differential equation. (The α_i are arbitrary coefficients.) This expression is not separable in x and t. In this way it is possible to build very many non-separable functions out of terms that are themselves separable.

Thus, for example, the standing wave

$$z_1 = A \sin kx \cos ckt \qquad (11.31)$$

is a separable solution of the wave equation. So is

$$z_2 = A \cos kx \sin ckt \qquad (11.31')$$

which is another standing wave, slightly displaced from the previous one in both space and time. By adding these, we obtain

$$z = z_1 + z_2$$
$$= A \sin k(x + ct) \tag{11.32}$$

which is a moving wave. This is also a solution of the wave equation, but is not separable in x and t, i.e. it cannot be written in the form of a single product $F(x)G(t)$.

We now insert our trial solution (11.29) in the wave equation (11.10). When we differentiate $F(x)G(t)$ partially with respect to x (i.e. at fixed t), the factor $G(t)$ is just a constant so we obtain

$$\frac{\partial^2 z}{\partial x^2} = G(t) \frac{d^2 F(x)}{dx^2} \tag{11.33}$$

where the derivative of F has been written as an ordinary one (rather than as a partial derivative) since F depends only on one variable x. Similarly

$$\frac{\partial^2 z}{\partial t^2} = F(x) \frac{d^2 G(t)}{dt^2} \tag{11.33'}$$

Then eqn (11.10) becomes

$$G(t) \frac{d^2 F(x)}{dx^2} = \frac{1}{c^2} F(x) \frac{d^2 G(t)}{dt^2} \tag{11.34}$$

It is now useful to divide each side of the equation by the solution $z = F(x)G(t)$; this step is required for most problems of this type. We then obtain

$$\frac{1}{F} \frac{d^2 F}{dx^2} = \frac{1}{c^2} \frac{1}{G} \frac{d^2 G}{dt^2} \tag{11.34'}$$

Now comes the crucial step in the argument. The left-hand side of this equation contains F and its second derivative; it is thus a function of x only. In contrast, the right-hand side involves only G (and a constant), and so depends only on t. It is in this sense that we have separated the variables. Now eqn (11.34') must be true for all x and t. The only way this can be so is if each side of the equation is equal to a constant,† which for reasons that will soon become clear we write as $-k^2$.

---

† If this is not obvious, it can be seen as follows. Eqn (11.34') applies at every possible x and t. Thus we can choose to consider a specific position x_0, but any time t. Then the left-hand side reduces to a constant, and hence so too must be the right-hand side. By interchanging the rôle of x and t, we can similarly show that the left-hand side is constant too. (Alternatively, since we have just shown that the right-hand side is constant, so too must the left-hand side, even for arbitrary x.)

Thus equation (11.34′) is equivalent to

$$\left.\begin{aligned}
\frac{1}{F}\frac{d^2 F}{dx^2} &= -k^2 \\
\text{and}\quad \frac{1}{G}\frac{d^2 G}{dt^2} &= -k^2 c^2
\end{aligned}\right\} \tag{11.35}$$

A very significant change has occurred between eqns (11.10) and (11.35). The former was the partial differential equation we were trying to solve, while the latter are two separate ordinary differential equations. This simplification is a direct result of our choice of a separable function as a trial solution. At this stage we no longer have a partial differential equation to deal with, and we can regard the problem as essentially solved.

In fact the ordinary differential equations that we obtain for the separable functions can sometimes be tricky to solve. It is thus well worth while to become familiar with the methods for extracting their solutions for the few partial differential equations that we regularly have to solve (see below). In this case, however, the ordinary differential equations (11.35) present no problems, and their solutions are

$$\left.\begin{aligned}
F(x) &= \sin k(x + \beta) \\
\text{and}\quad G(t) &= b \sin k(ct + \gamma)
\end{aligned}\right\} \tag{11.36}$$

where b, β and γ are arbitrary constants.† With suitable choices for these, we can reproduce the previously stated separable solutions (11.31) and (11.31′).

We now see why the separation constant in eqn (11.35) was chosen to be negative. This was to obtain harmonic solutions for the x and for the t behaviour of the solutions. A positive constant would have resulted in exponentials, which are physically not useful for most wave problems. In general, the choice of the constant of separation is dictated by the boundary conditions for the particular problem (see also Section 11.3.2).

It is worth reemphasising that the most important step in the argument is that which results in each side of eqn (11.34′) being set equal to a constant. In answering problems, and even more so for examination questions, it is absolutely necessary to include a couple of sentences explaining how this occurs. Not only does this enable you to convince the examiner that you understand what you are doing, it can also help you avoid making mistakes.

---

† A possible constant a multiplying the $\sin k(x + \beta)$ of $F(x)$ would result in a factor ab in the solution $F(x)G(t)$ of the differential equation. Hence only the combination ab is of any significance. It is thus simplest arbitrarily to take a as unity.

11.3.2 Application to the heat conduction equation

We now use the method of the separation of variables to find solutions of the heat conduction equation (11.5). As before, the trial solution is written as a product of separate functions of the independent variables

$$T(x,t) = C(x)D(t) \tag{11.37}$$

Then the procedure of differentiating with respect to x twice and with respect to t once, substituting in the partial differential equation and dividing by the solution T yields

$$\frac{1}{C}\frac{d^2C}{dx^2} = \frac{s\rho}{\kappa D}\frac{dD}{dt} \tag{11.38}$$

The usual form of words then results in each side being equal to a constant. Here there are several interesting possibilities for how it is chosen, corresponding to different types of solutions, as determined by the boundary conditions of the problem.

11.3.2.1 Constant $= -m^2$

We first investigate setting the constant equal to $-m^2$. Then the solutions of

$$\left. \begin{aligned} \frac{d^2C}{dx^2} &= -Cm^2 \\ \text{and} \quad \frac{dD}{dt} &= -D\frac{\kappa m^2}{s\rho} \end{aligned} \right\} \tag{11.39}$$

are

$$\left. \begin{aligned} C(x) &= a\sin(mx + \beta) \\ \text{and} \quad D(t) &= e^{(-\kappa m^2/s\rho)t} \end{aligned} \right\} \tag{11.40}$$

The harmonic form of C as compared with the exponential for D arises from the fact that the ordinary differential equations for C and D are second and first order, respectively. Another feature of the solutions, common to that for other partial differential equations, is that the scale of the x distribution (whose wavelength is $2\pi/m$) is related to that of the t distribution (whose exponential decay time is $s\rho/\kappa m^2$); this arises because the same constant m occurs in both the ordinary differential equations of (11.39).

Thus we have a solution of the form

$$T(x,t) = a\sin(mx + \beta)e^{-(\kappa m^2/s\rho)t} \tag{11.41}$$

If, for example, $\beta = 0$, this gives the temperature at $t = 0$ as

$$T(x,0) = a\sin(mx) \tag{11.42}$$

Then (11.41) (with $\beta = 0$) is the suitable solution for the temperature of an infinite rod for all values of x and t, when its initial distribution is specified by (11.42). Alternatively we could have a rod of length L, with ends at $x = 0$ and $x = L$ maintained at 0 °C, for example by being in contact with iced water, and with an initial temperature distribution (11.42), where $m = \pi/L$. Then according to eqn (11.41), at all subsequent times the x dependence of the temperature remains the same except that its amplitude is reduced everywhere by the exponential factor, with time constant $s\rho L^2/\pi^2\kappa$.

A sinusoidal distribution for the initial temperature sounds fairly unrealistic. How do we proceed if the temperature distribution is something else? For example, for the rod of length L,

$$T(x,0) = \begin{cases} \alpha x & \text{for } x < L/2 \\ \alpha(L-x) & \text{for } x \geq L/2 \end{cases} \tag{11.43}$$

The solution now is no longer separable. However, we need not give up this approach in despair and look for a completely different method of solution. All that is necessary is to make use of the linearity of the heat conduction equation to write

$$T(x,t) = \sum_m a_m \sin(mx + \beta_m)e^{-\left(\kappa m^2/s\rho\right)t} \tag{11.44}$$

where a_m and β_m are arbitrary constants which can differ for each possible value of m.

For a rod with ends at $x = 0$ and $x = L$ kept at $T = 0$ °C, the required values for m are just $n\pi/L$, where n is any positive integer. Also the β_m values are all zero. Finally the a_m are chosen so that the initial temperature distribution of eqn (11.43) is reproduced by eqn (11.44) with t set to zero, i.e.

$$T(x,0) = \Sigma a_m \sin mx \tag{11.44'}$$

The extraction of the coefficients a_m from eqn (11.44') is discussed in Chapter 12 on Fourier series.

The case where the ends of the rod are maintained at different temperatures requires also the term corresponding to $m = 0$ (see the next section).

11.3.2.2 Constant = 0

At first sight this appears to be just a special case of the above. However, the solution cannot be obtained simply by substituting $m = 0$ in eqn (11.44). Instead it is necessary to go back to the ordinary differential eqns (11.39), with $m = 0$:

$$\left. \begin{aligned} \frac{d^2C}{dx^2} &= 0 \\ \text{and} \quad \frac{dD}{dt} &= 0 \end{aligned} \right\} \tag{11.39'}$$

and solve them anew.† This is completely trivial and gives

$$\left. \begin{aligned} C &= ax + b \\ \text{and} \quad D &= 1 \end{aligned} \right\}$$

This is linear in x, but completely independent of t. It thus corresponds to an equilibrium (time-independent) solution of the heat conduction equation, with a temperature distribution of constant gradient along the rod. In order to maintain this temperature distribution along the rod, it is necessary to provide a source of heat at the hot end, and to remove heat from the cool end.

For a rod with a specified initial temperature distribution (see Section 11.3.2.1) but with its ends maintained at different temperatures, the $m = 0$ term (i.e. $C(x)D(t) = ax + b$) must be added to the right-hand side of eqn (11.44). Then a and b are obtained from the temperatures at the two ends, and the Fourier coefficients a_m and β_m are determined for the function $T(x, 0) - ax - b$.

At this point, it is worth pausing to consider this method of approach to solving partial differential equations. What we are doing is to choose a constant, and then discover for what sort of problem this provides a solution. (There will be other examples of this in Sections 11.3.3 and 11.3.4.) Of course, in a standard problem, things happen the other way round, in that some boundary conditions are specified, and we have to find the appropriate solution. However, by first studying which type of constant is useful for which problem, we should be capable of choosing the appropriate constant in any given case.

---

† In solving the separated equations derived from a partial differential equation, it is a common feature that the case of $m = 0$ needs to be investigated separately.

$$11.3.2.3 \quad Constant = +m^2$$

With the constant as $+m^2$, the solutions are

$$\left.\begin{array}{c} C = ae^{mx} + be^{-mx} \\ \text{and} \quad D = e^{(\kappa m^2/s\rho)t} \end{array}\right\} \qquad (11.45)$$

Thus the x dependence involves growing and decaying exponentials (or equivalently sinh and cosh), and the time dependence is a growing exponential. In order to maintain these growing temperatures at the ends of the rod, it is necessary to input heat at the hotter end (and either to input or to extract heat at the cooler end, depending on the sign of the temperature's spatial gradient there). Furthermore, these heat transfers at the ends must also increase in magnitude exponentially with time. Clearly this cannot continue indefinitely, since the rod will melt. Indeed the physical situation is rather artificial, and the choice of the constant as $+m^2$ is useful only very rarely.

11.3.2.4 Harmonic time variations

If a long conducting rod has one of its ends heated, the region of higher temperature gradually extends as the heat is conducted from that end into the body of the rod. If instead the end is heated and cooled periodically, thermal waves are conducted away from the end. This situation is not too different from the way large rocks on the earth have their surface temperatures varying with a period of 24 hours (they are heated by the sun during the day, but not at night), and also with a period of 1 year (the sun is more effective in summer than in winter); then we may be interested in how the temperature varies within the rock.

Clearly the heat conduction equation should provide the answer to this problem. What is needed is to find a solution of the equations

$$\left.\begin{array}{c} \dfrac{d^2C}{dx^2} = C \times \text{constant} \\ \text{and} \quad \dfrac{dD}{dt} = D\dfrac{\kappa}{s\rho} \times \text{constant} \end{array}\right\} \qquad (11.39'')$$

where $D(t)$ is harmonic.

Since the differential equation for D is only first order, this requires the constant to be chosen as imaginary — say im. Then

$$\frac{dD}{dt} = im\frac{D\kappa}{s\rho} \qquad (11.46)$$

yields

$$D(t) = e^{[(i\kappa m/s\rho)(t+\gamma)]} \qquad (11.47)$$

where γ is an arbitrary time offset; and there is the clear implication that we eventually need to extract the real (or the imaginary) part of the final answer.† The constant m is chosen to yield the correct frequency for the harmonic temperature variation as specified in the boundary condition describing how the temperature at the end of the bar is changing with time.

Of course it is necessary to insert the same constant into the first differential equation of (11.39″). Thus

$$\frac{d^2C}{dx^2} = Cim \qquad (11.48)$$

with solution

$$C = ae^{(\sqrt{im}\,x)} \qquad (11.49)$$

This involves the square root of i, but this presents no special problem as

$$\sqrt{i} = \pm(i+1)/\sqrt{2}$$

(see Section 4.2.5). Thus

$$C(x) = be^{(i+1)\sqrt{m/2}x} + ce^{-(i+1)\sqrt{m/2}x}$$

where b and c are arbitrary constants.

On multiplying by $D(t)$ from (11.47) with γ set equal to zero, we obtain

$$
\begin{aligned}
T(x,t) &= C(x)D(t) \\
&= be^{+\sqrt{m/2}x}e^{i(\frac{\kappa m}{s\rho}t+\sqrt{m/2}x)} \\
&\quad + ce^{-\sqrt{m/2}x}e^{i(\frac{\kappa m}{s\rho}t-\sqrt{m/2}x)}
\end{aligned}
\qquad (11.50)
$$

Now the imaginary exponentials will give cosines or sines when we take the real or the imaginary part of this expression, while the real exponentials give terms which grow or die with increasing x (for the positive and negative exponents, respectively). If the rod extends along positive x and the end being heated is at $x = 0$, the solution which grows

---

† There is actually some degree of subtlety in going from the complex solution of our partial differential equation to the real one appropriate to an actual physical problem. The simplest approach is to regard the complex function $C(x)D(t)$ that is obtained in (11.50) as a solution to the heat conduction equation. However, since this is complex but the differential equation is real, then the real and the imaginary parts of the product $C(x)D(t)$ are both also valid solutions. Which we choose depends on the appropriate boundary conditions.

with x is unphysical, and so we set $b = 0$ in eqn (11.50). Finally on taking the real part, we obtain the solution as

$$T(x, t) = ce^{-\sqrt{m/2}x} \cos\left(\frac{\kappa m}{s\rho}t - \sqrt{\frac{m}{2}}x\right) \qquad (11.50')$$

$$= ce^{-\beta x} \cos(\beta x - \omega t)$$

where $\omega = \kappa m / s\rho$ is the angular frequency of the temperature variations, and

$$\beta^2 = m/2 = s\rho\omega/2\kappa$$

Alternatively the imaginary part gives the same expression but with the cosine replaced by a sine. Which combination we select would depend simply on the choice of the time origin for the temperature variations at $x = 0$ (i.e. on the value of ϕ when the harmonically changing temperature at the end is written as $\cos(\omega t + \phi)$).

The form of the solution (11.50′) is worthy of some study. Just as for the solutions of the wave equation, the cosine term corresponds to a wave, here of temperature variations. Because of the relative minus sign between the t and the x terms, this wave propagates to the right, which is fortunate because the rod extends from $x = 0$ (where it was being heated and cooled) to the right. The speed v of the waves is given by the ratio of the relative changes Δx and Δt required to keep the phase of the wave constant, i.e.

$$v = \frac{\Delta x}{\Delta t} = \frac{\omega}{\beta} = \sqrt{\frac{2\kappa\omega}{s\rho}} \qquad (11.51)$$

Thus in contrast to the solution of the wave equation (11.10), here the waves travel with a speed which increases with the frequency of the temperature variations.

Another difference from the waves on a string is that here the waves have an amplitude which decreases exponentially with distance x (see fig. 11.3). Indeed the x-dependent factors of the exponential decrease and of the sinusoidal variation are the same, which results in the amplitude of the wave being decreased after one wavelength by a factor of $e^{-2\pi} \approx 0.002$. This is indeed severe damping.

An interesting feature is apparent from fig. 11.3. Even when the temperature at the end of the rod is a maximum† as in fig. 11.3(c), there

---

† The maximum is for the temperature considered as a function of time. At $t = 0$, the temperature as given by eqn (11.50′) is largest at $x = 0$, but is not a mathematical maximum there, in the sense that $\partial T / \partial x \neq 0$. This is because of the effect of the exponential decrease in x caused by the $\exp(-\sqrt{m/2}x)$ term (see also the remarks below eqn (11.52)).

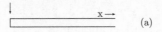

(a)

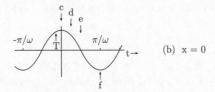

(b) x = 0

(c) ωt = 0

(d) ωt = π/4

(e) ωt = π/2

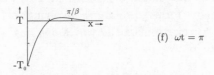

(f) ωt = π

Fig. 11.3. (a) A thermally insulated rod, extending from $x = 0$ along the x axis to infinity. The end of the rod at the origin is subjected to harmonic temperature variations. The arrow shows the position at which the time dependence of the temperature is plotted in (b). (b) The temperature T at $x = 0$ plotted as a function of time. The letters c—f show the times at which the temperature distributions in the rod are plotted below. (c) The temperature T at $t = 0$ plotted as a function of position x along the rod. The temperature variations at $x > 0$ are a result of the variations at $x = 0$ and $t < 0$ (see (b)) being conducted into the rod. The temperature at $x = 3\pi/4\beta$ is a minimum when the temperature at $x = 0$ is at its greatest. (d)–(f) The temperature variation along the rod at subsequent times. As compared with earlier times, the thermal wave has moved to the right. The temperature peak in (d) ($T_0/\sqrt{2}$ at $x = 0$) has been conducted in as far as $x = 3\pi/4\beta$ by $t = \pi/\omega$ (see (f)) in a very attenuated form.

is a position at positive x where the temperature is below the ambient temperature. This is where the minimum at $x = 0$ (from the earlier time $t = -\pi/\omega$) has been conducted to at time $t = 0$. Differentiation of the expression (11.50′) (with t set equal to zero) shows that this occurs at

$$\beta x = \sqrt{s\rho\omega/2\kappa}\,x = 3\pi/4 \qquad (11.52)$$

The naive guess that expression (11.50′) minimises as a function of x when the cosine term is -1 is incorrect; the exponential decrease is so fast that it pulls the minimum towards a significantly smaller x. Otherwise the right-hand side of eqn (11.52) would have been π, rather than $3\pi/4$.

Eqn (11.52) shows that the distance at which this "temperature inversion" takes place is inversely proportional to the square root of the frequency; it is thus larger for a slow variation (e.g. yearly) as compared with a faster one (e.g. daily). This is because, although the speed of the high frequency waves is larger (see eqn (11.51)), they have had less time to travel.

Because of the exponential damping factor, the temperature at this minimum is reduced by a factor of $e^{-3\pi/4} \approx 0.095$ as compared with the amplitude of the temperature variations at $x = 0$. The minimum is thus not very deep.

This type of solution to the heat conduction equation is used by kangaroos sitting down on the sand in the Australian desert (see Problem 11.5). Completely analogous forms of solutions involving very strongly damped waves are found for electromagnetic waves entering a medium of high electrical conductivity.

11.3.2.5 Discrete or continuous constants

In Sections 11.3.2.1–11.3.2.4, we have discussed the choice of the nature of the arbitrary constant of separation m, i.e. whether it is negative, zero, positive or imaginary. Having made that choice, we still have to select the particular values of m that are relevant for our particular problem.†️ Again this is determined by the boundary conditions.

If the boundary conditions are particularly favourable, a single value of m provides the complete solution. Thus in Section 11.3.2.1 we found that eqn (11.41) satisfies the heat conduction equation. Then if at $t = 0$, the temperature distribution is

$$T(x,0) = T_0 \sin(2\pi x/L) \qquad (11.53)$$

---

† Unless $m = 0$, in which case the choice is already made.

and the ends of the rod at $x = 0$ and $x = L$ are maintained at a temperature of 0 °C, a choice of $m = 2\pi/L$ and $\beta = 0$ in (11.41) provides the unique solution to the heat conduction equation which satisfies the boundary conditions. We end up with just a single value of m because the initial condition (11.53) was especially simple (in relation to the functional form of the separated solution of the partial differential equation).

If the same rod with its ends maintained at 0 °C has a non-sinusoidal initial temperature distribution $T(x, t = 0)$, it will be necessary to use not just one m but an infinity of terms with discrete values of m (and which have a constant separation). This was discussed briefly in Section 11.3.2.1, and is returned to at length in Chapter 12.

If instead of being of length L, the rod is infinite with an initial temperature distribution which repeats after a fixed distance d (i.e. $T(x, t = 0) = T(x + d, t = 0)$), the required values of m are again discrete. However, if $T(x, t = 0)$ does not repeat *ad infinitum*, it is necessary to use a continuous distribution of m values to solve the problem. Thus instead of eqn (11.44), we need to replace the summation by an integral, so

$$T(x, t) = \int a(m) \sin[mx + \beta(m)] e^{-(\kappa m^2/s\rho)t} dm \qquad (11.54)$$

where $a(m)$ is the function of m which specifies the amplitude of each of the basic spatial sine waves required in the solution, and $\beta(m)$ gives the phase for each value of m. This then is Fourier analysis (see Section 12.12), rather than Fourier series.

Before leaving the heat conduction equation for the time being, it is worth recalling that in this section we have been considering problems with only one spatial dimension x. In more realistic situations, it may well be necessary to consider the full three-dimensional equation (11.6). This then will have an even richer variety of solutions.

11.3.3 Application to Laplace's Equation

We now apply the method of separating the variables to Laplace's Equation. We shall do this twice, first using the variables x, y and z, and then in polar coordinates. These are not merely useful exercises, but bring out interesting features of the mathematics and of the physics.

Since, in the special case where the time derivative is zero, the heat diffusion equation reduces to Laplace's Equation, the following analysis is also relevant for the equilibrium temperature distribution within a semi-infinite thermally conducting bar with rectangular cross-section, when the temperatures on the boundaries are as specified below.

11.3.3.1 Cartesian coordinates

Laplace's Equation in Cartesian coordinates is

$$\frac{\partial^2 V}{\partial x^2} + \frac{\partial^2 V}{\partial y^2} + \frac{\partial^2 V}{\partial z^2} = 0 \tag{11.15'}$$

We now look for a solution which is separable in the variables x, y and z, i.e.

$$V(x, y, z) = X(x)Y(y)Z(z) \tag{11.55}$$

where X, Y and Z are each functions of only one variable. After inserting eqn (11.55) into (11.15') and dividing through by V, we obtain

$$\frac{1}{X}\frac{d^2 X}{dx^2} = -\left(\frac{1}{Y}\frac{d^2 Y}{dy^2} + \frac{1}{Z}\frac{d^2 Z}{dz^2}\right) \tag{11.56}$$

Since there are now three functions to be determined and only one equation relating them, we have to invoke the special argument (about functions of different variables being equal implying that each is a constant) twice. Thus the left-hand side of equation (11.56) is a function of x only, while the rest is a function of y and z; we set them equal to a constant,† say $-m^2$. Then

$$\left. \begin{aligned} \frac{d^2 X}{dx^2} &= -m^2 X \\ \text{and} \quad \frac{1}{Y}\frac{d^2 Y}{dy^2} &= m^2 - \frac{1}{Z}\frac{d^2 Z}{dz^2} \end{aligned} \right\} \tag{11.57}$$

Finally the second equation of (11.57) involves a function of y being equal to a function of z; we put each of them equal to $-n^2$. This results in the following set of ordinary differential equations:

$$\left. \begin{aligned} \frac{d^2 X}{dx^2} &= -m^2 X \\ \frac{d^2 Y}{dy^2} &= -n^2 Y \\ \text{and} \quad \frac{d^2 Z}{dz^2} &= (m^2 + n^2)Z \end{aligned} \right\} \tag{11.58}$$

† Here too we can choose positive or negative constants depending on the form of the particular type of function that we need in order to satisfy the boundary conditions of the specific problem.

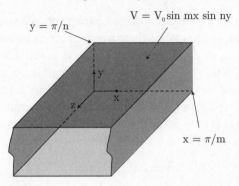

Fig. 11.4. The potentials on the semi-infinite walls of the box are maintained at zero, while that on the end at $z = 0$ is $V_0 \sin(mx) \sin(ny)$. The potential throughout the box is then given by eqn (11.59′).

The solutions of these equations can be written down by inspection, so that we obtain

$$V = XYZ$$
$$= V_0 [a \sin mx + (1 - a) \cos mx][b \sin ny + (1 - b) \cos ny]$$
$$\times \left[c e^{-\sqrt{m^2+n^2}z} + (1 - c)e^{+\sqrt{m^2+n^2}z} \right] \qquad (11.59)$$

With the choice $a = b = c = 1$, this thus yields

$$V(x, y, z) = V_0 \sin(mx) \sin(ny) e^{-\sqrt{m^2+n^2}z} \qquad (11.59')$$

This could be, for example, the potential within an infinitely long rectangular box (see fig. 11.4) whose walls at $x = 0$ and π/m, and at $y = 0$ and π/n, are all kept at zero potential, while the potential on the wall at $z = 0$ is maintained as $V_0 \sin mx \sin ny$. The term with the positive exponential in z has been set to zero, so that V remains finite in the whole of the physically relevant region (in particular, as z tends to $+\infty$).

A characteristic feature of the solution is that the x and the y dependences of the potential determine the way it varies with z. This is completely analogous to the fact that in solutions of the wave equation, the space dependence (e.g. the wavelength) is related to that in time (e.g. the frequency) by the speed of the waves. Also for the heat conduction equation, the spatial variation of the temperature distribution on a rod determines how quickly it dies away in time (see eqn (11.41)).

Eqn (11.59) is the general separable solution of this type. However, we can produce even more general solutions by taking linear combinations of separable solutions. (Similar remarks apply to solutions of the wave equation in Section 11.3.1. See also the paragraph below eqn (11.75) for a fuller discussion.) Thus since there are two independent functions for each of x, y and z in eqn (11.59), we can have eight independent terms (e.g. $\sin(mx)\cos(ny)e^{+\sqrt{m^2+n^2}z}$), each with its own arbitrary amplitude. Any linear combination of these is a solution of Laplace's Equation; it is not separable in x, y and z, but is made out of separable terms. The separable function V of eqn (11.59) is more restrictive in that it contains only four arbitrary coefficients (V_0, a, b and c). Thus an acceptable solution

$$V = \sin(mx)e^{mz} + \cos(mx)e^{-mz}$$

cannot be obtained from (11.59) for any choice of the constants.

11.3.3.2 Two-dimensional polar coordinates

Because of the boundary conditions of a specific problem, it may be useful to solve Laplace's Equation in polar coordinates. For example, the potential V might be specified along lines where θ is constant and circles where r is constant. We thus start with Laplace's Equation

$$\frac{1}{r}\frac{\partial}{\partial r}\left(r\frac{\partial V}{\partial r}\right) + \frac{1}{r^2}\frac{\partial^2 V}{\partial \theta^2} = 0 \tag{11.25'}$$

in terms of r and θ, and look for a solution which is separable in these variables

$$V(r,\theta) = A(r)B(\theta) \tag{11.60}$$

Substitution of this form of solution into (11.25') yields

$$\frac{1}{Ar}\frac{d}{dr}\left(r\frac{dA}{dr}\right) + \frac{1}{Br^2}\frac{d^2B}{d\theta^2} = 0 \tag{11.61}$$

Before using the magic words which result in each term being set equal to a constant, it is essential to multiply the equation throughout by r^2, since as it stands the second term contains a $1/r^2$ factor, and so is not yet a function only of θ. Then

$$\frac{r}{A}\frac{d}{dr}\left(r\frac{dA}{dr}\right) = -\frac{1}{B}\frac{d^2B}{d\theta^2} = +m^2 \tag{11.62}$$

where the constant of separation is chosen as positive in order to provide the required form of solutions for θ (see immediately below).

The ordinary differential equation for B is readily solved to give

$$B(\theta) = a[b\cos m\theta + (1-b)\sin m\theta] \tag{11.63}$$

If this solution is to apply for all values of θ, the requirement that it is single valued then forces m to be integral, since the function at a given θ must be identical with that at $2\pi + \theta$, i.e.

$$V(r,\theta) = V(r, 2\pi + \theta) \tag{11.64}$$

However, if we are interested in a restricted region, say for θ between 10° and 152°, then m does not necessarily have to be integral. This is because the solution has meaning only in the restricted region of angle, and it cannot be continued all the way round in θ to return to the starting point.

The ordinary differential equation for A can be rewritten as

$$r^2 \frac{d^2 A}{dr^2} + r \frac{dA}{dr} - m^2 A = 0 \tag{11.62'}$$

This is not the standard second order differential equation discussed at length in Chapter 5, in that the coefficients multiplying A and its derivatives in (11.62') are not constants, but are functions of r. We then need some other method of solution, which should be remembered.

Most textbooks recommend changing the independent variable from r to

$$w = \ln r \tag{11.65}$$

Then eqn (11.62') simplifies to

$$\frac{d^2 A}{dw^2} = m^2 A \tag{11.66}$$

with solutions

$$A = e^{\pm mw} = r^{\pm m} \tag{11.67}$$

However, it requires several lines of algebra to derive (11.66) from (11.62'). Rather than the seemingly arbitrary (but successful) change of variable (11.65), it is equally valid to look (again arbitrarily, but successfully) for solutions of the form

$$A(r) = r^\alpha \tag{11.68}$$

Substitution of this into (11.62') immediately gives

$$\alpha^2 = m^2 \tag{11.69}$$

so that the solutions are as stated in (11.67).

Then a solution of Laplace's Equation is given by the r dependence as specified in (11.67), multiplied by the function of θ given in (11.63), with the same value of m used in each.

We still have to consider the special case $m = 0$. The solution of eqn (11.62) for B now yields

$$B(\theta) = d + e\theta \tag{11.70}$$

where d and e are constants. For the case where $B(\theta)$ is required to be single-valued, e must be zero.

For the r dependence, we solve (11.62') with $m = 0$. This is achieved by noting that, with the substitution

$$h = \frac{dA}{dr} \tag{11.71}$$

eqn (11.62') becomes

$$\frac{dh}{dr} = -\frac{h}{r} \tag{11.72}$$

which can be solved as the variables are separable (see Section 5.3). Then this solution is substituted in (11.71), which yields

$$A(r) = f + g \ln r \tag{11.73}$$

Alternatively, if the substitution (11.65) has been used, (11.66) gives

$$A = f + gw \tag{11.74}$$

which results in the same solution (11.73). This then has to be combined with (11.70) to give $V(r, \theta)$ for the case where $m = 0$.

At last we can write the general solution of the problem (for the case where $V(r, \theta)$ is required to be single-valued for all θ) as

$$V(r, \theta) = (f + g \ln r) + \Sigma\{d_m r^m [e_m \cos m\theta + (1 - e_m) \sin m\theta]$$
$$+ f_m r^{-m} [g_m \cos m\theta + (1 - g_m) \sin m\theta]\} \tag{11.75}$$

where the summation extends over positive integral values of m.

A rather inelegant subtlety has been introduced at this stage (compare the discussion at the end of Section 11.3.3.1). If solutions of the type (11.63) and (11.67) for $B(\theta)$ and for $A(r)$ are multiplied and then summed over m, we obtain

$$V(r, \theta) = (f + g \ln r)$$
$$+ \Sigma a_m [b_m \cos m\theta + (1 - b_m) \sin m\theta][c_m r^m + (1 - c_m) r^{-m}] \tag{11.75'}$$

This functional form, however, is not completely adequate; for example, it forces the ratio of the $\sin m\theta$ and the $\cos m\theta$ factors to be the same $((1 - b_m)/b_m)$ for the r^m and the r^{-m} terms. In contrast eqn (11.75) has these ratios as $(1 - e_m)/e_m$ and $(1 - g_m)/g_m$, which are completely independent. There is absolutely no reason mathematically for these ratios to be identical, as each term of (11.75) itself satisfies Laplace's Equation.†

Furthermore, as mentioned later in this section, it turns out that physically the negative powers of r correspond to various arrangements of electrical charges near the origin (e.g. dipoles), while the positive powers are suitable for describing the potential due to configurations of charges at infinity (e.g. constant fields); in each case the angular factors describe the orientation of these charge arrangements. There is again no need for these angular dependences to be related. Once again we conclude that we need the more general form of eqn (11.75).

For example, the solution of Laplace's Equation in two dimensions which behaves like a dipole term $(\mu \sin \theta)/r$ at small r, and as a constant electric field $Ex = Er \cos \theta$ at large r is simply

$$V = \frac{\mu}{r} \sin \theta + Er \cos \theta$$

There is no way the coefficients of the incomplete form (11.75′) can be chosen to reproduce this. However, with the correct solution (11.75), we merely need

$$\left.\begin{array}{l} e_1 = 1 \\ d_1 = E \\ g_1 = 0 \\ f_1 = \mu \end{array}\right\}$$

and all other coefficients are zero.

Similar care is needed in constructing the general solution for Laplace's Equation in other coordinates (e.g. x, y and z as discussed in Section 11.3.3.1; and spherical polar coordinates in Section 11.3.3.3) and for the wave equation. The one-dimensional heat conduction equation (11.5) does not have this complication, because it is only first order in t.

Eqn (11.75) provides a general solution of Laplace's Equation, but contains many arbitrary constants, whose values are determined by the

---

† There is an alternative argument for obtaining the more general solution (11.75) from the more restrictive form (11.75′). Because the differential equation is linear, the individual separated terms satisfying Laplace's Equation can be summed to give a valid solution. The sum extends over integral values of m, if we want the solution to be single-valued in θ. However, there is no particular reason why each integral value of m should occur only once in the summation. Repeated values would not be interesting if they produced only the same functional form. However, a pair of solutions with the same m but with different values for the coefficients enables us to write the solution in the more general form (11.75).

specific details of each individual problem. Thus if we are looking for a solution within a finite region containing the origin, we cannot have terms that diverge as r tends to zero, and so those that behave like $\ln r$ or r^{-m} must be omitted.† Thus g and all the f_m have to be zero. Alternatively, if the region over which the solution must be valid excludes the origin but extends out to infinity, then g and the d_m must be zero in order to avoid divergence at large r.

Finally we look at the physical significance of some of the terms in eqn (11.74). Since this is the solution of Laplace's Equation, we expect to find functional dependences corresponding to familiar electrostatic phenomena, such as a point charge, an electric dipole or quadrupole or higher multipole moments, or a constant electric field (such as exists between the plates of a large capacitor).

The well-known potential due to a point charge q is

$$V = \frac{1}{4\pi\epsilon_0}\frac{q}{r} \tag{11.76}$$

No term in eqn (11.75) corresponds to this, since r^{-1} occurs with a $g_1\cos\theta + (1 - g_1)\sin\theta$ dependence, while (11.76) is independent of θ. Similarly the potentials for other electrostatic multipoles do not correspond to the other terms in (11.75). This is at first disconcerting. However, it is important to remember that (11.75) is the solution of Laplace's Equation in two dimensions, while all the well-known forms apply in three dimensions.

Thus, for example, in two dimensions, the electric field **E** from a point charge falls off like r^{-1} (rather than r^{-2}), simply because the lines of force spread out over circular regions rather than over spheres. Then the potential, whose derivative is $-\mathbf{E}$, is $-q \ln r/4\pi\epsilon_0$, and corresponds to the logarithmic term in (11.75).

Similarly a dipole has an $r^{-1}[g_1\cos\theta + (1 - g_1)\sin\theta]$ dependence, and the $(n + 1)$-fold multipole gives rise to the term

$$V(r,\theta) = f_n r^{-n}[g_n\cos n\theta + (1 - g_n)\sin n\theta] \tag{11.77}$$

Of the other terms in (11.75), the constant is fairly innocuous, in that it just affects the overall level of the potential everywhere; it is

---

† A charge at the origin produces a potential that diverges there, which appears to violate this statement. However, this situation is excluded because we are considering cases where Laplace's Equation applies throughout the whole region of interest (Laplace's Equation is valid only in regions where there are no charges — compare eqns (11.14) and (11.15)).

the derivative of V which gives the electric field, which is more directly related to physically measurable effects.

The terms involving r^{+1} are $r \cos \theta = x$ and $r \sin \theta = y$. The potential has these forms for a uniform field in the x and in the y directions respectively. With the constant g_1 different from zero or unity, the field can be in an arbitrary direction in the two-dimensional space.

The terms with higher positive powers of r correspond to other configurations of charges at large distances, which become more and more artificial from a physical viewpoint.

11.3.3.3 Spherical polar coordinates

In this section, we merely sketch the solutions of Laplace's Equation in three-dimensional spherical polar coordinates.

We look for a solution of eqn (11.15) (with $\nabla^2 V$ being given by (11.28)) of the form

$$V(r, \theta, \phi) = A(r)B(\theta)C(\phi) \tag{11.78}$$

After performing the separation of variables in two stages, we obtain the ordinary differential equations

$$\left. \begin{aligned} r^2 \frac{d^2 A}{dr^2} + 2r \frac{dA}{dr} - l(l+1)A &= 0 \\ \sin^2 \theta \frac{d^2 B}{d\theta^2} + \sin \theta \cos \theta \frac{dB}{d\theta} + B[l(l+1)\sin^2 \theta - m^2] &= 0 \\ \text{and} \quad \frac{d^2 C}{d\phi^2} &= -m^2 C \end{aligned} \right\} \tag{11.79}$$

where the constants of separation have (somewhat surprisingly) been chosen as $l(l+1)$ and $-m^2$. The first and third of these equations readily give

$$A = ar^l + br^{-(l+1)}$$

and

$$C = \begin{cases} c \cos m\phi + (1-c) \sin m\phi & \text{if} \quad m \neq 0 \\ d + (1-d)\phi & \text{if} \quad m = 0 \end{cases}$$

The requirement of C being single valued as a function of ϕ results in $d = 1$ (for $m = 0$), or m being integral otherwise.

The equation for B is not trivial, but its solutions are known. For $m = 0$ they are called Legendre Polynomials $P_l(\cos \theta)$, and for $m \neq 0$

they are associated Legendre functions† $Y_l^m(\cos\theta)$. In either case, for the solution to be meaningful, l is required to be integral (hence the rather curious choice of $l(l+1)$ for the separation constant), and $l \geq |m|$. The P_l and Y_l^m are polynomial functions of $\cos\theta$ and/or $\sin\theta$ with powers not higher than l.

Then the solution of Laplace's Equation that is well behaved for all angles‡ is

$$V(r,\theta,\phi) = \sum_l \sum_m \{a_{l,m}r^l[c_{l,m}\cos m\phi + (1 - c_{l,m})\sin m\phi]$$
$$+ b_{l,m}r^{-(l+1)}[c'_{l,m}\cos m\phi + (1 - c'_{l,m})\sin m\phi]\}Y_l^m(\cos\theta)$$
$$(11.80)$$

where the summations extend over integral values ($l \geq 0$ and $|m| \leq l$). This has application to a very wide range of physical problems. As just one example, the terms with negative powers of r in the above expression are identifiable with the potential produced by the various electrostatic multipoles, situated at the origin.

11.3.4 Application to the wave equation

The wave equation (11.10) with one space dimension was already used in Section 11.3.1 to illustrate the separation of variables. The typical solution obtained there was of the form

$$z(x,t) = a\sin kx \cos kct \qquad (11.81)$$

This is a standing wave in that there are positions ($x = n\pi/k$) at which the displacement z is zero at all times.

For boundary conditions (e.g. $z(x)$ and $\partial z/\partial t$ at $t = 0$) that are suitably simple, a separated solution can be chosen so that the boundary conditions are satisfied. Otherwise it is necessary to use a combination of terms of different wavelengths. As with Laplace's Equation, it is necessary to write this combination in a sufficiently general way (compare the remarks below eqn (11.75)). The amplitude of each of the components is determined by the method of Fourier series or Fourier analysis as

---

† Because the differential equation for B in (11.79) is second order, there are two independent solutions for each l and m. However, the other ones ($Q_l^m(\cos\theta)$, Legendre Polynomials of the second kind) diverge for $\cos\theta = 1$.

‡ In going from the separable function $A(r)B(\theta)C(\phi)$ to the sum of separable terms, we have again written the summation to be as general as possible.

appropriate (compare the use of these methods for heat conduction problems).

The wave equation also has solutions of travelling waves. These are obtained by combining two different standing waves with different phases. Thus

$$z = a \sin kx \cos kct - a \cos kx \sin kct$$
$$= a \sin k(x - ct) \tag{11.82}$$

is a wave travelling to the right (see also Chapter 14).

The solutions of the wave equation become more interesting when we go to more than one spatial dimension, e.g. eqn (11.11) for longitudinal waves. Because of the separation in the variables, however, each single solution will be a stationary wave.

If ∇^2 in eqn (11.11) is expressed in terms of x, y and z, then there are completely separable solutions of the form

$$w = a \sin k_x x \sin k_y y \sin k_z z \sin \omega t$$

where

$$c^2(k_x^2 + k_y^2 + k_z^2) = \omega^2$$

and where any of the sine terms can be replaced by a cosine. Similarly there are partially separated solutions (which can be obtained as suitable combinations of the completely separated ones) of the type

$$w = a \cos(k_x x + k_y y + k_z z) \cos \omega t$$

This is a stationary plane wave, with w being constant at any given t along a plane perpendicular to the direction (k_x, k_y, k_z).

In spherical polar coordinates, the separated types of solutions correspond to waves centred on the origin (but still stationary). Their angular dependence is given by the associated Legendre functions $P_l^m(\cos\theta)\, e^{\pm im\phi}$, just as for Laplace's Equation in spherical polar coordinates. Here, however, the radial dependence is determined by the equation

$$r^2 \frac{d^2 A}{dr^2} + 2r \frac{dA}{dr} + \left[\frac{\omega^2 r^2}{c^2} - l(l+1) \right] A = 0$$

whose solutions are related to Bessel functions. Further discussion of this is beyond the scope of this text.

11.3.5 *Schrödinger's Equation*

This is an exceedingly important equation with applications to a wide range of physical problems. It determines the behaviour of non-relativistic electrons (and some other fundamental particles too) in a potential $V(x, y, z)$. It is

$$-\frac{\hbar^2}{2m}\nabla^2\psi + V\psi = i\hbar\frac{\partial\psi}{\partial t} \qquad (11.83)$$

where $\hbar$ is Planck's constant divided by 2π, m is the mass of the electron, and the solution of the equation ψ is known as the wave function. It describes the electron, in that $|\psi(x, y, z, t)|^2$ gives the probability of finding the electron at the specified coordinates. It is a basic concept of quantum mechanics that ψ contains all the information about the electron that it is in principle possible to know. It is not that the probability is just temporarily covering up information that could perhaps be revealed by more detailed investigation.

What defines a particular problem is the potential V, and the boundary conditions that ψ must obey. Furthermore, both ψ and its gradient must be continuous and single-valued, and it should not be infinite.

An important class of solutions is when the time dependence of ψ is separable from the spatial part, and is given by

$$e^{-i\omega t} = e^{-iEt/\hbar} \qquad (11.84)$$

where $E = \hbar\omega$ is according to Planck the total energy (potential plus kinetic) of the electron. Then Schrödinger's Equation becomes

$$-\frac{\hbar^2}{2m}\nabla^2\psi + V\psi = E\psi \qquad (11.85)$$

This is an eigenvalue equation (see Section 16.1.2), with ψ_i being the ith eigenfunction, and E_i the corresponding energy. This means that acceptable wavefunctions ψ exist only for those specific values of E for which eqn (11.85) yields eigenfunction solutions. Other values of E are unphysical.

As an example, an electron in a hydrogen atom experiences an electrostatic potential

$$V = -\frac{1}{4\pi\varepsilon_0}\frac{e^2}{r} \qquad (11.86)$$

due to the attraction of the nucleus. Here e is the magnitude of the negative charge on the electron and of the positive charge on the nucleus.

With eqn (11.86) substituted into (11.85), Schrödinger's Equation has solutions separable in spherical polar coordinates of the form

$$\psi(r, \theta, \phi) = R(r)Y_l^m(\theta)e^{im\phi} \qquad (11.87)$$

where $R(r)$ satisfies the ordinary differential equation

$$-\frac{\hbar^2}{2mr}\frac{d^2}{dr^2}(rR) + \left[V + \frac{\hbar^2 l(l+1)}{2mr^2}\right]R = ER \qquad (11.88)$$

The angular dependence of ψ ($Y_l^m(\theta)e^{im\phi}$) is the usual form obtained for this type of partial differential equation, where θ and ϕ occur implicitly in the ∇^2 operator, but nowhere else (compare Section 11.3.3.3).

There is an infinity of possible solutions. The simplest one is independent of θ and ϕ:

$$\psi = e^{-r/r_0} \qquad (11.89)$$

where

$$r_0 = \frac{4\pi\varepsilon_0\hbar^2}{me^2} \qquad (11.90)$$

and characterises the radial scale of the wave function. It is the Bohr radius of the hydrogen atom.

The energy eigenvalue for this state is given by

$$E = -\frac{me^4}{2(4\pi\varepsilon_0\hbar)^2} \qquad (11.91)$$

This also agrees with the simple Bohr Theory for the atom in its lowest possible energy state. The negative sign shows that the electron is bound within the atom.

All other ψ_i which satisfy Schrödinger's Equation correspond to the hydrogen atom being in an excited state, which will sooner or later decay to the ground state by emitting a photon. The fact that only the eigenvalues E_i are allowed immediately explains why the hydrogen atom has a discrete spectrum of bound states, rather than a continuous one.

This is thus a good example of applying the method of separation of variables to obtain solutions of a partial differential equation, which provide a physical description of a very interesting problem.

11.4 d'Alembert's method

A very neat method of solving the wave equation (11.10) is due to d'Alembert. It makes use of new variables

$$\left. \begin{array}{r} u = x + ct \\ \text{and} \quad v = x - ct \end{array} \right\} \tag{11.16}$$

As shown in Section 11.2.1, the wave equation now becomes

$$\frac{\partial^2 z}{\partial u \partial v} = 0 \tag{11.21}$$

This can be readily integrated with respect to u, to give the result that $\partial z / \partial v$ is equal to a "constant". This "constant", however, is simply something which, when differentiated partially with respect to u, gives zero. It can hence be a function $f(v)$ of the other variable v, i.e.

$$\frac{\partial z}{\partial v} = f(v) \tag{11.92}$$

(compare the discussion of Section 5.7, especially around eqns (5.59) and (5.59′)). Similarly this new equation can be integrated to give

$$z = g(v) + h(u) \tag{11.93}$$

where

$$g(v) = \int f(v) dv$$

and $h(u)$ is the "constant" of integration with respect to v.

The solution (11.93) can now be written in terms of the original variables x and t as

$$z(x,t) = g(x - ct) + h(x + ct) \tag{11.93′}$$

In contrast to the method of separating variables which naturally produces standing waves, the d'Alembert approach results in travelling wave solutions, with $g(x - ct)$ being a wave travelling in the direction of increasing x, and $f(x + ct)$ moving to the left.

That the above statement is so can be seen by drawing $g(x-ct)$, first at $t = 0$. Then $g(x)$ is some completely arbitrary function (see fig. 11.5(a)). At some slightly later time Δt_1, the function is now $g(x - c\Delta t_1)$. This is identical to $g(x)$, except that the curve has been shifted by an amount $\Delta x_1 = c\Delta t_1$. Similarly at another somewhat later time Δt_2, the curve is shifted ever further, by $\Delta x_2 = c\Delta t_2$. If we thus consider all possible times,

$$z = g\,(x - ct)$$

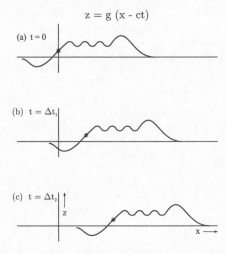

Fig. 11.5. $g(x - ct)$ as a travelling wave: (a) An arbitrary displacement $z = g(x)$ at $t = 0$. The star marks the place on the wave initially at $x = 0$. (b) The wave $g(x - c\Delta t_1)$ at a slightly later time $t = \Delta t_1$. The whole wave pattern has been displaced *en masse* to the right by a distance $\Delta x_1 = c\Delta t_1$. The stars here and in (c) correspond to the same point on the wave as in (a). (c) At a somewhat later time $t = \Delta t_2$, the wave has moved even further to the right. The speed with which it moves is c.

the general displacement moves at speed c to the right, while maintaining exactly the same shape.†

By analogy, $h(x + ct)$ corresponds to a general displacement moving to the left with speed c.

If the above discussion was a little too general, it is possible to reconsider the argument in terms of a specific function, e.g.

$$g(x - ct) = a\sin(x - ct)$$

Then the sine wave can be drawn as a function of x for various t values, and again it follows that this is a wave which moves in the direction of increasing x.

In a typical problem, the specified boundary conditions are likely to be the initial displacement and speed of the string‡ as functions of x. Then

† The reason that any arbitrary-shaped displacement keeps the same shape as it moves is due to the fact that the speed c of the waves is independent of their frequency. If this were not so, it is only harmonic waves which would remain the same shape as they travel.

‡ It is important to distinguish the constant speed c with which the waves propagate along the length of the string, and the speeds with which individual segments of the string are moving in the transverse direction, and which vary along the string and in time.

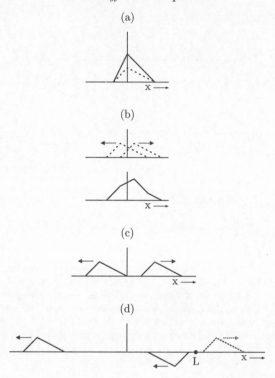

Fig. 11.6. d'Alembert's solution for a string with an initial displacement: (a) At
$t = 0$, the string is at rest with the displacement as shown. It is considered to be
made up of two functions $g(x)$ and $h(x)$, each half of the total displacement. They
are shown as the dashed lines. (b) At a slightly later time, the waves have moved
apart, since $g(x - ct)$ and $h(x + ct)$ travel in opposite directions. The arrows show
the directions in which the components of the wave are travelling. The total
displacement, shown as the solid curve, is the sum of g and h. (c) Somewhat later,
the waves have moved even further apart. (d) If one end of the string at $x = +L$
is fixed, the part of the wave that reaches it is reflected from it, with a change
in the sign of its amplitude. The dashed line shows where $g(x - ct)$ would have
been had it not been reflected.

all that is required is to choose the functions g and h to match these
boundary conditions, and the problem is solved.

Thus if the string is initially at rest with displacement as shown in
fig. 11.6, $g(x)$ and $f(x)$ are each chosen as half this displacement, and
then as time goes by, each travels at speed c in its appropriate direction.

Why $g(x)$ and $h(x)$ were chosen each to be equal to half the initial
displacement (rather than having, say, $h(x) = 0$, or $g(x) = -2h(x)$) was
glossed over in the above paragraph. The reason is that the initial velocity

of the string was zero everywhere. For a solution of the type (11.93')

$$\frac{\partial z}{\partial t} = -cg'(x - ct) + ch'(x + ct) \tag{11.94}$$

and so

$$g'(x) = h'(x) \tag{11.95}$$

ensures zero initial velocity at all x. This in turn implies that

$$g(x) = h(x) + \text{ constant} \tag{11.96}$$

where the constant of integration can be set to zero, since the solution for z of eqn (11.93') is completely independent of its value. This justifies what was done above.

If the initial velocity of the string were non-zero, then eqns (11.93') and (11.94) (but with t set equal to zero, assuming that is when z and $\partial z/\partial t$ are specified) provide two simultaneous equations for z at each point x, so that z can be uniquely determined.

If the string on which the waves are travelling is not infinite but has a fixed end at $x = +L$, the above type of analysis can still be used. What happens is that, as the component $g(x - ct)$ reaches this fixed point, it is reflected with a change in sign of its amplitude (see fig. 11.6(d)). This is discussed further in Sections 12.11 and 14.6.

Thus the d'Alembert approach is a most flexible one for finding travelling wave solutions of very general shape.

11.5 Worked examples

To illustrate the techniques described earlier in this chapter, this final section contains a few worked examples.

11.5.1 Laplace's Equation

Find the solution of the two-dimensional Laplace's Equation for $r > a$, given that

$$V(a, \theta) = V_1 \cos \theta + V_2 \sin 2\theta \tag{11.97}$$

Here it is appropriate to use two-dimensional polar coordinates (rather than x and y), so Laplace's equation is

$$\frac{1}{r} \frac{\partial}{\partial r} \left(r \frac{\partial V}{\partial r} \right) + \frac{1}{r^2} \frac{\partial^2 V}{\partial \theta^2} = 0 \tag{11.25'}$$

The general solution for this is as given in (11.75).

In order for the solution to be valid for all values of r larger than a, it must not diverge as r tends to infinity. Thus g and all the d_m must be zero.

At $r = a$, the solution is thus of the form

$$V(a, \theta) = f + \sum a^{-m}[f_m g_m \cos m\theta + f_m(1 - g_m)\sin n\theta] \qquad (11.98)$$

which must match (11.97). Then the only non-zero coefficients in (11.98) are $f_1 g_1$ and $f_2(1 - g_2)$, which must be set equal to V_1 and V_2 respectively.†

Thus the complete solution is

$$V(r, \theta) = V_1(a/r)\cos\theta + V_2(a/r)^2 \sin 2\theta \qquad (11.99)$$

11.5.2 Thermal conduction equation

A thin uniform conducting rod of length l initially has the end at $x = 0$ at 0 °C, and the other at 100 °C. Subsequently the end at $x = 0$ is kept of 0 °C, while the other end is thermally insulated. Find the temperature $T(x, t)$ along the rod, for $t > 0$.

Here we need to find the solution of the heat conduction equation (11.5) that satisfies the initial condition $T(x, 0)$, and the boundary conditions for $T(0, t)$ and $T(l, t)$. They are

$$T(x, 0) = 100x/l \qquad (11.100a)$$

$$T(0, t) = 0 \qquad (11.100b)$$

$$\frac{\partial T}{\partial x}(l, t) = 0 \qquad (11.100c)$$

The first of these is the equilibrium temperature distribution along the rod whose ends are at 0 °C and 100 °C‡ (see fig. 11.7(a)); and the third says that, if the end at $x = l$ is insulated, there cannot be any heat conducted in or out of that end, and hence the temperature gradient must be zero there.

The form of solution suitable for problems with a specified initial temperature distribution is with the constant of separation being negative (see Section 11.3.2.1). The general solution is as given in (11.44), with

---

† Either this is obvious, or else the values of all the coefficients can be determined by the method of Fourier series. The latter is very heavy handed for this simple problem, but would be necessary if $V(a, \theta)$ were not given as the sum of a few harmonic terms (e.g. $V(a, \theta) = 0$ for θ in the range 0–π, and $V = V_0$ otherwise).

‡ Again if this is not obvious it can be derived by solving the time-independent heat conduction equation $\partial^2 T/\partial x^2 = 0$, with the given temperatures at the ends.

the values of m and the coefficients being chosen to match the boundary conditions (11.100).

Because eqns (11.100b) and (11.100c) apply *at all times*, it is necessary to ensure that each of the terms of (11.44) describing the temperature variation satisfies them too. Now (11.100b) implies that sines (rather than cosines) are required for the spatial dependence, and so all the β_m are zero.

Then (11.100c) says that we must use only those sine waves that have zero gradient at $x = l$. Thus the argument ml of the sine waves must be an odd multiple of $\pi/2$, i.e. the allowed values of m are

$$m = (2n - 1)\pi/2l \quad (n = 1, 2, 3, ...) \tag{11.101}$$

Finally we have to choose the coefficients a_m in order to satisfy the initial condition (11.100a). Since all the exponential factors in (11.44) are simply unity at $t = 0$,

$$T(x, 0) = 100x/l$$
$$= \sum_{n=1}^{\infty} a_m \sin mx \quad (0 \le x \le l)$$

The coefficients a_m are found by the method of Fourier series.

A little care is now necessary. The temperature distribution (11.100a) has meaning only for $0 \le x \le l$, and so it might be thought that this is the repeat distance for the Fourier series. This is incorrect for two reasons. Because the Fourier series must consist only of sine waves (as a result of (11.100b)), for $x < 0$ it satisfies

$$T(x) = -T(-x)$$

This has the effect of extending the repeat distance of the series. Secondly, because of the boundary condition (11.100c), we are forced to use sine terms with zero gradient at $x = l$, and hence the Fourier series for $x > l$ is related to that in the physical region by

$$T(x) = T(2l - x)$$

(see fig. 11.7(b)). Then $T(x, 0)$ is forced to return to zero only at $x = 2l$. The net result is that the basic wavelength for the series is $4l$, rather than l.†

---

† It is not completely necessary to go through the above argument, as the repeat wavelength of $4l$ is already implied in the boundary conditions (11.100). However, it is very easy to make errors in deriving the values of the Fourier coefficients. By looking at fig. 11.7(b), we can have much better intuition about whether the Fourier series we have derived is physically reasonable, and also save ourselves time by not bothering to calculate coefficients which must be zero.

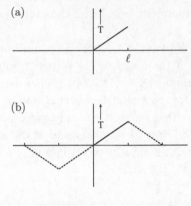

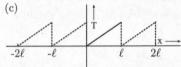

Fig. 11.7. (a) Temperature distribution at $t = 0$, for $0 < x < l$. (b) Temperature distribution at $t = 0$, extended over one full cycle of the Fourier series. The continuation in the region $l < x < 2l$ is determined by the boundary condition that the end at $x = l$ is insulated; and that at $x < 0$ by the fact that the end at $x = 0$ is kept at zero temperature. (c) The incorrect form of $T(x, 0)$ that would result from the wrong assumption that the repeat distance of the Fourier series is l. At $t > 0$, the boundary conditions at $x = 0$ and $x = l$ would be violated.

By applying the standard methods of Fourier series (see Section 12.3), we obtain

$$a_m = \frac{1}{2l} \int_{-2l}^{+2l} f(x) \sin mx\, dx \qquad (11.102)$$

where $f(x)$ is the function drawn in fig. 11.7(b), and m is as given in eqn (11.101). This is equivalent to

$$a_m = \frac{2}{l} \int_0^l (100x/l) \sin mx\, dx$$

$$= \pm \frac{200}{m^2 l^2}$$

$$= \pm \frac{800}{\pi^2 (2n - 1)^2} \qquad (11.103)$$

where the sign alternates, and is positive for $m = \pi/2l$ $(n = 1)$.

Thus our solution is

$$T(x,t) = \sum_m \pm \frac{200}{m^2 l^2} \sin mx e^{-\left(\kappa m^2/s\rho\right)t}$$

with the values of m as given in (11.101). As t tends to infinity, all the exponential terms tend to zero, and so does $T(x,\infty)$. This is consistent with the fact that heat is conducted out of the rod from the $x = 0$ end, until all the rod is at 0 °C.

This is an example of the topic discussed in Section 12.10, where the Fourier series is required for a function that is defined only over a finite range. Of all the possible extensions of the function over the full infinite range, we choose the one which is physically meaningful in terms of the boundary conditions.

A similar problem can be found in Section 12.10.1.

11.5.3 Waves on a string

A string of density ρ and under tension T is stretched between two rigid supports at $x = \pm l$. A mass m is attached to the mid-point of the string at $x = 0$. Show that there are two sorts of normal modes of vibration of the string:

(a) those with frequencies

$$\omega = \frac{n\pi}{l} \sqrt{\frac{T}{\rho}}$$

which are independent of the mass m; and

(b) those with frequencies which depend on m, and are given by

$$\cot kl = \frac{mk}{2\rho}$$

where

$$k = \omega \sqrt{\rho/T}$$

As in most problems involving waves on strings, gravity is to be neglected. For example, we can imagine that the mass is supported by a frictionless table, and the string undergoes transverse horizontal displacements.

This problem is, however, somewhat unusual in that there is a mass in the middle of the string. This results in our having to consider separately the waves on the two parts of the string ($y_-(x,t)$ for $x < 0$; and $y_+(x,t)$

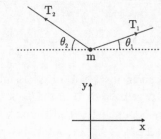

Fig. 11.8. The region around $x = 0$ for the string with a mass on it. In order for the wave to be transverse, the longitudinal force on the mass must vanish. i.e. $T_1 \cos\theta_1 = T_2 \cos\theta_2$. For small angles θ, $\cos\theta \approx 1$, so the tensions on the two sides are equal.

The net transverse force on the mass results in its acceleration, and is given by $T_1 \sin\theta_1 + T_2 \sin\theta_2$. For small angles, $\sin\theta_1 \approx dy_+/dx$ and $\sin\theta_2 \approx -dy_-/dx$, and so Newton's Second Law reduces to eqn (11.105).

In order not to have a discontinuity in the string at $x = 0$, we also require $y_+(0,t) = y_-(0,t)$. This is our other boundary condition at $x = 0$.

for $x > 0$). We must then worry about the boundary conditions on the displacement y at $x = 0$, as well as those at the ends $(x = \pm l)$.

For the displacement itself, we require

$$y_-(0,t) = y_+(0,t) \qquad (11.104)$$

In the absence of the mass, we would require also the gradients to match at $x = 0$, because otherwise there would be a transverse force on an infinitesimal mass of string, which would cause an undesirable infinite acceleration. Here, however, that is not the case, because of the presence of the mass. Newton's Second Law, equating the net transverse force to the mass m times its acceleration, gives†

$$T\left(\frac{\partial y_+}{\partial x} - \frac{\partial y_-}{\partial x}\right) = m\frac{\partial^2 y}{\partial t^2} \qquad (11.105)$$

(see fig. 11.8), where all the derivatives in (11.105) refer to $x = 0$. Thus (11.104) and (11.105), together with the requirement that the ends of the string at $x = \pm l$ have $y = 0$, are the required boundary conditions.

We now write the waves as

$$\left.\begin{array}{l} y_-(x,t) = A\sin[k(x+l)]\cos\omega t \\ \text{and} \quad y_+(x,t) = B\sin[k(x-l)]\cos\omega t \end{array}\right\} \qquad (11.106)$$

† The fact that the tension is the same in both parts of the string follows because we are allowing only transverse vibrations of the string. Then if $T_+ \neq T_-$ (and given that the gradients of the string are small), the mass would move longitudinally.

where ω and k are related by the speed of the waves c:

$$\frac{\omega^2}{k^2} = c^2 = \frac{T}{\rho} \tag{11.107}$$

The form of eqns (11.106) deserves some comment. First, we have written down separated functions of x and t, which are suitable for describing standing waves. The same factor $\cos \omega t$ occurs for the time dependence of y_- and y_+. This implies that they have the same frequency and the same phase. This is necessary in order to satisfy the boundary condition (11.104) at all values of t.

For the spatial dependence, we have chosen a harmonic form such that y_- is zero at $x = -l$; $\sin[k(x+l)]$ is suitable, with k and ω related by the speed of the waves (eqn (11.107)). Similarly for y_+, we write $\sin k[(x-l)]$. Because the values of ω are the same for y_+ and y_-, so too must be those for k.

All that remains is to implement the boundary conditions at $x = 0$. From (11.104) and (11.105) respectively, we obtain†

$$A \sin kl = -B \sin kl \tag{11.108}$$

and

$$Tk(B - A) \cos kl = -m\omega^2 A \sin kl \tag{11.109}$$

From (11.108), we deduce that

$$\sin kl = 0 \tag{11.110}$$

or

$$A = -B \tag{11.111}$$

Then from (11.109), if $\sin kl = 0$,

$$A = B \tag{11.112}$$

while if $A = -B$

$$2Tk \cot kl = m\omega^2 \tag{11.113}$$

We have thus solved the problem, and have found the two different types of solutions. The first have $\sin kl = 0$, and hence $y = 0$ at $x = 0$. The allowed values of k are simply $n\pi/l$, and are independent of m, because the mass at $x = 0$ is never displaced in these modes. From (11.110) and (11.112), it follows that the two parts of the string at $x < 0$ and at $x > 0$

---

† On the right-hand side of (11.109), we have written $-m\omega^2 A \sin kl$, as derived from $\partial^2 y_-/\partial t^2$. If instead we had used $\partial^2 y_+/\partial t^2$, we would have obtained $m\omega^2 B \sin kl$. From (11.108), these alternatives are identical.

join smoothly together at the origin; the functions, defined in the two different regions, are exact continuations of each other, since

$$y_+(x,t)/\cos\omega t = B\sin[k(x-l)]$$
$$= A\sin\left[\frac{n\pi}{l}(x-l)\right]$$
$$= A\sin[k(x+l)]$$
$$= y_-(x,t)/\cos\omega t$$

The other solutions have $A = -B$, and hence

$$y_+(x,t)/\cos\omega t = B\sin[k(x-l)]$$
$$= -A\sin[k(x-l)]$$
$$= A\sin[k(-x+l)]$$
$$= y_-(-x,t)/\cos\omega t$$

i.e. the displacement of the whole string is an even function of x.

The boundary condition (11.113) must also be satisfied. Since ω and k are always related by (11.107), (11.113) can be rewritten as

$$\cot kl = mk/2\rho \qquad (11.114)$$

The solutions of this equation provide the allowed values of k for these modes. They are most easily found graphically (see fig. 11.9(a)). It is seen that these values of k are interleaved between the ones at $n\pi/l$ for the other type of modes. Here the mass at $x = 0$ is displaced, and the allowed wavelengths and frequencies do depend on m.

The displacements of the string for the first few allowed frequencies are shown in fig. 11.9(b).

11.5.4 Schrödinger's Equation for harmonic oscillators

Show that the wave function for a two-dimensional harmonic oscillator (in x and y) is simply the product of two one-dimensional harmonic oscillator wave functions (in x and in y separately); and that the allowed energy in two dimensions is the sum of the energies in the two one-dimensional cases.

The Schrödinger Equation for a one-dimensional oscillator is

$$-\frac{\hbar^2}{2m}\frac{d^2\psi}{dx^2} + \frac{1}{2}kx^2\psi = E\psi \qquad (11.115)$$

where the potential $\frac{1}{2}kx^2$ is such as to produce a restoring force $-kx$

(a)

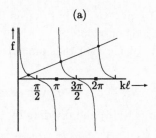

(b)

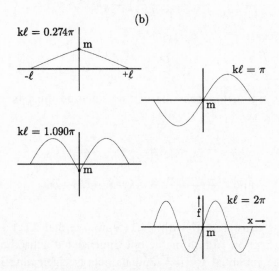

Fig. 11.9. (a) The functions $f = \cot kl$ and $f = mk/2\rho$, plotted against kl. Their intersections (small circles) give the solutions of eqn (11.114), and hence the allowed values of kl. These are for the modes in which the mass at $x = 0$ oscillates. These values of kl depend on the gradient of the line, and hence on $m/2\rho l$ (i.e. on the mass). In contrast, the modes in which the mass does not move have $kl = n\pi$ (small squares), and are independent of m.

The allowed values of k are such that the two types of solutions alternate. For example, for $m/2\rho l = 1$ (i.e. the mass m and the mass of the string are equal), the solutions are $kl = 0.274\pi$, 1.090π, 2.049π, ... for the modes where the mass moves; and π, 2π, 3π, ... for the ones where it does not. (b) The displacement of the string at $t = 0$, for the first four modes. Those on the left (including the lowest possible k and ω) have the mass moving; for those on the right, it is always stationary.

(and an angular frequency of oscillation $\omega = \sqrt{k/m}$ in the corresponding classical problem). For the two-dimensional situation, Schrödinger's Equation is

$$-\frac{\hbar^2}{2m}\left(\frac{\partial^2\psi}{\partial x^2} + \frac{\partial^2\psi}{\partial y^2}\right) + \frac{1}{2}k(x^2 + y^2)\psi = E\psi \qquad (11.116)$$

The derivatives in (11.116) are partial, but that in (11.115) is a total one, since in that case ψ depends only on x.

We look for solutions of (11.116) that are separable in x and y by writing

$$\psi(x, y) = X(x)Y(y)$$

On substituting this into (11.116) and dividing by ψ, we obtain

$$-\frac{\hbar^2}{2m}\left(\frac{1}{X}\frac{d^2X}{dx^2} + \frac{1}{Y}\frac{d^2Y}{dy^2}\right) + \frac{1}{2}k(x^2 + y^2) = E$$

This can be recast as

$$-\frac{\hbar^2}{2m}\frac{1}{X}\frac{d^2X}{dx^2} + \frac{1}{2}kx^2 = -\left(-\frac{\hbar^2}{2m}\frac{1}{Y}\frac{d^2Y}{dy^2} + \frac{1}{2}ky^2 - E\right) \qquad (11.117)$$

By the usual argument, each side of this equation must be equal to a constant, say λ, i.e.

$$-\frac{\hbar^2}{2m}\frac{d^2X}{dx^2} + \frac{1}{2}kx^2X = \lambda X \qquad (11.118a)$$

$$\text{and} \quad -\frac{\hbar^2}{2m}\frac{d^2Y}{dy^2} + \frac{1}{2}ky^2Y = (E - \lambda)Y \qquad (11.118b)$$

On comparing eqns (11.118) with (11.115), we see that $X(x)$ is an eigenfunction (see Section 16.1.2) for the one-dimensional Schrödinger Equation when the potential is $\frac{1}{2}kx^2$, and its energy eigenvalue is $E_x = \lambda$. Similarly $Y(y)$ is an eigenfunction for the corresponding y equation, with eigenvalue $E_y = E - \lambda$. Then the solution of the two-dimensional harmonic oscillator equation is simply $X(x)Y(y)$, with energy $E = E_x + E_y$.

For example, as can be verified by direct substitution, an eigenfunction of eqn (11.118a) is

$$X(x) = e^{-\alpha x^2}$$

with

$$\alpha = \sqrt{km}/\hbar$$

and eigenenergy

$$E_x = \frac{\hbar}{2}\sqrt{\frac{k}{m}}$$

(This is, in fact, the solution with the lowest possible energy.) Then for the two-dimensional case, a solution is

$$\psi = X(x)Y(y) = e^{-\alpha(x^2 + y^2)}$$

with α as above, and with energy

$$E = E_x + E_y = \hbar\sqrt{k/m}$$

Thus the lowest energies of the quantum mechanical harmonic oscillator are $\frac{1}{2}\hbar\omega$ and $\hbar\omega$ for the one- and two-dimensional cases, respectively (and $\frac{3}{2}\hbar\omega$ for the three-dimensional case).

A whole series of excited states $X_i(x)$ (and correspondingly $Y_j(y)$) exists for the one-dimensional situation. Then other solutions for the two-dimensional case can be constructed as $\psi = X_i(x)Y_j(y)$, for any combination of i and j.

Problems

11.1 Use the transformation

$$u = xe^y$$
$$\text{and} \quad v = xe^{-y}$$

to show that

$$x^2\frac{\partial^2\phi}{\partial x^2} - \frac{\partial^2\phi}{\partial y^2} + x\frac{\partial\phi}{\partial x} = 4uv\frac{\partial^2\phi}{\partial u\partial v}$$

This transformation enables solutions of

$$x^2\frac{\partial^2\phi}{\partial x^2} - \frac{\partial^2\phi}{\partial y^2} + x\frac{\partial\phi}{\partial x} = 0$$

to be easily found, since this equation is equivalent to

$$\frac{\partial^2\phi}{\partial u\partial v} = 0$$

Then a solution is

$$\phi = f(xe^y) + g(xe^{-y})$$

where f and g are arbitrary functions. (Compare Section 11.4.)

11.2 The transverse displacement y of a thin rod of length l obeys the differential equation

$$k^4\frac{\partial^4 y}{\partial x^4} + \frac{\partial^2 y}{\partial t^2} = 0$$

Find solutions for y which are separable in the variables x and t, and whose time dependence is oscillatory.

The rod is clamped at each end, such that at $x = 0$ and at $x = l$, both y and $\partial y/\partial x$ are zero at all times. Show that the natural frequencies ω satisfy

$$\cos\alpha\cosh\alpha = 1$$

where

$$\alpha = \sqrt{\omega}l/k$$

11.3 Laplace's Equation in two dimensions, expressed in plane polar coordinates, has the form

$$\frac{\partial^2 V}{\partial r^2} + \frac{1}{r}\frac{\partial V}{\partial r} + \frac{1}{r^2}\frac{\partial^2 V}{\partial \theta^2} = 0$$

Determine separated solutions of this equation which are sinusoidal in θ, but otherwise as general as possible. How do you ensure that the solutions are single-valued functions of position?

Determine the solution $V(r,\theta)$, valid within the annulus defined by $1 \le r \le 2$, which is zero for all θ when $r = 1$ and takes the form $63\sin 3\theta$ when $r = 2$.

11.4 An infinite string is stretched along the x axis. At $t = 0$, it is given a transverse velocity distribution

$$\dot{y}(x,0) = \begin{cases} A\cos kx & \text{for } |x| < 2\pi/k \\ 0 & \text{otherwise} \end{cases}$$

Use d'Alembert's method to find the displacement of the string at any subsequent time. Make sketches of these displacements at $t = \varepsilon$, $\pi/2kc$, π/kc and $2\pi/kc$; ε is a small positive constant.

11.5 It is the hottest time of day in the Australian desert. Kate the kangaroo is tired, but realises that if she sits down, the hot sand would burn a certain part of her anatomy. However, she reasons as follows.

Maybe it is possible to treat the sand as a continuous medium with thermal conductivity κ, density ρ and heat capacity per unit mass s. Its surface is the horizontal plane $z = 0$, and it surely obeys the heat conduction equation

$$\frac{\partial T}{\partial t} = \frac{\kappa}{\rho s}\frac{\partial^2 T}{\partial z^2}$$

where T is the difference between the actual temperature and the time-averaged temperature, t the time of day and z the distance below the surface. The surface temperature at $z = 0$ is given by

$$T = \Delta\cos[2\pi(t - t_0)/t_d] \tag{1}$$

where Δ is a constant, t_d is 24 hours and t_0 is the time at the hottest part of the day. Kate then wonders whether it would be more comfortable to use her tail to flip away the top layer of sand before sitting down.

By finding a solution for T that is separable in t and in z, and by suitably choosing the relevant constant in order to match the boundary condition (1), find how the temperature varies with z when $t = t_0$. Hence determine the thickness of the layer of sand to be removed, in order for Kate to have the coolest surface to sit on.

($s = 800$ J kg^{-1} K^{-1}; $\rho = 1600$ kg m^{-3}; $\kappa = 0.3$ J s^{-1} m^{-1} K^{-1}. You may assume that the time-averaged temperature is independent of z.)

11.6 (i) Solve Schrödinger's Equation for a particle of energy E in a region of constant potential V, where (1) $E > V$, and (2) $E < V$. Show that the solutions can be written as (1) $\psi = e^{\pm ikx}$, and (2) $\psi = e^{\pm vx}$, where k and v are constants. The solutions in (1) correspond to waves travelling to the right or left; in (2), they are exponential standing waves.

(ii) Solve Schrödinger's Equation for the case of a wave $\psi = e^{\pm ikx}$, corresponding to an energy E, travelling in a region $x < 0$ where the potential is zero, and reaching a step where the potential for $x > 0$ is again a constant V_0, where $V_0 > E$. (The solution consists of the incident wave and a reflected one in the region $x < 0$, and a decaying exponential at $x > 0$. The arbitrary constants in front of the reflected wave and the decaying exponential are determined by the boundary conditions at $x = 0$:

$$\psi_- = \psi_+$$

and

$$\frac{\partial \psi_-}{\partial x} = \frac{\partial \psi_+}{\partial x},$$

where the $\pm$ signs refer to the solutions beyond and before $x = 0$.)

(iii) The above can be regarded as an extremely simplified model of a nucleus containing an alpha particle. The alpha particle is held in by the Coulomb barrier, such that it sees a potential roughly as shown in (b). The alpha particle inside the nucleus approaches the barrier, with energy E_α below the height of the barrier. The wave function inside the barrier (a region which classically is forbidden to the alpha particle) is not zero, but falls off exponentially like

(a)

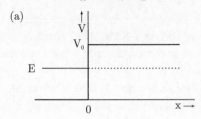

(b)

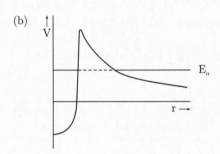

(c)

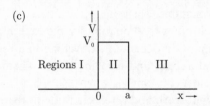

(a) A particle of energy E in the region where the potential is constant at zero, for $x < 0$; and at $V_0 > E$, for $x > 0$. (b) The potential seen by an alpha particle as a function of its distance r from the centre of a nucleus. For large r, the Coulomb repulsion results in a potential $\sim 1/r$, while at small r, the attractive nuclear force produces a well. The energy E_α of the alpha particle is such that it is below the top of the potential barrier. (c) An approximation to the barrier seen by the alpha particle in (b). Region I is supposed to correspond to the central potential well inside the nucleus; region II is where $E_\alpha < V$; and in region III the alpha particle is free.

$e^{-\nu x}$. At the far side of the barrier, the wave function ψ is not zero but merely very small, and hence so is ψ outside the barrier, where the alpha particle can appear as a free particle. Because of the approximately exponential decrease of ψ within the barrier, the probability of the alpha emerging is very small indeed, and so the lifetime for alpha decay is very large compared with the time it takes an alpha particle to travel across the nucleus.

(iv) The potential step (a) was a very crude model of the potential shown in (b). A little better is that in (c). Find the amplitude of the wave function ψ in region III when a wave e^{ikx} is incident from $x < 0$. (The solution will be of the form

$$\psi_- = e^{ikx} + Re^{-ikx} \quad \text{for} \ \ x < 0$$

$$\psi_B = Ae^{-vx} + Be^{+vx} \quad \text{for} \ \ 0 < x < a$$

and

$$\psi_+ = Te^{ikx} \quad \text{for} \ \ x > a$$

Appropriate boundary conditions need to be applied to ψ and $\partial\psi/\partial x$ at both boundaries.) Show that, for large va, $|T|$ is smaller than the incident amplitude (unity) by a factor of order e^{-va}, as in the simpler picture.

If the algebra of this general situation is too tedious for you, examine the special case where $E = V_0/2$.

More problems involving partial differential equations are to be found in the chapter on Fourier series.

12

Fourier series

12.1 Functions that repeat, and repeat, and repeat...

Sine waves and cosine waves are beautiful. They don't have any jumps or kinks. We can differentiate them as many times as we want to and they still don't develop discontinuities. And when we differentiate them, they stay the same shape. We can even integrate them, and they still remain a cosine or sine wave.

In physics we may come across functions that repeat, but which are not necessarily sinusoidal in shape. For example, we could have an electric waveform generator, whose output voltage V drawn as a function of time could look like any of the graphs of fig. 12.1.

Since harmonic waves† are so pleasant, we can see the extent to which we can produce a non-sinusoidal waveform by the sum of a few sine waves. Thus in fig. 12.2, we show the waveform of fig. 12.1(d), together with successive approximations:

(a) $y_1 = \dfrac{4}{\pi} \cos t$

(b) $y_2 = y_1 + \dfrac{4}{9\pi} \cos 3t$

(c) $y_3 = y_2 + \dfrac{4}{25\pi} \cos 5t$

It is clear from the diagram that, as we take more and more terms, our approximation to the original function becomes better and better. The Fourier series for a function $f(t)$ consists of an infinite number of such terms. Mathematicians would worry about whether the series converges, but for almost any repetitive function that we are likely to encounter in

† We use the term "harmonic" to apply to waves that have a sine or cosine shape, without worrying about where the origin is. Thus $y = A \sin t + B \cos t$ is harmonic, but $y = 2 \sin t + 3 \cos 2t$ is not.

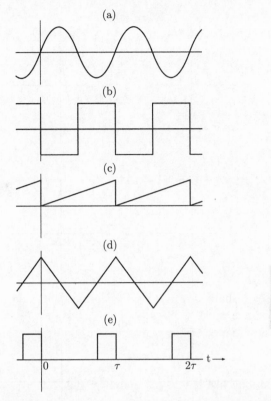

Fig. 12.1. A set of functions that repeat themselves as t increases i.e $f(t+\tau) = f(t)$, where τ is the period.

physical situations, we can assume that the series provides a very good representation of the function almost everywhere.

In general, we will need both sine and cosine terms for our function $f(t)$. For functions that repeat themselves after a time $t = 2\pi$, we write the Fourier series as

$$f(t) = a_0/2 + \sum_{n=1}^{\infty} a_n \cos nt + \sum_{n=1}^{\infty} b_n \sin nt \qquad (12.1)$$

Here, a_0, a_n and b_n are the Fourier coefficients to be determined for the given function $f(t)$. Thus for the case depicted in fig. 12.2, the first few coefficients are

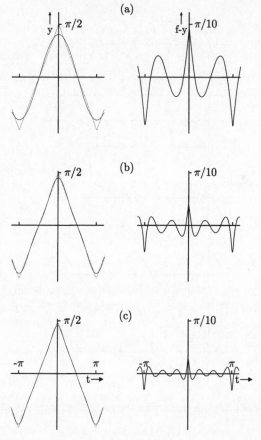

Fig. 12.2. The dashed lines on the left-hand side are the function $f(t)$ of fig. 12.1(d), with the period chosen as 2π. The solid lines show successive approximations of the form

$$y = \frac{4}{\pi} \sum_{n=1}^{N} \frac{1}{(2n-1)^2} \cos(2n-1)t$$

As the number N of terms used is increased from 1 in (a) to 3 in (c), the approximation of the series to the original function improves. This is seen even more clearly on the right-hand side, where $f - y$ (i.e. the actual function minus the first few terms of its Fourier series) are plotted as functions of t, with an expanded vertical scale.

$$\left. \begin{aligned} a_1 &= 4/\pi \\ a_3 &= 4/9\pi \\ a_5 &= 4/25\pi \\ a_7 &= 4/49\pi \end{aligned} \right\} \tag{12.2}$$

and all the bs as well as the even as are zero.

The fact that the Fourier series starts off with the term $a_0/2$ deserves some comment. If we had written the summations as extending from zero (rather than from 1) to infinity, the extra contribution would have been

$$a_0 \cos 0 + b_0 \sin 0 = a_0$$

Apart from the factor of $1/2$, this corresponds to our separate term $a_0/2$. It is simply a convention to include this factor of $1/2$; the motivation will become clear in Section 12.3.

For functions that repeat with a period τ different from 2π, formula (12.1) becomes modified to

$$f(t) = a_0/2 + \sum_{n=1}^{\infty} a_n \cos(2\pi n t/\tau) + \sum_{n=1}^{\infty} b_n \sin(2\pi n t/\tau) \qquad (12.3)$$

We can see from (12.3) that if t increases to $t + \tau$, every term in the series representation of $f(t)$ keeps its original value; this is as required by the fact that $f(t)$ has a period of τ.

Since (12.1) is a little neater than (12.3), we shall try to work with functions whose period is 2π. In other cases, we can make this so by changing the variable to $t' = 2\pi t/\tau$. In fact this is a way of deducing eqn (12.3) from eqn (12.1).

Of course it is not necessary to restrict ourselves to functions that are repetitive in t, rather than in any other variable. Thus for a function $f(x)$ that has a wavelength $\lambda = 2\pi$, we could write

$$f(x) = a_0/2 + \sum_{n=1}^{\infty} a_n \cos nx + \sum_{n=1}^{\infty} b_n \sin nx \qquad (12.4)$$

For wavelengths different from 2π, this would become

$$f(x) = a_0/2 + \sum_{n=1}^{\infty} a_n \cos nkx + \sum_{n=1}^{\infty} b_n \sin nkx \qquad (12.5)$$

where the wave number $k = 2\pi/\lambda$. Thus fig. 12.3(a) could represent a rather bumpy road along which a car was being driven; or fig. 12.3(b) could be the transparency of a diffraction grating as we scan in a direction perpendicular to the grating lines.

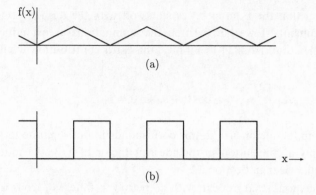

Fig. 12.3. Repeating functions of possible physical interest: (a) $f(x)$ could be an uneven road along which a car is being driven. (b) Here $f(x)$ could be the transparency of a diffraction grating, where x is the direction perpendicular to the grating's lines.

12.2 Why is it useful?

The question that immediately arises is "What is the point of replacing a function $f(t)$ such as one of those drawn in fig. 12.1 by an infinite series of terms?"

The answer depends on the fact that, in many circumstances, the response of a physical system to a harmonic driving force (e.g. the current in an electric circuit when a harmonic voltage is applied) is very much simpler than for other input waveforms. In particular, the response itself may be harmonic (perhaps shifted in phase and probably of a different amplitude from the input harmonic waveform, but with the same frequency). Thus, using the complex number notation of Section 4.4, we can write†

$$R(\omega) = I(\omega)H(\omega) \tag{12.6}$$

where $I(\omega)$ is the input of frequency $v = \omega/2\pi = 1/\tau$; $R(\omega)$ is the corresponding output at the same frequency; and $H(\omega)$ is a factor which relates the two, which can depend on ω, and which can be complex, in order to represent a phase shift of the output relative to the input.

Eqn (12.6) could, for example, describe the output voltage from a given electric circuit when a sinusoidal input is applied. Thus for the circuit of

---

† We are, of course, implicitly assuming that our system is linear; this means that if, for example, we double the input, the output is doubled in magnitude as well. This is, of course, not an essential restriction to the application of the Fourier method. It merely provides a simple situation in which to study its use.

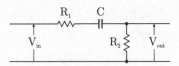

Fig. 12.4. A simple circuit consisting of two resistors R_1 and R_2 and a capacitor C. When the input voltage V_{in} is harmonic of angular frequency ω, the output voltage V_{out} is also harmonic with the same frequency, and is given by equation (12.7). For a non-harmonic input (see, for example, fig. 12.1(b)), the output can be deduced by first determining the Fourier series for V_{in}.

fig. 12.4, the input and output voltages V_{in} and V_{out} are related by

$$V_{out} = V_{in} \frac{j\omega C R_2}{1 + (R_1 + R_2)j\omega C} \qquad (12.7)$$

Later we shall consider other situations in which the behaviour for harmonic disturbances is particularly straightforward. These will include the vibrations of strings, and the way in which an initial temperature distribution in a thermally conducting material changes with time.

The great advantage of the Fourier series is that it now enables us to discover what is the effect of applying to our system *any* periodic waveform. All that is required is to:

(i) decompose the waveform into its Fourier series;
(ii) apply eqn (12.6) to each Fourier term separately, in order to find the response of the system to each harmonic component of the input; and
(iii) add up all the components of the response.

Thus if the input is a function whose Fourier series is given in complex notation† by

$$f(t) = C_0/2 + \sum_{n=1}^{\infty} \text{Re}[C_n e^{i2\pi nt/\tau}] \qquad (12.8)$$

and the response of the system to a harmonic wave is as defined in (12.6), then the output $g(t)$ is determined as

$$g(t) = C_0 H(0)/2 + \sum_{n=1}^{\infty} \text{Re}[C_n H(2\pi n/\tau)e^{i2\pi nt/\tau}] \qquad (12.9)$$

From this solution, we note that in the special case where H is a

† That is, the real and imaginary parts of C_n in eqn (12.8) correspond to a_n and $-b_n$ respectively in eqn (12.3). A more useful convention for writing the complex notation, with more similarity to the Fourier transforms of Section 12.12, is given in Problem 12.9.

constant (i.e. independent of frequency), it can be factorised out of the right-hand side, in which case

$$g(t) = Hf(t)$$

That is, the output is just a constant multiple of the input. In the more general, and probably more typical, case where H varies with frequency, the output will usually differ in shape from the input. An exception to this would occur if $f(t)$ were such that only one of the Fourier coefficients C in eqn (12.8) were non-zero, i.e. the input is harmonic. This agrees with our introductory comments that sine waves are beautifully simple, whereas other waveforms result in more complicated behaviour.

An example that illustrates these ideas is provided by an electrical circuit. In the special case where all the circuit elements are resistors, the transmission of the circuit (i.e. the ratio of the output to the input voltages) is frequency-independent, and the output waveform is a scaled down version of the input. If, however, the circuit elements include capacitors and/or inductors, the transmission varies with frequency, and unless the input is harmonic, the circuit will change the shape of the applied voltage waveform. Thus, for example, the square wave of fig. 12.1(b) applied to the circuit will not result in a square wave output.

It is worth emphasising what we have achieved. Eqn (12.6) represents the response of the system we are considering to a *harmonic input*. By the use of the Fourier series, we can now extend our understanding of the way the system behaves to *any periodic input*.

The only problem is that our solution (12.9) is again an infinite series. It is, however, a formal solution to our problem. Furthermore, as we shall see, the Fourier series for any function of interest in physical applications has coefficients C_n which decrease as n increases, sometimes rapidly. This means that a reasonably accurate answer can be obtained by considering at most a few terms of the series. Then we can use a computer to add up as many terms as we want in order to obtain a good approximation to the solution in any particular case.

12.3 Determining the Fourier coefficients

So far we have been describing how useful the Fourier technique can be. In order to use it in practice, we need to determine the Fourier coefficients (e.g. the as and bs in eqn (12.1)) for our specified function $f(t)$. For simplicity of notation, we have chosen $f(t)$ to have a period of 2π.

The trick we use is the general one of "multiplying" both sides of eqn (12.1) by something clever, in order to make all the terms but one on the right-hand side of the equation disappear. If the only remaining term is, for example, the one involving b_{15}, then this immediately determines the value of this coefficient. Next we make other cunning choices in order to determine each one of the as and bs. Then our problem is solved. All we have to do is to discover the nature of the clever trick.

The procedure utilises the fact that the functions $\sin mt$ and $\cos nt$ are orthogonal over the range 0–2π, provided that m and n are integers. This means that

$$\left. \begin{array}{ll} \displaystyle\int_0^{2\pi} \sin mt \sin lt\, dt = 0 & m \neq l \\[2ex] \displaystyle\int_0^{2\pi} \cos nt \cos kt\, dt = 0 & n \neq k \\[2ex] \text{and} \quad \displaystyle\int_0^{2\pi} \sin mt \cos nt\, dt = 0 & \text{any } m, n \end{array} \right\} \qquad (12.10)$$

When $m = l \neq 0$ or when $n = k \neq 0$, the first two integrals are clearly non-zero, and in fact

$$\int_0^{2\pi} \sin^2 mt\, dt = \int_0^{2\pi} \cos^2 nt\, dt = \pi \qquad (12.11)$$

(Most of us remember that the average value of $\sin^2 \theta$ over a whole number of cycles is $1/2$. For those who do not, think of: (i) how the average values of $\sin^2 \theta$ and $\cos^2 \theta$ compare; and (ii) the value of the sum of these averages.)

By now it should be obvious what is needed. In order to determine the coefficient b_{15} in eqn (12.1), we simply multiply the equation by $\sin 15t$, and then integrate both sides over the range 0–2π in t. The right-hand side is integrated term by term; because of the orthogonality conditions (12.10), almost all the terms disappear and we are simply left with

$$\int_0^{2\pi} f(t) \sin 15t\, dt = \int_0^{2\pi} b_{15} \sin^2 15t\, dt$$

Then from (12.11) we find

$$b_{15} = \frac{1}{\pi} \int_0^{2\pi} f(t) \sin 15t\, dt$$

Of course, we could have been a bit more intelligent, and multiplied by $\sin lt$, rather than $\sin 15t$. Then we would have found

$$b_l = \frac{1}{\pi} \int_0^{2\pi} f(t) \sin lt\, dt \qquad (12.12)$$

Similarly had we instead multiplied by $\cos kt$ before integration, we would have obtained

$$a_k = \frac{1}{\pi} \int_0^{2\pi} f(t) \cos kt\, dt \qquad (12.12')$$

At this stage, we discover why the rather curious factor of $1/2$ appears in the first term of the eqn (12.1). If we integrate this equation as it stands, we find that

$$a_0 = \frac{1}{\pi} \int_0^{2\pi} f(t)\, dt \qquad (12.12'')$$

This looks just like equation (12.12′), with k set equal to zero. This means that we do not have to remember the solution (12.12″) separately. Without the factor of $1/2$ in eqn (12.1), we would have needed to replace the π in (12.12″) by 2π. The reason for this slight difference with the a_0 term is due to the fact that the normalisation conditions (12.11) apply for all positive integral values of m and n, but not for $n = 0$ (or $m = 0$).

We have now achieved what we set out to do — the coefficients a_k and b_l are determined for any specific repeating function f (such as the one of fig. 12.1(b)) by eqns (12.12) and (12.12′) respectively. Of course, in many cases, the integrations on the right-hand sides of these equations may be difficult or impossible to perform analytically. It will then be necessary to evaluate the integrals, and hence the a_k and b_l, numerically.

12.4 Least squares approach to Fourier series

An alternative derivation which leads to a series with the above Fourier coefficients for a periodic function $f(t)$ is as follows. We consider a *finite* series with only N cosine and M sine terms, i.e.

$$s(t) = a_0/2 + \sum_{n=1}^{N} a_n \cos nt + \sum_{m=1}^{M} b_m \sin mt \qquad (12.13)$$

We then ask what are the values of the as and the bs such that the series gives the best approximation to $f(t)$ in the least squares sense, i.e.

$$D = \int_0^{2\pi} [s(t) - f(t)]^2 dt \quad \text{is a minimum}$$

Thus we must minimise D with respect to the $M + N + 1$ coefficients. On performing the relevant partial differentiations, we obtain

$$\frac{1}{2}\frac{\partial D}{\partial b_l} = \int_0^{2\pi} [s(t) - f(t)] \sin ltdt = 0$$

Making use of the orthogonality and normalisation conditions (12.10) and (12.11) results in

$$b_l = \frac{1}{\pi}\int_0^{2\pi} f(t)\sin ltdt$$

which is identical to (12.12). Similarly the a_n and a_0 coefficients are as determined earlier.

This then provides us with extra insight into the Fourier series. We see that, if we take only a finite number of terms rather than the complete infinite series, the Fourier coefficients are such as to ensure that we have the best least squares approximation possible to $f(t)$ for the given number of terms we have chosen to use. This helps our conscience a bit when we decide to stop calculating in some problem after just a few terms.†

12.5 An actual example

Let us see how we find the Fourier coefficients in a specific case. As an example, we choose the graph shown in fig. 12.5(a), i.e.

$$f(t) = \begin{cases} -1 & 0 < t < \pi \\ +1 & \pi < t < 2\pi \end{cases} \tag{12.14}$$

As before, we assume that the period of $f(t)$ is 2π.

Because $f(t)$ is defined separately in the two different t ranges, it is necessary to split up the integrals (12.12) similarly. Thus we write

$$b_l = \frac{1}{\pi}\int_0^{2\pi} f(t)\sin ltdt$$

$$= \frac{1}{\pi}\left\{-\int_0^{\pi}\sin ltdt + \int_\pi^{2\pi}\sin ltdt\right\}$$

$$= \frac{1}{\pi l}\left\{[\cos lt]_0^{\pi} - [\cos lt]_\pi^{2\pi}\right\} \tag{12.15}$$

We now need to be careful, as it is important to distinguish between the cases where l is odd and even. In the former, $\cos l\pi = -1$ but

---

† We should remember, however, that this result shows us only that our approximation is the best possible of this type. It does *not* tell how good (or bad) it is.

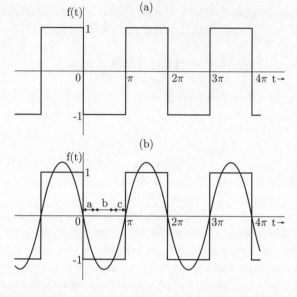

Fig. 12.5. (a) The function $f(t)$ of eqn (12.14), for which we wish to obtain the Fourier series. (b) The curve shows the best approximation to $f(t)$ of a single sine wave of period 2π (equal to that of $f(t)$); its amplitude is negative, with magnitude slightly larger than unity. By considering the signs of the discrepancies between $f(t)$ and the sine curve in the regions a, b and c, we find that a correction of the form $\sin 3t$ with a negative amplitude is needed; this is the next term of the Fourier series.

$\cos 2l\pi = +1$, whence

$$b_l = -4/\pi l$$

In contrast for l even, both $\cos l\pi$ and $\cos 2l\pi$ are $+1$, and hence b_l is zero.

We can similarly determine the a coefficients as

$$
\begin{aligned}
a_k &= \frac{1}{\pi} \int_0^{2\pi} f(t)\cos kt\, dt \\
&= \frac{1}{\pi}\left\{ -\int_0^{\pi} \cos kt\, dt + \int_{\pi}^{2\pi} \cos kt\, dt \right\} \\
&= \frac{1}{\pi k}\left\{ -[\sin lt]_0^{\pi} + [\sin lt]_{\pi}^{2\pi} \right\} \\
&= 0
\end{aligned}
\tag{12.15'}
$$

for all values of k, and similarly for a_0.

Having calculated that the a coefficients are all zero, we can ask whether we ought to have been able to see this without bothering to do the integrations. The answer is "yes". In the top line of eqn (12.15'), we have written the range of integration as 0–2π, but, because the function is periodic, any range encompassing one period would have been equally suitable. Thus we could have chosen $-\pi$ to $+\pi$, which is symmetric about $t = 0$. Now we concentrate on the integrand. We see that we have the product of two factors. The first, $f(t)$, is *odd* as a function of t (i.e. $f(t) = -f(t)$, — see fig. 12.5), while the second, $\cos kt$, is *even* (i.e. $\cos(kt) = +\cos(-kt)$). Thus the product of the two is *odd*, and hence its integral between symmetric limits is zero (compare Appendix A8).

It is always worth while to spend a few seconds looking at integrals in this way to see if it is "obvious" whether they vanish. Not only is this quicker than doing the actual integration explicitly (which in some cases may even be impossible to perform analytically), it avoids the danger of making some trivial error which could result in our getting the wrong answer.

Returning to our problem, we see that we have now determined all the coefficients, with the result that we now have the Fourier series as

$$f(t) = -\frac{4}{\pi}(\sin t + \frac{1}{3}\sin 3t + \frac{1}{5}\sin 5t + \frac{1}{7}\sin 7t + \cdots) \qquad (12.16)$$

Several points are worth noting:

(i) As promised, the coefficients decrease as we go to the higher terms of the series.

(ii) Across $t = 0$, the function jumps from $+1$ at negative t, to -1 at positive t. On the other hand, the series (12.16) for $f(t)$ is continuous across $t = 0$, and yields $f(0) = 0$. The fact that the series gives the average of the limits of the function on the two sides of the discontinuity is a general property of Fourier series for discontinuous functions.

(iii) According to Section 12.4, the first term of the series (i.e. $-(4/\pi)\sin t$) should be the best approximation of the type "$b_1 \sin t$" to our function. As we can see from fig. 12.5(b), this looks plausible. Certainly we need the minus sign, and the value of the coefficient $(4/\pi \sim 1.3)$ should be a bit bigger than 1. So if we had made a numerical error by, say, a factor of 2 or more, we should have been aware of this from the rough diagram.

(iv) If we look a bit harder at fig. 12.5, we see that our single $b_1 \sin t$ term lies:

above $f(t)$ in the region (a);

below it in region (b);

above it in region (c);

etc.

Thus we need a correction whose frequency is three times that of the first term, and of the same sign. This agrees with the absence of a $\sin 2t$ term in our Fourier series, and we also see that we have the correct sign for the $\sin 3t$ term.

At this stage, it is not worth drawing increasingly more complicated diagrams to see what further corrections are needed, but our Fourier series has passed the simple tests we have made. In the next section we will discover how to check whether the coefficients of the various terms decrease in a reasonable manner.

12.6 n dependence of Fourier coefficients

For the Fourier series we have considered, the magnitudes of the coefficients a_n and b_n decrease as we proceed to the higher frequency components, i.e. as n increases. In this section we are going to investigate experimentally how fast the coefficients decrease for large n.

First we are going to draw three functions over the range -2π to $+4\pi$ in the variable t. These functions are

$$f_1(t) = t \text{ for } 0 < t < 2\pi \tag{12.17}$$

$$f_2(t) = \begin{cases} \frac{1}{2}\pi - t & \text{for } 0 < t < \pi \\ \frac{1}{2}\pi + t & \text{for } -\pi < t < 0 \end{cases} \tag{12.18}$$

$$f_3(t) = \begin{cases} t(\pi - t) & \text{for } 0 < t < \pi \\ t(\pi + t) & \text{for } -\pi < t < 0 \end{cases} \tag{12.19}$$

and all of them have period 2π.

The next stage is for the reader (i.e. *you*) to take a pen and paper and actually calculate the Fourier coefficients for each of these three functions. This will take about an hour, especially as we want to get the correct coefficients, and we are then going to stare at them and try to understand their n dependence. We therefore want to check our calculations carefully. So, see you again later.

* * * * * * * * * * * * * * * * * * *

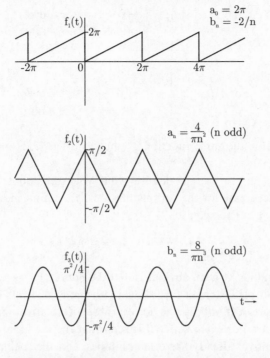

Fig. 12.6. The functions $f(t)$ of eqns (12.17)–(12.19), whose Fourier series have coefficients as given in eqns (12.20)–(12.22).

Welcome back. Now we are going to look at the diagrams of the three functions f_1, f_2 and f_3 that we drew earlier (see fig. 12.6). We can see that $f_2(t)$ is "nicer" than $f_1(t)$, in that f_1 has discontinuities at $t = 2n\pi$. This means that the function has a jump in value there, i.e.

$$\lim_{\varepsilon \to 0}[f_1(2n\pi + \varepsilon)] \neq \lim_{\varepsilon \to 0}[f_1(2n\pi - \varepsilon)]$$

In a similar way, $f_3(t)$ is "nicer" than $f_2(t)$, in that the gradient of f_2 is discontinuous at $t = n\pi$, whereas f_3 has been constructed so that the gradients match at the boundaries of the different regions in which it is defined.

Just to make sure that we calculated the coefficients correctly, the values should be:

$$\text{For } f_1(t) : \quad \begin{cases} a_0/2 = \pi \\ a_n = 0 \\ b_n = -2/n \end{cases} \tag{12.20}$$

$$\text{For } f_2(t) : \quad \begin{cases} a_n \begin{cases} = 0 & \text{for } n \text{ even} \\ = 4/\pi n^2 & \text{for } n \text{ odd} \end{cases} \\ b_n \quad = 0 \end{cases} \qquad (12.21)$$

$$\text{For } f_3(t) : \quad \begin{cases} a_n \quad = 0 \\ b_n \begin{cases} = 0 & \text{for } n \text{ even} \\ = 8/\pi n^3 & \text{for } n \text{ odd} \end{cases} \end{cases} \qquad (12.22)$$

In fact, $f_2(t)$ is the same function as shown in fig. 12.2, and considered in Section 12.1.

Now it is up to you to deduce what you can about the way the coefficients decrease for the various funtions. Take a bit more time off to try to work it out for yourself.

$$* \; * \; * \; * \; * \; * \; * \; * \; * \; * \; * \; * \; * \; * \; * \; * \; * \; * \; *$$

What should have been obvious is that the Fourier terms for the "nicer" functions fall off faster. In fact, for large n the coefficients decrease like $n^{-(m+1)}$, where $d^m f / dt^m$ is the lowest order derivative which exhibits discontinuities. For example, $f_1(t)$ is such that the function itself (i.e. the $m = 0$ derivative) already has discontinuities, and the coefficients fall off like the first power of $1/n$. However, $f_3(t)$ is continuous and so is its gradient. It is only $d^2 f_3 / dt^2$ which has discontinuities at $t = n\pi$, and so we expect the coefficients to fall off as n^{-3}, as in fact they do.

As another example of the validity of the above, you should now obtain the Fourier series for the function

$$f_4(t) = \sin t \sin 2t \qquad (12.23)$$

Does the behaviour of the coefficients agree with our expectations?

Since we can differentiate $f_4(t)$ as many times as we want without ever finding any discontinuities over the whole range of t, we would expect the coefficients to decrease very quickly indeed. We ought to have found that a_1 and a_3 are equal to $1/2$ and $-1/2$ respectively, and all other coefficients are zero. This then agrees with our prediction that the coefficients disappear rapidly for larger n. It also may make us suspect that, if we had used eqns (12.12) and (12.12') to calculate the Fourier coefficients, perhaps there would have been a better way of doing so. Indeed this is the case, since we may make use of the identity

$$2 \sin A \sin B = \cos(A - B) - \cos(A + B) \qquad (12.24)$$

in order to determine the complete Fourier series by inspection.

12.7 Checking Fourier coefficients

We are now ready to describe the steps we ought to take when determining Fourier coefficients for a function $f(t)$, in order to check that our calculated answer is plausible.

Step 1: Draw the function for a few complete cycles. This enables us to see what the repeat period τ is, and hence also the fundamental frequency† ω. We can then write the series as in (12.3), or in the simpler form (12.1) if $\tau = 2\pi$.

Step 2: Look at the function to decide whether it is odd or even, in which case we can delete all the a_n terms or all the b_n terms respectively. If the function $f(t)$ is neither odd nor even, we need both a_n and b_n terms.

Step 3: It is almost always possible to draw a very good approximation to the constant $a_0/2$ and to lowest frequency term of the series. This enables us to check the signs and the approximate magnitudes of these coefficients in our calculations.

Step 4: In some cases we can also see from our diagram what the next term must be in order to correct for the deficiencies of the leading term. For example, we may well be able to check its frequency. This in general will be 2ω, but for some functions this term will be missing and we will have only a 3ω term as the next one. If this is so, it would not be too surprising if all higher terms of the type $2n\omega$ were zero as well.

We would also hope to check the sign of the second term. Again it will often be the case that if the second term has the same sign as the first, then all the terms have this sign, while if they are opposite, then the signs along the complete series may alternate.

Step 5: We next look at the n dependence of our coefficients, and see if they agree with our expectations based on the function's behaviour (see Section 12.6).

Step 6: As a final test we can substitute some specific value t' for the variable t (e.g. $t' = \pi$ or $\pi/2$), such that all the sine and cosine terms take on simple values. We should then be able to use a few terms of our series to check numerically that the sum gives a plausible approximation to $f(t')$ (we need to sum a few terms, since if we use only the first term, we are doing no more than the

---

† As in Volume 1, ω is often referred to simply as the frequency, rather than the more accurate but cumbersome "angular frequency".

check of Step 3). It is important to remember that, if the function displays a discontinuity at $t = t'$, the Fourier series should give the average of the two values on either side of the discontinuity, e.g. for f_1, it yields an answer of π at $t = 2n\pi$.

If our solution passes all these tests, it may well even be correct.

As an example of Step 6, we can take the function $f_2(t)$ in Section 12.6, and set $t = 0$. Then $f_2(0) = \pi/2$, while the series solution is

$$\frac{4}{\pi}\left(\cos t + \frac{1}{3^2}\cos 3t + \frac{1}{5^2}\cos 5t + \cdots\right) = \frac{4}{\pi}\left(1 + \frac{1}{9} + \frac{1}{25} + \cdots\right) \quad (12.25)$$

from which we deduce π^2 is somewhat larger than 9.2. This looks reasonable.

12.8 Half range series

In physical situations, a function $f(x)$ is often meaningful only for a finite range of the independent variable x. Thus an elastic string held between two nails at $x = 0$ and $x = L$ may be given some specific displacement $y(x)$ for $0 < x < L$. Similarly a bar of length L may have a temperature distribution $\theta(x)$ at some particular time. In both cases, the value of the function $y(x)$ or $\theta(x)$ is not too relevant outside the range of interest (but see later).

In this type of problem, we shall determine Fourier series for functions which are defined only over a limited range, which we take to be 0–π for simplicity. This contrasts with the standard situation where the function must repeat for ever.

In order to convert the current case to the more standard one, we extend the definition of the function outside the region $0 < x < \pi$, so that it does repeat. For example, for the function shown in fig. 12.7(a), we could choose the extension to be as shown in 12.7(b) where the repeat distance is π. Of course, this is not the only possibility, and some others are shown in the rest of fig. 12.7. All of these are identical to our original $f(x)$ in the range 0–π, but their Fourier series will be very different. Indeed the periods of the various functions shown in fig. 12.7 are not all the same.

Of all the infinite possible choices, two are particularly simple. They are shown in figs. 12.7(e) and (f). They have the property of being even and odd functions respectively which repeat over the range 0–2π.† Thus their

---

† The phrase "half range series" derives from the fact that the function was originally defined over a range that is half the repeat distance for the Fourier series.

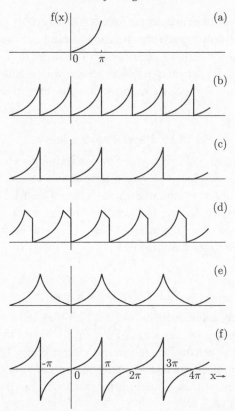

Fig. 12.7. The function $f(x)$ shown in (a) is defined only over the range $0 < x \leq \pi$. In (b)–(f), functions are displayed which extend from $-\infty$ to $+\infty$ in a repetitive manner, and which are identical with $f(x)$ over the region in which it is defined. In (b) the repeat distance is π, while in (c) it is 2π, with the function being set equal to zero in the x range from $-\pi$ to zero. An extension of arbitrary shape is chosen in (d), resulting in a repeat distance rather larger than π. Finally (e) and (f) both repeat after an x interval of 2π, and are respectively even and odd as functions of x; their Fourier series will contain only cosine or only sine terms respectively.

Which continuation of $f(x)$ is sensible in order to derive a Fourier series depends on the physics of the problem, as specified by the boundary conditions that have to apply.

Fourier series will consist only of a_n and of b_n terms, respectively. They are even neater (and also tend to be more relevant in physics applications) than if the repeat distance has been chosen as in fig. 12.7(b).

Of course, as a purely mathematical problem of finding a representation of the function $f(x)$ over the range 0–π, any of our choices is satisfactory, since they are all designed to reproduce $f(x)$ for $0 < x < \pi$, and differ from each other only outside the meaningful range. However, in an actual physical situation where we need the Fourier series in order to make use of it in solving a particular problem, there will usually be one choice which is relevant. We will see examples of this later in Section 12.10.

12.9 Physics applications

In this section, we will apply the Fourier technique to some simplified physical situations. This is to give a flavour of the use of Fourier series in practice. In order to minimise the number of Fourier coefficients that we have to determine, we will assume that in each of the examples considered below the "disturbance" is as shown in fig. 12.1(d), with Fourier coefficients as given in eqns (12.2).

12.9.1 Electrical circuit

We have already mentioned in Section 12.2 that the Fourier method is useful for analysing the response of electrical circuits to repetitive input waveforms. Here we consider the simple circuit of fig. 12.8.† Fig. 12.1(d) represents the applied voltage V_{in}, with maximum amplitude $\pm\pi/2$, and with variable repeat time τ; its Fourier coefficients are as in eqn (12.21).

Our circuit is such that

$$V_{out} = V_{in}\frac{\omega^2 LC}{\omega^2 LC - 1} \tag{12.26}$$

or in the notation of eqn (12.6)

$$H(\omega) = \frac{\omega^2 LC}{\omega^2 LC - 1}$$

$$= \frac{\omega^2}{\omega^2 - \omega_0^2} \tag{12.27}$$

where $\omega_0^2 = 1/LC$, a constant for a given circuit; and ω is the frequency of the applied voltage if it is harmonic, or of a Fourier component if it is not.

Depending on the value of $\tau\omega_0$, the output can have many different waveforms (see fig. 12.9). If we change the period τ and hence the

---

† Although the impedances of both L and C are imaginary (see Sections 4.4 and 4.5), by excluding resistors from the circuit, we obtain an $H(\omega)$ (see eqn 12.6) which is frequency-dependent, but real. This simplifies the appearance of the formulae as the relative phase of the output with respect to the input is always 0 or π.

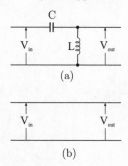

(a)

(b)

Fig. 12.8. (a) An electric circuit consisting of a capacitor C and inductor L in series. For a harmonically varying input voltage V_{in}, the output V_{out} is given by eqn (12.26). (b) In the high frequency limit, the impedances of C and L become zero and infinity, respectively, so the circuit simplifies as shown, and $V_{out} = V_{in}$. Because this relationship is frequency-independent, the output is identical to the input, regardless of the latter's waveform. For finite frequencies, this will not in general be so.

fundamental frequency ω_1 ($= 2\pi/\tau$) of the input, we will obtain a very large output when ω_1 is close to the circuit's natural frequency ω_0 (see fig. 12.9(d)). As τ is increased, ω_1 decreases below the resonance frequency, but when $\tau \approx 3(2\pi/\omega_0)$, the third harmonic is near the circuit's natural frequency and we again obtain a large output, as in fig. 12.9(b). Thus whenever τ is such that an odd multiple $2n-1$ of ω_1 is close to ω_0, the denominator of $H(\omega)$ becomes very small for that particular harmonic, and hence $H(\omega)$ will be much larger than for any other harmonic.

For example, if $\omega_1 = 0.201\omega_0$, $H(5\omega_1) \approx 101$, as compared with $H(3\omega_1) \approx -0.6$ and $H(7\omega_1) \approx 2$. This means that the output voltage

$$V_{out} = \sum_{n=1}^{\infty} H((2n-1)\omega_1)a_{2n-1} \cos(2n-1)\omega_1 t$$

$$\approx 101\frac{4}{25\pi} \cos 5\omega_1 t \tag{12.28}$$

The output, being almost purely harmonic with $\omega = 5\omega_1$, is a very different shape from the input (see fig. 12.9(a)).

Thus whenever τ approaches $(2n-1)2\pi/\omega_0$, the output is nearly harmonic in shape and its amplitude rises rapidly; it then suddenly changes sign and decreases in magnitude as τ increases beyond this value. For τ not too close to one of these values, the output waveform is smaller and more complicated in shape (see fig. 12.9(c)).

In fact the relative amplitudes of the outputs, for inputs close to the

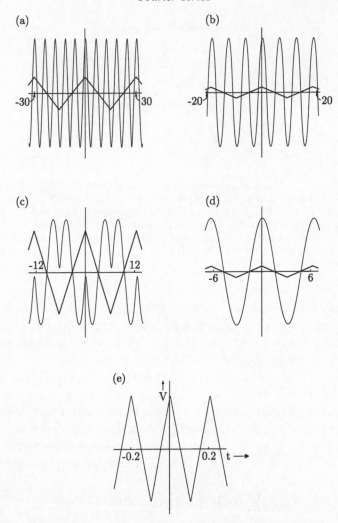

Fig. 12.9. Response of the circuit of fig. 12.8(a), with $\omega_0 = 1$, to the triangular-shape input voltage of fig. 12.1(d), with decreasing repeat time τ (and hence increasing fundamental frequency $\omega_1 = 2\pi/\tau$). The curves are the output voltages V_{out}, and the lines show the triangular-shaped inputs V_{in}, in the same units. As ω_1 is increased, the time scale of the plots is made correspondingly smaller. The vertical scales are adjusted according to the magnitude of the output. (a) $\omega_1 = 0.201$. Here the repeat time is such that the fifth harmonic $5\omega_1$ is close to the natural frequency ω_0 of the circuit. The output is to a good approximation a pure cosine wave, even though the input is triangular at one fifth of the frequency. For $\omega_1 = 0.199$, the output is inverted with respect to that shown here, but of very closely the same magnitude.

various resonant frequencies, give information on the Fourier coefficients of the input voltage waveform. Because the frequencies used in figs. 12.9(a), (b) and (d) satisfy†

$$5\omega_a = 3\omega_b = \omega_d \approx \omega_0$$

the amplitudes are in the ratios of approximately

$$a_5 : a_3 : a_1 = 1/25 : 1/9 : 1$$

In contrast, if the fundamental frequency of the input is very much larger than ω_0, all the relevant $H(\omega) \approx 1$, and so V_{out} will consist of almost exactly the Fourier sum into which we decomposed V_{in}, i.e. $V_{out} \approx V_{in}$ (compare fig 12.9(e)). This is of course reasonable, since at high frequency the impedences of the capacitor and inductor tend to zero and infinity respectively, and so the circuit approximates to that of fig. 12.8(b). Here clearly $V_{out} = V_{in}$, independent of the shape of V_{in}.

12.9.2 Car on bumpy road

A moving car will be jolted by any bumps on the road along which it is travelling. If the surface is repetitive, we can use its Fourier series to analyse what happens to the car. Thus we take fig. 12.2 as representing the shape of the road's surface $h(x)$, and we neglect the fact that the car may temporarily lose contact with the road as it passes over the top of a bump.

We model the vertical motion of the car's body as follows. The car's

---

† Because the circuit has no resistance, the amplitude of the response at precisely a resonant frequency is infinite. Thus it is necessary for ω_a, ω_b and ω_c to be not exactly at the resonance frequency, but close to it, and in the specified ratios. However, in a realistic circuit with $R \neq 0$, the responses at the different resonances are finite and in the ratios stated. To a good approximation, the ratios of the first few Fourier coefficients are thus readily determined by observing the maximum responses of the circuit as the frequency of the input voltage is tuned through the various harmonic resonances.

Fig. 12.9. (contd) (b) $\omega_1 = 0.335$. Now the third harmonic is close to ω_0, and the output is again large. (c) $\omega_1 = 0.5$. None of the harmonics is close to the circuit's natural frequency. The output is of a more complicated shape, and is small. (Hence the amplitude of the input has been reduced by a factor of 4 to show it on the same scale.) (d) $\omega_1 = 1.005$. This time the fundamental frequency is close to resonance, and the output is very large. (To make it visible on the same scale, the input has been multiplied by a factor of 10.) (e) $\omega_1 = 30$. The frequencies of all the harmonics are large compared with ω_0, and the output is almost identical to the input. (Hence the input is not shown separately.)

suspension transmits a force to the body of the car which is equal to $k(h - y)$, where y is the vertical displacement of the driver's seat from its position when the car is on a flat surface, and k is the constant of proportionality for the suspension. If we neglect dissipative effects, we derive the differential equation for y as

$$m\frac{d^2y}{dt^2} + ky = kh(x)$$
$$= kh(vt) \qquad (12.29)$$

where m is the mass of the car and v is its speed.

After we decompose h into its Fourier components as

$$h(vt) = \Sigma a_n \cos nvt \qquad (12.30)$$

with the a_n as before given by (12.2), we obtain the response of the car as

$$y = \sum_n \frac{ka_n}{k - mn^2v^2} \cos nvt \qquad (12.31)$$

As with the electrical circuit, we see that the response of the system will be large when nv (where n is an odd integer) is close to the natural frequency†, in this case $\sqrt{k/m}$. Thus as the car is driven along the road, there will be a series of speeds which will be most uncomfortable for the passengers.

In this and the previous problem, our solution is such that the response would be infinite for certain values of the car's speed here, or of the frequency of the applied voltage in Section 12.9.1. This is clearly unphysical, and hence undesirable. We stress that this is *not* connected with using Fourier methods, but arises simply from our neglect of dissipative effects in the car's suspension, or of resistive elements in the electrical circuit. With their inclusion, our solution would remain finite under all conditions (compare eqns (5.137) and (5.135)).

12.9.3 Wave on string

If we establish some periodic transverse displacement along an infinite elastic string and then release it, it will vibrate. Thus fig. 12.2 could represent the transverse displacement $y(x, 0)$ at time zero. It is tempting (but incorrect) to imagine that as the string vibrates, it will always look

---

† This looks to be dimensionally incorrect. The reason that all is in fact well is that we have taken the repeat distance of $h(x)$ to be 2π, and hence the speed v has dimensions $(\text{time})^{-1}$.

like the original form, but multiplied by a factor that varies continuously and harmonically from $+1$ at $t = 0$, through zero, -1, zero again and then back to $+1$ after one period.

The displacement obeys the one-dimensional wave equation

$$\frac{\partial^2 y}{\partial x^2} = \frac{1}{c^2}\frac{\partial^2 y}{\partial t^2} \tag{12.32}$$

where c is the propagation speed of moving waves along the string. For initial displacements whose spatial dependence is harmonic with wavelength $\lambda = 2\pi/n$, a solution is

$$y(x,t) = A\cos(nx + \phi)\cos(nct + \psi) \tag{12.33}$$

(See Section 14.3 for a derivation of the wave equation, and Section 11.3.1 for a justification of the above solution by the method of separation of variables.)

For a string initially at rest, $\psi = 0$. Then $A\cos(nx + \phi)$ gives the initial displacement at $t = 0$. The really important feature of eqn (12.33) is that the frequency $\nu = nc/2\pi$ of the wave depends on its wavelength λ, such that $\nu = c/\lambda$ (as expected for a wave of speed c). As the wave equation is linear in y, we can generalise this to our non-harmonic displacement of fig. 12.2 by considering each Fourier component separately, and summing the contributions. Thus

$$y(x,t) = \sum_n a_n \cos nx \cos nct \tag{12.34}$$

where the a_n are as given in eqn (12.2).†

At $t = 0$, the right-hand side of eqn (12.34) is simply $\Sigma a_n \cos nx$ which is our Fourier series for $y(x,0)$, as required. The crucial point is that at subsequent times, each term in the series is multiplied by its relevant $\cos nct$. For some arbitrary time, these will in general be different for $n = 1, 3, 5, \ldots$, as each Fourier component is oscillating at a different frequency. Hence the summation does *not* reduce to $\Sigma a_n \cos nx$ multiplied by a common overall reduction factor. The shorter wavelength components oscillate faster, and hence the various components will get

---

† If the string had been initially undeflected, but each part of it had been given a transverse velocity $\dot{y}(x, t = 0)$, we would have needed a $\sin nct$ time dependence rather than $\cos nct$, with the Fourier coefficients being determined by $\dot{y}(x, t = 0)$. If the string were both displaced and moving transversely at $t = 0$, we would have needed to keep both $\cos nct$ and $\sin nct$ terms.

out of phase with each other. It is this which causes the shape of the string to change as time goes by.

Of course, at special times this will not be true. For example, at $t = 2r\pi/c$ (where r is any integer), all the $\cos nct$ terms will be unity, and so the string will be back at its initial position; this is because the fundamental harmonic displacement will have completed a whole number of oscillations, and the higher frequencies even more. Similarly at $t = 2(r \pm \frac{1}{4})\pi/c$, all the relevant $\cos nct$ terms (i.e. those with n odd) are zero, so the string is in its undeflected position. The other simple situation is at $2(r + \frac{1}{2})\pi/c$, when $\cos nct = -1$; this can be factored out to give

$$y(x, 2(r + \tfrac{1}{2})\pi/c) = -\sum_n a_n \cos nx$$
$$= -y(x, 0) \qquad (12.35)$$

i.e. the string is simply inverted from its original position. Thus the string does oscillate between:

$$\text{its initial position at } t = 2r\pi/c$$
$$\text{zero at } t = 2(r + \tfrac{1}{4})\pi/c$$
$$\text{the initial position inverted at } t = 2(r + \tfrac{1}{2})\pi/2$$
$$\text{zero at } t = 2(r + \tfrac{3}{4})\pi/c$$
$$\text{and its initial position again at } t = 2r'\pi/c$$

At intermediate times the shape is not simply a multiple of the initial shape, but can be obtained from eqn (12.34). It is, however, easier to derive the solution by d'Alembert's method (see Section 11.4). Thus we imagine the initial pattern as consisting of two components, each of half its amplitude, and moving in opposite directions with velocities $\pm c$. At time t, we see how far the components have separated, and then add them. The net displacements at some specific times are shown in fig. 12.10. When we look at them, we realise that maybe we should have seen from the beginning that this is what the solution should look like. Initially most of the string is straight, and so individual elements have no restoring force on them, and hence do not move. The only places for which this does not hold at $t = 0$ are the tops and bottoms of the waveform; these points are pulled towards the $y = 0$ line. Gradually the regions that move towards the undisplaced position extend until at

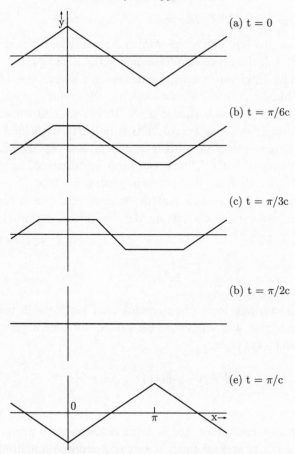

Fig. 12.10. The transverse deflection of an infinite length of string shown at different times: (a) The initial position at $t = 0$. (b) $t = \pi/6c$, where c is the speed of waves along the string. (c) $t = \pi/3c$. (d) $t = \pi/2c$, and the string instantaneously lies along the x axis. (e) $t = \pi/c$, and the string is inverted with respect to its initial displacement.

For even later times, the string collapses back towards the x axis, and then, as t approaches $2\pi/c$, builds up to become identical to the initial displacement. After this the whole cycle repeats.

$t = \pi/2c$, the whole string lies along the x axis (but is moving). And so the process continues.

This problem is thus another example where different approaches and intuition can be compared in order to check that our solution is reasonable.

12.9.4 *Thermally conducting bar*

This is in fact the type of problem which originally led Fourier to devise his method of dealing with non-harmonic periodic functions.

We consider an infinite conducting bar along which a periodic temperature distribution $f(x)$ has been established. We want to find what the distribution will be at subsequent times. Before we had considered the previous example, we might have been tempted to think that the temperature differences would gradually disappear, and that the distribution would maintain its original functional form, with reduced amplitude as time goes by. We shall see that this is in general not true.

In Section 11.1.2.1, we saw that the temperature θ as a function of position and time in a conducting bar obeys the partial differential equation

$$\frac{\partial^2 \theta}{\partial x^2} = k \frac{\partial \theta}{\partial t} \tag{12.36}$$

where k is a constant for a given conductor. Then if the distribution in x is harmonic with wavelength 2π, the subsequent time dependence will be exponential decay, i.e.

$$\theta(x, t) = A \cos(nx + \phi) e^{-n^2 t / k} \tag{12.37}$$

is a possible solution (see Section 11.3.2.1). It is only for a spatial dependence of temperature that is a single harmonic function that the temperature $\theta(x, t)$ at later times is simply a reduced amplitude replica of the original pattern $\theta(x, 0)$. It is also worth noting that the wavelength $2\pi/n$ of the spatial dependence of the temperature in this case determines the time constant k/n^2 of the time variation.

If the initial temperature distribution is of the form of fig. 12.2, we can thus write down the solution by inspection as

$$\theta(x, t) = \Sigma a_n \cos nx e^{-n^2 t / k} \tag{12.38}$$

with the a_n as given by eqn (12.2).

At any fixed time $t \neq 0$, the factors $e^{-n^2 t / k}$ act as a set of coefficients which, because of the n^2, are different for each Fourier component. Since the higher terms are more severely damped than the lower ones, the temperature distribution will become more like a harmonic one of the fundamental wavelength as time goes by. Eventually of course, the effect

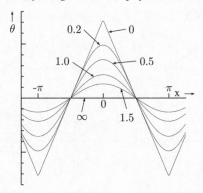

Fig. 12.11. Temperature distributions along an infinite conducting bar, shown at different times.

The initial temperature is labelled 0. The other plots are for subsequent times, as specified in units of t/k; k is the parameter occurring in the heat conduction equation (its dimensions are time/(distance)^2, but as the repeat distance of the temperature distribution is 2π (i.e. dimensionless), t can be expressed in terms of k). As t tends to infinity, the temperature variations die away, and θ tends to zero everywhere.

of the conduction is to reduce to zero the temperature variations along the bar.

The temperature distributions at various times are shown in fig. 12.11.

12.10 Half range series in physical situations

The examples of the previous section were a little unrealistic in that they assumed, for example, that the temperature distribution was established on a bar of infinite length. Real bars are shorter than this, and the temperature distribution of a thermal conductor is meaningful only within its physical extent. Thus we are forced to consider the "half range series" problems that we discussed from a mathematical point of view in Section 12.8. Here we are particularly interested in the question of which of the possible infinite extensions is the relevant one required by the physics of our particular problem.

12.10.1 Temperature of a finite bar

We assume that we are provided with a bar extending from the origin to $x = \pi$, and that initially its temperature distribution $f(x)$ is given. As a

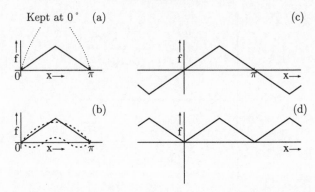

Fig. 12.12. (a) Initial temperature distribution along a finite bar, whose ends at $x = 0$ and $x = \pi$ are kept at $0\,^{\circ}\mathrm{C}$ at all times. (b) First two sine components of the Fourier series for $f(x)$. Because of the boundary conditions on the temperatures at the ends of the bar, only components which themselves are zero at $x = 0$ and $x = \pi$ are satisfactory. (c) The Fourier series used to extend the definition of $f(x)$ outside the range 0 to π in x. The resulting function is odd in x. (d) An alternative extension of $f(x)$ in (a), which repeats over the infinite range. This is unsatisfactory in that, although the boundary conditions are satisfied at $t = 0$, they will not be at subsequent times. In particular, the temperature at $x = 0$ and $x = \pi$ would begin to rise.

This figure can alternatively represent the transverse displacement of a string, described in Section 12.10.2.

specific example, we take $f(x)$ to be as shown in fig. 12.12. We want to find the temperature at subsequent times.

As indicated in Section 12.8, we imagine the bar to be continued to $\pm\infty$, and that the definition of $f(x)$ is extended in some repetitive manner for the infinite range. We then find the solution $\theta(x, t)$ by the Fourier method of Section 12.9.4. The idea is that the temperature behaviour on our limited bar will be identical at all times with that at the same position for the infinite bar.

The crucial question is how to extend $f(x)$, or equivalently what sort of terms we are going to allow in our Fourier series for $f(x)$. The answer depends on what boundary conditions are imposed on $f(x)$.

Thus if the ends of the bar at $x = 0$ and π are held at $0\,^{\circ}\mathrm{C}$ (e.g. by having them in thermal contact with baths of an ice–water mixture), then in order to ensure that at all times the temperature remains zero at the ends, it is essential to choose only those Fourier terms which individually satisfy the constraints. This implies that we use terms of the type $b_n \sin nx$, since $\sin nx = 0$ at both $x = 0$ and $x = \pi$, for any integer n. As time goes by, each of these terms will decay exponentially at its own rate, but the temperature at the ends is guaranteed to remain always at the correct value. In effect we are assuming that the infinite extension of $f(x)$ is as shown in fig. 12.12(c).

If we had ignored the boundary conditions, we could have found an equally acceptable solution for $f(x)$ at $t = 0$ by choosing, for example, a cosine series, e.g.

$$f(x) = a_0/2 + \Sigma a_n \cos nx \qquad (12.39)$$

The way this achieves $f(0) = 0$, however, is by a cancellation among all the terms; the a_n are such that

$$a_0/2 + \Sigma a_n = 0 \qquad (12.40)$$

(A similar effect occurs at $x = \pi$.) The problem is that, as time passes by, the amplitude of each of the cosine terms decays exponentially, and each at a different rate (given by $e^{-n^2t/k}$ — see Section 12.9.4). Thus the right-hand side of eqn (12.39) is modified in that, at $t \neq 0$, each a_n is multiplied by a different factor, so the special cancellation feature at the bar's ends is lost, and the temperature will no longer remain 0 °C; this contradicts the conditions imposed on the problem.

Another way of understanding why the cosine series is unsatisfactory in this case is by considering the extended initial temperature distribution $f(x)$ that it implies. This is shown in fig. 12.12(d). We see that in the neighbourhood of $x = 0$, the temperature is positive on both sides. Thus a short time later, heat will be conducted into the region at $x = 0$, and the temperature will rise there. This is forbidden, and hence the solution is unsatisfactory. Thus we conclude by repeating that in order to ensure that we satisfy a boundary condition that something is zero at all times, we choose only those Fourier terms which do so individually.

Of course, keeping both ends at 0 °C is not the only possible set of boundary conditions for a conducting bar. Another simple case would be to have the $x = 0$ end kept at 0 °C, and the other end insulated. This means that no heat can flow in or out of that end, and hence, from the definition of thermal conductivity, $\partial \theta / \partial x = 0$ at all times there.

For this situation, we must therefore choose Fourier terms which are zero at $x = 0$ and have zero gradient at $x = \pi$. Thus we write

$$f(x) = \Sigma b_n \sin(nx/2), \quad \text{for } n \text{ odd}$$

In fig. 12.13(b), we show the first couple of Fourier terms used in the expansion, and in (c) the implied extension of $f(x)$. The boundary conditions are such that the wavelength of the extended $f(x)$ is 4π (which

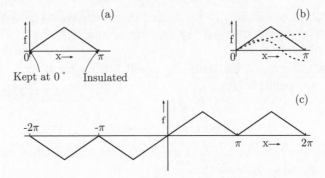

Fig. 12.13. (a) Initial temperature distribution along a finite bar, whose end at $x = 0$ is kept at $0\,°C$ and whose end at $x = \pi$ is thermally insulated. This initial temperature distribution is identical to that of fig. 12.12(a). (b) The first two sine components of the Fourier series for $f(x)$. Because of the boundary conditions, the only satisfactory components are those which individually are zero at $x = 0$, and have zero gradient at $x = \pi$. (c) The Fourier series, used to extend the definition of $f(x)$ outside the range $0–\pi$ in x. The function is symmetric about $x = \pi$, as a result of the bar being insulated there; it is odd in x because the end at $x = 0$ is kept at $0\,°C$. The repeat distance of the pattern is now 4π.

explains why the lowest term in our series is $b_1 \sin(x/2)$). We also see that $f(x)$ is such that in the neighbourhood of $x = \pi$, the temperature there will initially rise as t increases from zero. For this problem, however, that is satisfactory since that end of the bar is not fixed in temperature, but is insulated. Indeed, since the region to the left of $x = \pi$ is at a higher temperature, heat conducted from there will indeed cause the temperature of that end to rise.

If both ends of the bar were insulated, we would need Fourier terms with zero gradient at $x = 0$ and at $x = \pi$ (see fig. 12.14). Thus we would write†

$$f(x) = a_0/2 + \sum_n a_n \cos nx, \quad \text{for } n \text{ even}$$

Because the individual terms decay at rates given by $e^{-n^2 t/k}$, after an infinite time the cosine terms will all have disappeared while the $a_0/2$ term will remain. Thus in contrast to the two previous examples, the steady state condition of the bar is not that of zero temperature. Again this is sensible: the ends of the bar are insulated, and so the heat merely

† The boundary conditions merely require n to be integral, but the symmetry of the initial temperature distribution results in all the odd a_n being zero.

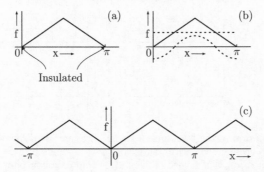

Fig. 12.14. As in fig. 12.13, but here both ends of the bar are insulated. The allowed Fourier terms in (b) must now have zero derivatives at both ends of the bar. Since all components except for the constant in x (i.e. the $a_0/2$ term) are multiplied by an exponentially decreasing time dependence, as t becomes large the temperature tends to $a_0/2$, the average of the initial temperature distribution. The extension of the Fourier series to all x (see (c)) has repeat distance π; this is due to the specific symmetry of the initial temperature distribution.

redistributes itself in such a way that the average temperature at any time remains constant.

Of course, other sets of boundary conditions are possible, but can be treated in a similar manner. For example, we leave as an exercise for the reader the case where the two ends of the bar are kept at different temperatures (see Problem 12.7(i)). Section 11.5.2 contains a worked example of the heat conduction equation where the boundary conditions are discussed in detail.

We end by emphasising that the Fourier series and hence also the complete solution of a problem of this sort depend crucially on the boundary conditions imposed. You cannot expect to obtain the correct solution by ignoring them.

12.10.2 *Waves on a finite string*

In Section 12.9.3, we considered waves on an infinite string. With the more realistic case of a finite string (extending from $x = 0$ to $x = L$), there are strong similarities with the corresponding problems of the temperature distribution along a finite bar (except that the individual Fourier components of the string oscillate, but decay exponentially in the thermal conduction situation).

As a specific example, we can take fig. 12.12(a) as representing the transverse displacement $y(x, 0)$ at $t = 0$ of a string whose ends are rigidly

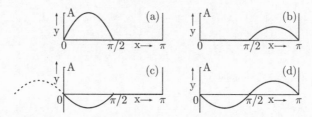

Fig. 12.15. Solution to the "wave on a string" problem by d'Alembert's method:
(a) Initial displacement $y(x, 0)$ of the string with its ends at $x = 0$ and π fixed.
The d'Alembert approach regards this as two identical disturbances each of half
the original amplitude, and which will subsequently travel in opposite directions.
(b) After a time $t = \pi/2c$, the right-moving disturbance is as shown. (c) The left
moving part would appear as shown dashed, were it not for the fact that the
string extends only as far as $x = 0$, where it is fixed. This results in the wave being
reflected, as shown by the solid line. (d) The total disturbance at time $t = \pi/2c$
is the sum of the two contributions from (b) and (c), and hence is $-\frac{1}{2} A \sin 2x$.

held such that $y(0, t)$ and $y(\pi, t)$ are both zero. Once again we must
choose a Fourier series whose individual terms all obey the boundary
conditions, i.e. we require a sine series. Thus the behaviour of the finite
string will be identical to that of the corresponding portion of the infinite
string depicted in fig. 12.12(c).

Once again, had we chosen the wrong type of Fourier series (e.g. cosine
terms), the two ends of the string would not have remained with zero
displacement as time went by. Since they are assumed to be fixed, that
type of solution is clearly incorrect.

12.11 Worked example

We consider here the problem of a string which is initially displaced in
the form shown in fig. 12.15(a), i.e.

$$y(x, 0) = \begin{cases} A \sin 2x & \text{for} \quad 0 < x < \dfrac{\pi}{2} \\ 0 & \text{for} \quad \dfrac{\pi}{2} < x < \pi \end{cases} \tag{12.41}$$

The string is held securely at its ends at $x = 0$ and $x = \pi$ with zero
displacement. We want to find the subsequent displacement of the string,
and in particular $y(x, t)$ at $t = \pi/2c$.

We are going to use the Fourier technique to solve the problem and
will compare our answer with that obtained (rather more easily) by
d'Alembert's method. Readers are advised that we are going to make a

mistake in the Fourier approach, for which they should look out. All will be revealed at the end of the problem.

Because the ends of the string are not allowed to move, we need the sine series

$$y(x,0) = \Sigma b_n \sin nx$$

since the sines are all zero at $x = 0$, and the fact that we use $\sin nx$ ensures that the end at $x = \pi$ also has $y = 0$. Then after we have determined the coefficients b_n, our solution will be

$$y(x,t) = \sum_n b_n \sin nx \cos nct \qquad (12.42)$$

where c is the speed of moving waves on the string.

The b_n are as usual determined by

$$b_n = \frac{2}{\pi} \int_0^\pi y(x,0) \sin nx \, dx$$

$$= \frac{2}{\pi} \int_0^{\pi/2} A \sin 2x \sin nx \, dx \qquad (12.43)$$

To perform this integration we make use of the identity (12.24) in order to rewrite

$$b_n = \frac{A}{\pi} \int_0^{\pi/2} \{\cos(n-2)x - \cos(n+2)x\} \, dx$$

$$= \frac{A}{\pi} \left\{ \left[\frac{\sin(n-2)x}{n-2} \right]_0^{\pi/2} - \left[\frac{\sin(n+2)x}{n+2} \right]_0^{\pi/2} \right\} \qquad (12.44)$$

Both contributions from the lower limit are zero, while at $x = \pi/2$, the value of $\sin(n \pm 2)x$ depends on whether n is even or odd. In the former case, it is again zero, while for n odd, it is -1 for $n = 1, 5, 9, \ldots$ and $+1$ for $n = 3, 7, 11, \ldots$, i.e.

$$\sin(n \pm 2)\pi/2 = (-1)^{(n+1)/2} \quad (n \text{ odd}).$$

Thus

$$b_n = (-1)^{(n+1)/2} \frac{A}{\pi} \left(\frac{1}{n-2} - \frac{1}{n+2} \right)$$

$$= (-1)^{(n+1)/2} 4A/\pi(n^2 - 4) \quad (n \text{ odd}) \qquad (12.45)$$

and

$$b_n = 0 \quad (n \text{ even}) \qquad (12.45')$$

We then obtain the time dependence of the displacement from eqn

(12.42), with the b_n as given above. In particular, if we substitute $t = \pi/2c$, we find that $y(x, \pi/2c)$ is zero, since for odd n the $\cos nct$ factor vanishes, while for even n the b_n are zero. We thus conclude that at this instant, the string is undisplaced.

This result is incorrect, as we shall now demonstrate.

An alternative approach to this problem is to use d'Alembert's method (see Section 11.4). The initial displacement is regarded as the sum of two displacements, each of amplitude $A/2$. At $t = 0$ they are superimposed, but are moving in opposite directions along the string with velocities $\pm c$. After a time $\pi/2c$, each component will have travelled a distance $\pi/2$. The one moving to the right thus appears as shown in Fig. 12.15(b). The other component would have moved off the string to the left, which is impossible; the effect of the string being fixed to the origin is that this wave is reflected and inverted. This results in the contribution shown in fig. 12.15(c), and so the net displacement is as drawn in fig. 12.15(d), i.e.

$$y(x, \pi/2c) = -\frac{A}{2} \sin 2x \qquad (12.46)$$

This is indeed the correct answer.

With this in mind, we can go back to our Fourier approach and see if we can find the error. We note that the correct solution (12.46) consists of a $\sin nx$ term with n even; these we "found" were all zero.

Rather than simply pointing out the error in the algebra of the Fourier method above, we shall now use an alternative and somewhat more intelligent Fourier approach. We first move our origin to the centre of the string, and then write $y(x', t)$ as an even plus an odd component; these are shown in fig. 12.16(a). Again we want to decompose these into Fourier series which have each term vanishing at the ends of the string, i.e. at $x' = \pm\pi/2$. For the odd part, this is particularly easy; $\frac{1}{2}A \sin 2x$ already is harmonic with the correct boundary conditions, and so is its own Fourier series. The even part can be decomposed into a $\cos mx'$ series, and the boundary conditions at $x' = \pm\pi/2$ require m to be odd.[†]

As before, the frequencies at which the $\cos mx'$ components oscillate are such that at $t = \pi/2c$, they are all zero. The new feature is that we have a $\frac{1}{2}A \sin 2x$ term, whose time dependence is $\cos 2ct$, and hence at $t = \pi/2c$, this will give a contribution (which in fact is the complete

---

[†] Because of the change of origin, the $\cos mx'$ terms with odd m correspond directly to our $\sin mx$ terms before.

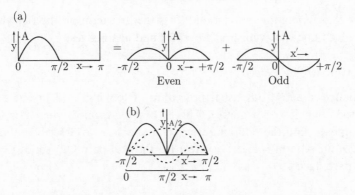

Fig. 12.16. Solution to the "wave on a string" problem using Fourier series: (a) The initial displacement $y(x,0)$ is expressed as the sum of an even and an odd function, with respect to a new variable x' whose origin is at the centre of the string. The odd part is the single harmonic term $\frac{1}{2}A\sin 2x$, but the even one needs to be Fourier analysed. (b) The first two terms in the Fourier series for the even part of $y(x',0)$. At $t = \pi/2c$, all the even components will become zero, while the odd part will have gone through half a complete oscillation. The total displacement will thus be $-\frac{1}{2}A\sin 2x$, in agreement with the d'Alembert solution.

solution) of

$$y(x, \pi/2c) = -\frac{1}{2}A\sin 2x$$

This agrees with our d'Alembert solution.

We thus need to find a non-zero b_2 coefficient for our original Fourier series. Now that we know where to look, we should be able to see the error of our ways. Our integration in eqn (12.44) was correct for all values of n, except for $n = 2$. In that case, the $[\sin(n-2)x]/(n-2)$ term is meaningless. It is probably easiest to start again for b_2, and to write

$$b_2 = \frac{2}{\pi}\int_0^{\pi/2} A\sin^2 2x\,dx$$
$$= A/2 \tag{12.47}$$

This agrees with what we found by performing the Fourier analysis in the variable $x' = x - \pi/2$.

Thus our full solution is now

$$y(x,t) = A\left[\frac{1}{2}\sin 2x\cos 2ct + \Sigma(-1)^{(n+1)/2}\frac{4}{\pi(n^2-4)}\sin nx\cos nct\right]$$

$$\tag{12.48}$$

where the summation is for n odd. It is the first term on the right-hand side which does not vanish at $t = \pi/2c$, and again gives

$$y(x, \pi/2c) = -\frac{1}{2}A \sin 2x$$

Finally we adopt our usual procedure of looking hard at our series to check that it is reasonable. This is best performed when $y(x', 0)$ is decomposed into the odd and even components x' as described above.† The $\frac{1}{2}A \sin 2x$ term is clearly correct for the odd part. Our series for the even part starts‡

$$\frac{4A}{\pi}\left(\frac{1}{3}\sin x + \frac{1}{5}\sin 3x - \frac{1}{21}\sin 5x + \frac{1}{45}\sin 7x + \cdots\right) \qquad (12.49)$$

The first term is shown in fig. 12.16(b), and its amplitude $+4A/3\pi \approx A/2$ is clearly of the correct sign and reasonable in magnitude. Because the function is even in x', the series will consists only of odd n, and it looks plausible that the b_3 coefficient is also positive, in order to enhance the first term near $x = 0$ and $x = \pi$, and to compensate for it near $x = \pi/2$. (The only mistake we might have made in our rough check would be to assume that as the first two signs were positive, all the terms would also be.)

Even when extended beyond the range $0–\pi$, the function $y(x, 0)$ has no discontinuities, but the gradient does at $x = r\pi/2$. We are thus pleased to see that our coefficients decrease at large n like $1/n^2$.

Thus there is nothing obviously wrong with our Fourier series, and it is in fact correct.

Once again, we see the value of looking at a problem in different ways. This helps us detect possible errors, and gives us additional insight.

12.12 Fourier transforms

12.12.1 Motivation

As was mentioned at the very beginning of this chapter, harmonic waves display particularly simple behaviour. This led us to consider Fourier series, for functions that repeated themselves but which were not harmonic. In this final section, we acknowledge the fact that physical phenomena do not continue out to $\pm\infty$, and so the concept of a function

---

† In what follows we think about the odd or even character of the function in the variable x', but write down our Fourier series in x.

‡ The signs are a little surprising. Basically they alternate, but because of the $1/(n^2 - 4)$ factor, the first term with $n = 1$ has the opposite sign from the regular sequence.

that repeats for ever is somewhat idealistic. We thus consider in what way the previous analysis needs modifying for functions that are not infinitely repetitive, and we are led to the subject of Fourier transforms.

As a specific example, consider a beam of light passing through a glass block, whose absorption and refractive index are frequency-dependent. Then any given harmonic wave would have its phase shifted and its amplitude reduced. However, for a combination of waves of different frequencies, the situation would be more complicated, as each component would now have its own phase shift and amplitude reduction factor.

Thus if our light source were monochromatic, with its wave train being infinitely long in both directions, i.e.

$$y = A \cos k_0 x, \quad \text{for } -\infty < x < \infty \tag{12.50}$$

we would just use the absorption and refractive index for the given unique wavelength to decide what happens. But real wave trains are finite in extent, typically being tens of centimetres long for a non-laser source, e.g.

$$y = A \cos k_0 x, \quad \text{for } x_1 < x < x_2 \tag{12.50'}$$

Since this is not identical to (12.50), it cannot be truly monochromatic. We in fact will need lots of pure harmonic waves of neighbouring frequencies to ensure that the resultant sum builds up to $A \cos k_0 x$ in the range x_1–x_2, and cancels to zero elsewhere. Then we will have to take into account how each component is affected by the block, and finally add all of these together to discover what happens to the pulse of light after it passes through the block.

In many other situations, we need to find out how a non-repeating disturbance can be composed out of harmonic waves. This subject is called Fourier transforms. It resembles Fourier series in that we build something up out of harmonic components, but differs from it in that Fourier series deal with functions that repeat for ever.

In the next subsection, we see how it is necessary to modify our results for Fourier analysis, to cope with non-repeating functions. The main difference will turn out to be that we will need harmonic components that are infinitesimally close together in wave number or frequency, whereas in the Fourier series situation, they were separated by constant finite intervals.

12.12.2 From Fourier series to Fourier transforms

The easiest way to make plausible the relation between Fourier series and transforms is to start with a repeating function, and then to modify it so that the repeat distance increases, until at last the separation between corresponding parts of the pattern increases to infinity.

For simplicity, we consider the function of fig. 12.17(a):

$$y = \begin{cases} A & \text{for } -\dfrac{L}{2} \le x < \dfrac{L}{2} \\ 0 & \text{for } \dfrac{L}{2} \le x < \dfrac{3L}{2} \end{cases} \tag{12.51}$$

and repeats with a period $2L$. We write this even function of x as a Fourier series

$$y = a_0/2 + \Sigma a_n \cos(n\pi x/L)$$
$$= a_0/2 + \Sigma a_n \cos kx$$

where $k = n\pi/L = 2\pi/\lambda$, k being the wave number and λ the wavelength of each Fourier component. Simple application of the formulae for the coefficients (see Problem 12.10) yields

$$\left.\begin{array}{l} a_0 = A \\ \text{and} \quad a_n = A\sin(kL/2)/(kL/2) \\ \qquad = \begin{cases} \pm 2A/n\pi & \text{for } n \text{ odd} \\ 0 & \text{for } n \text{ even} \end{cases} \end{array}\right\} \tag{12.52}$$

where we have expressed the coefficients in terms of k rather than the more usual n (compare the Fourier series (12.16) for the similar function (12.14)).

We now go to the situation where the non-zero regions of y are still of length L, but are separated by a distance rL from each other, where r is a factor which is greater than unity (see fig. 12.17(b)). Because the repeat distance is now $(r + 1)L$, the Fourier series is

$$y = a_0/2 + \Sigma a_n \cos\left[2n\pi x/(r+1)L\right]$$
$$= a_0/2 + \Sigma a_n \cos kx \tag{12.53}$$

where

$$k = 2n\pi/(r+1)L$$

We can again use the formulae for the coefficients to obtain

$$\left.\begin{array}{l} a_0 = 2A/(r+1) \\ \text{and} \quad a_n = \dfrac{2A}{r+1}\sin(kL/2)/(kL/2) \end{array}\right\} \tag{12.54}$$

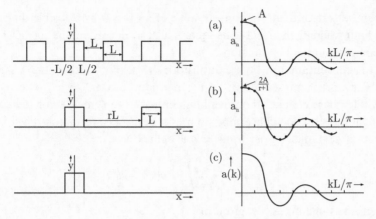

Fig. 12.17. (a) Repeating square wave, where the pulse is "on" for a distance
L, and "off" for an identical distance. Its Fourier coefficients a_n are shown as
a function of $kL/\pi = n$, where k is the wave number of the relevant Fourier
component. The curve is $A\sin(kL/2)/(kL/2)$. (b) The pulses are still of length L,
but are now separated by a greater distance rL. Its Fourier coefficients a_n are
shown as a function of $kL/\pi = 2n/(r + 1)$. The curve is as in (a), but reduced
by a factor of $(r + 1)/2$. When plotted against k, the Fourier points are closer
together than in (a). (The diagram here is for $r = 3$.)

As r becomes very large, the shape of the curve of a_n against k remains the
same, although its overall height decreases and the individual points at which the
Fourier coefficients exist become closer and closer. (c) The pulse at the origin is
still of length L, but now the rest of the function is identically zero. It is plausible
that this is the limit of case (b) as r tends to infinity. We then expect the k
values at which the amplitudes of the various harmonics are defined to become
infinitesimally close together, and to lie on a continuous curve. Since the shape
of the curve in (b) is $\sin(kL/2)/(kL/2)$, independent of r, we expect this to apply
also in this case. The amplitudes of the various components are now given by
the continuous function $a(k)$, rather than by the discrete coefficients a_n.

The first thing to note is that the values of the wave number k at
which the coefficients are defined are closer together than when $r = 1$, by
a factor $(r + 1)/2$. The second point is that the values of the coefficients
at the same values of k in the second case are simply equal to those in
the first example, but scaled down by the same factor of $(r + 1)/2$. This
is illustrated on the right-hand side of fig. 12.17. The Fourier coefficients
lie on the same shaped curve, independent of the value of r. Thus
as r increases towards infinity, all that happens is that the individual
coefficients lie closer and closer together on the same curve, but are
smaller in absolute magnitude. (Despite the coefficients being smaller,
they still manage to build up the same shape near $x = 0$ because there

are now more of them for a given range of k.) It is thus plausible that, in the limit where there is only one square pulse at the origin, the coefficients become a continuous function of k.

It is conventional in Fourier transforms to use e^{ikx} rather than $\cos kx$ and $\sin kx$ as representations of the harmonic waves, so as to include possible phase changes more readily (compare Problem 12.9 for the corresponding situation concerning Fourier series). Then for a non-repeating function $y(x)$, the Fourier transform is defined as

$$y(x) = \frac{1}{\sqrt{2\pi}} \int_{-\infty}^{+\infty} a(k)e^{ikx} dk \tag{12.55}$$

where the function $a(k)$ is given by

$$a(k) = \frac{1}{\sqrt{2\pi}} \int_{-\infty}^{+\infty} y(x)e^{-ikx} dx \tag{12.56}$$

Several features of these equations are noteworthy.

 (i) Because of the symmetry between the equations for $y(x)$ and $a(k)$, not only is $a(k)$ the Fourier transform of $y(x)$, but also $y(x)$ is the Fourier transform of $a(-k)$.

 (ii) Fourier transforms exist for pairs of variables other than x and k. Another example is t and ω. Thus a time-dependent function can alternatively be viewed as being made up of an infinite number of harmonic functions of the type $e^{i\omega t}$, with the required ω values being infinitesimally closely spaced.

 (iii) The way to think of eqn (12.55) is that it shows how a non-repeating function $y(x)$ can be composed out of a whole range of harmonic curves e^{ikx}, each with amplitude given by $a(k)$. Because a continuous infinity of wave numbers k is needed to reproduce a non-repeating function, the expression for $y(x)$ involves an integral over k, rather than the sum for the Fourier series. Also the coefficients a_n (and b_n) of the Fourier series had a subscript n which took on integral values from 0 or 1 up to infinity. Here, because the wave numbers are infinitesimally close together, we use $a(k)$, a function of the continuous variable k.

 (iv) An infinitely long monochromatic wave of the type (12.50) consists of course of a single harmonic component. Since

$$\cos k_0 x = \frac{1}{2}(e^{ik_0 x} + e^{-ik_0 x}) \tag{12.57}$$

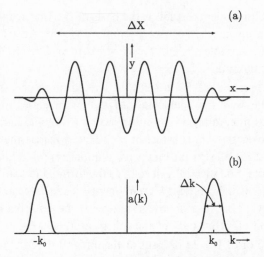

Fig. 12.18. The length Δx of the wave packet modulating an oscillation $\cos k_0 x$, and the width Δk of its Fourier transform are inversely related, such that their product is of order of magnitude unity. (A somewhat more precise statement can be made in terms of the standard deviations of these distributions.) Thus as the number of oscillations in the wave packet grows, there is a corresponding decrease in the width of the wave number distribution required to represent it.

As with all properties of Fourier transforms, (a) could alternatively be regarded as a pulse in time with duration Δt, and (b) would be the corresponding frequency spectrum to reproduce it. Then $\Delta t \Delta \omega \approx 1$.

its Fourier transform is simply two spikes† at $k = \pm k_0$. A long wave of the type (12.50′), where $x_2 - x_1$ is large enough to contain very many wavelengths, in some sense approximates to an infinite wave; its transform $a(k)$ will be sharply peaked around these values of $\pm k_0$.

(v) It is a basic feature of Fourier transforms that the width Δk of the distribution of $a(k)$ around k_0 for a truncated harmonic wave is inversely proportional to the size Δx of the wave packet. In fact, the product $\Delta k \Delta x$ is of order unity (see fig. 12.18). This is true for any shape envelope modulating the original harmonic wave; it is only the numerical factor which differs slightly as the shape of the envelope is changed.

† Another way of looking at the reason for the two spikes is that $\cos k_0 x$ represents a stationary wave, while the $e^{\pm ikx}$ representation is for moving waves; and a stationary wave can be made from two waves travelling in opposite directions. Thus had we instead found the Fourier transform of a moving wave rather than a stationary one, we would have avoided the extra peak.

This fundamental result is the basis of the Uncertainty Principle in quantum mechanics (see Section 12.12.4.3). An example using different variables ω and t was provided in Section 4.7.5, for an electric circuit.

(vi) The expression (12.56) for $a(k)$ is derived more or less in analogy with the way the coefficients of a Fourier series are found. It involves integrating $y(x)$, multiplied by a suitable harmonic wave $e^{-ik'x}$, over the whole range of x.† (Thus whereas the decomposition (12.55) of $y(x)$ into its wave number representation required us to integrate over all values of k, the formula (12.56) for the amplitudes of the different k components necessitates an integration over all x.) Because we integrate over x, the resulting $a(k)$ is not a function of x, but only of the other variable k.

(vii) The factor of $1/\sqrt{2\pi}$ in front of the integral in (12.55) is arbitrary, just as was the factor of $\frac{1}{2}$ in front of the a_0 term of the Fourier series. It then results in a factor $1/\sqrt{2\pi}$ in front of the integral in (12.56), so that the equations for $y(x)$ and $a(k)$ are symmetric (see also point (i)).

(viii) Even if $y(x)$ is real, $a(k)$ can be complex. Regarded as a function of k, $a(k)$ can be decomposed into an even and an odd part. These correspond respectively to the cosine and sine components of the original $y(x)$. Then for real $y(x)$, the even part of $a(k)$ must be real and its odd part imaginary (compare Section 12.12.3.2).

12.12.3 Examples

The above description of how to calculate a Fourier transform will now be made clearer by some specific examples of its application.

12.12.3.1 Square pulse

We require the Fourier transform of the square pulse of fig. 12.19(a), i.e.

$$y(x) = \begin{cases} A & \text{for } |x| < L/2 \\ 0 & \text{otherwise} \end{cases} \tag{12.58}$$

and we want

$$y(x) = \frac{1}{\sqrt{2\pi}} \int_{-\infty}^{+\infty} a(k)e^{ikx}\,dk \tag{12.55}$$

† This procedure actually yields eqn (12.56), but with k replaced by k'.

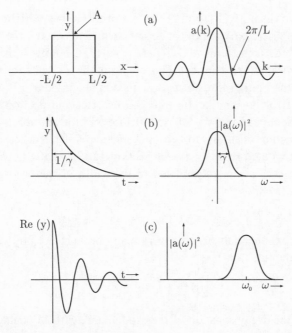

Fig. 12.19. Some Fourier transforms of simple functions: (a) A square pulse of length L. The first zeroes of its Fourier transform are at $k = \pm 2\pi/L$ (b) Decaying exponential $y(t) = e^{-\gamma t}$. The widths of $y(t)$ and of its Fourier transform are related to $1/\gamma$ and to γ, respectively. (c) $y(t) = e^{i\omega_0 t} e^{-\gamma t}$. Its Fourier transform is as in (b), but centred on ω_0 rather than at the origin. In (b) and (c), the Fourier transforms are complex, so we display $|a(\omega)|^2$.

Then our recipe (12.56) for $a(k)$ gives

$$
\begin{aligned}
a(k) &= \frac{A}{\sqrt{2\pi}} \int_{-L/2}^{+L/2} e^{-ikx} dx \\
&= \frac{A}{\sqrt{2\pi}} \frac{1}{ik} (e^{ikL/2} - e^{-ikL/2}) \\
&= \frac{AL}{\sqrt{2\pi}} \frac{\sin(kL/2)}{(kL/2)}
\end{aligned}
\tag{12.59}
$$

This is plotted on the right-hand side of fig. 12.19(a).

We immediately see that the shape of $a(k)$ is very closely related to the values of the coefficients of the Fourier series, for the problem described in Section 12.12.2 and in fig. 12.17. The fact that $a(k)$ here exists for negative values of k is simply a result of our expressing the transform in terms of e^{ikx} components, rather than $\cos kx$ and $\sin kx$ for the Fourier series. Also the normalisation of the two results appears slightly different, due to the discrete nature of the series' coefficients a_k and the continuous nature of the transforms $a(k)$. Apart from that, however, the shapes of the distributions are identical. This confirms that, at least for the present example, the Fourier transform as defined in (12.55) and (12.56) is indeed equivalent to the limit of the process, described in Section 12.12.2, of moving the repeating parts of $y(x)$ further and further apart.

The function $a(k)$ falls to zero at $k = \pm 2\pi/L$. This is consistent with the feature stated in point (v) of Section 12.12.2 that the width of the transform $a(k)$ is inversely related to the width of the original function $y(x)$.

12.12.3.2 Decaying exponential

Decaying exponentials occur in very many branches of physics. We thus consider the function

$$y(t) = \begin{cases} e^{-\gamma t} & \text{for } t \geq 0 \\ 0 & \text{for } t < 0 \end{cases} \tag{12.60}$$

Here γ is the decay constant of the exponential, with $1/\gamma$ being the mean lifetime.

We have thus changed our variable from x to t, so our Fourier transform is now expressed as

$$y(t) = \frac{1}{\sqrt{2\pi}} \int_{-\infty}^{+\infty} a(\omega)e^{i\omega t} d\omega \tag{12.61}$$

As usual, we obtain the frequency distribution $a(\omega)$ by

$$\begin{aligned} a(\omega) &= \frac{1}{\sqrt{2\pi}} \int_{-\infty}^{+\infty} y(t)e^{-i\omega t} dt \\ &= \frac{1}{\sqrt{2\pi}} \int_{0}^{\infty} e^{-\gamma t} e^{-i\omega t} dt \\ &= \frac{1}{\sqrt{2\pi}} \frac{1}{\gamma + i\omega} = \frac{1}{\sqrt{2\pi}} \frac{\gamma - i\omega}{\gamma^2 + \omega^2} \end{aligned} \tag{12.62}$$

The real part of $a(\omega)$ (i.e. $\gamma/[\sqrt{2\pi}(\gamma^2 + \omega^2)]$) is an even function of ω, while the imaginary part is odd. This is exactly what we expect for a real $y(t)$.

In fig. 12.19(b), we plot the intensity $|a(\omega)|^2$ of the Fourier transform as a function of ω. The half width at half height of this curve is γ, while for the $y(t)$ of eqn (12.60), $t = 1/\gamma$ is where the curve has fallen to $1/e$ of its initial height. Once again the widths of the function and its Fourier transform are inversely related.

12.12.3.3 Decaying oscillation

Even more interesting is the case of an oscillating source whose intensity is decaying. We represent this as

$$y(t) = \begin{cases} e^{i\omega_0 t}e^{-\gamma t} & \text{for } t \geq 0 \\ 0 & \text{for } t < 0 \end{cases} \tag{12.63}$$

where we are using the complex exponential form for the wave of angular frequency ω_0. Then

$$a(\omega) = \frac{1}{\sqrt{2\pi}} \int_0^{\infty} e^{(i\omega_0 - \gamma)t}e^{-i\omega t}dt$$

$$= \frac{1}{\sqrt{2\pi}} \frac{1}{\gamma + i(\omega - \omega_0)} \tag{12.64}$$

The intensity of the frequency spectrum is thus

$$|a(\omega)|^2 = \frac{1}{2\pi} \frac{1}{(\omega - \omega_0)^2 + \gamma^2} \tag{12.65}$$

This is plotted in fig. 12.19(c). It is a familiar shape, with the maximum occurring at $\omega = \omega_0$ and its half width at half maximum is again $\Delta\omega = \gamma$. This is the resonance curve, which was discussed at length in Chapter 4. Thus, for example, $y(t)$ of fig. 12.19(c) could represent the free oscillations of an electric circuit, with $|a(\omega)|^2$ giving the response of the circuit when an oscillating voltage of variable frequency is applied.

Alternatively $y(t)$ could represent the radiation emitted by a set of atoms in a specific state at excitation energy E as they decay back to the ground state. Because of Planck's relation between the energy of photons and their frequency, the radiation would be monochromatic with frequency $\omega_0 = 2\pi E/h$ were it not for the fact that the upper state is characterised by a finite lifetime. (Were the lifetime infinite, atoms in the upper energy level would not decay and there would be no radiation emitted.) Thus the expected oscillation $e^{i\omega_0 t}$ is modulated by the decaying exponential factor $e^{-\gamma t}$. Then $|a(\omega)|^2$ describes the intensity of the frequency spectrum of this radiation. It is not confined to the unique value ω_0 because the radiation is not given by $e^{i\omega_0 t}$ over the whole range of t from $-\infty$ to $+\infty$. The width $\Delta\omega$ of the frequency spectrum is determined by the time $t \approx 1/\gamma$ for which the radiation is emitted according to the relationship $\Delta\omega \cdot t \approx 1$. Multiplication of each side of this equation by Planck's constant $h/2\pi$ yields the Uncertainty Principle $\Delta E \cdot t \approx h/2\pi$, giving the ultimate accuracy $\Delta E (= h\Delta\omega/2\pi)$ with which an energy measurement is possible if there is a maximum time t available for performing it.

A consequence of the emitted radiation being composed of a range of frequencies is that, if we were to use incident radiation to try to excite these atoms from their ground state to the excited one, it would be possible to do so not only with an incident frequency of exactly ω_0, but for frequencies in the range $\omega \approx \omega_0 \pm \Delta\omega$. The curve of $|a(\omega)|^2$ shows how effective the radiation of the different frequencies would be in exciting the atoms.

Very similar ideas are relevant in the excitation of nuclei or of elementary particles (compare Sections 4.8.3 and 4.8.4).

12.12.4 More physical applications

Finally we examine examples where the ideas of Fourier transforms help us to understand physically what is happening. Explicit evaluation of Fourier transforms is not performed here.

12.12.4.1 Wave train

†Consider a wave packet passing through a dispersive medium. This could be, for example, a finite length light pulse travelling in a glass block. In such a medium, only exactly harmonic waves are transmitted without change of shape, with each particular wavelength λ travelling at a slightly different speed (as determined by the way the refractive index of the material depends on λ).

The wave train $y(x)$ is decomposed by Fourier analysis into its harmonic components, each with its own amplitude and phase, as determined by $a(k)$. We then find out what happens to each of these as they propagate through the glass, and add up these components to find the final answer $z(x)$. The result of dispersion is that $z(x)$ is not simply $y(x)$ shifted along in space; the shape of $z(x)$ will in general differ from that of $y(x)$.

If the pulse $y(x)$ is many wavelengths long, it may not look too different after travelling a short distance. This is because a long pulse has a narrow Fourier transform; thus not too wide a range of k values are relevant, and hence the speeds of the different significant components will not be too far apart. One consequence would be that if $y(x)$ started with fairly sharp leading and trailing edges, those of $z(x)$ would be somewhat more rounded (see fig. 12.20).

A more significant effect, however, is that the overall envelope of the pattern travels at a speed which can differ from that of an individual

† This was discussed briefly in the introduction to Fourier transforms of Section 12.12.1.

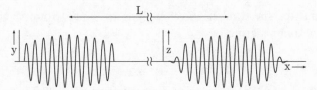

Fig. 12.20. Passage of a wave train of light through a large glass block. It is shown initially as $y(x)$ on the left. After it has travelled a certain distance, it appears as $z(x)$ on the right. In time T, the distance L the wave packet has travelled (as given for example by the separation of the maxima of the two envelopes) is determined by the group velocity $g = L/T = \partial\omega/\partial k$. Individual waves would travel at the phase velocity $v = \omega/k$. For a dispersive medium, these in general will not be equal. Thus for v greater than g, the individual ripples would appear to move forwards through the packet as it moves (except that for light the individual wavelets are not observable; this description is more appropriate for ripples within a wave on the sea).

Another feature of the dispersion is that the edges of the wave packet tend to be smeared out. For a short wave travelling a long distance, the distortion of the original shape may be so severe that it is impossible to ascribe a sensible speed to the wave packet.

wave. To investigate this, we express the packet† $f(x,t)$ in terms of its individual harmonic components $e^{i(kx-\omega t)}$:

$$f(x,t) = \frac{1}{\sqrt{2\pi}} \int a(k)e^{i(kx-\omega t)}dk \qquad (12.66)$$

Here ω is a function of k as determined by the refractive index

$$\mu(k) = ck/\omega \qquad (12.67)$$

and c is the speed of light in a vacuum. Then the maximum of the envelope of the wave packet will move such that the various components combine so as to reinforce each other there, i.e.

$$\frac{\partial\phi}{\partial k} = 0$$

where

$$\phi = kx - \omega t$$

is the phase of the wave. Thus

$$\frac{\partial\phi}{\partial k} = x - \frac{\partial\omega}{\partial k}t = 0$$

† $f(x,t)$ gives the disturbance as a function of both x and t. Then $y(x) = f(x,0)$ and $z(x) = f(x,T)$.

and hence the speed of the envelope, known as the group velocity g of the wave train, is given by

$$g = \frac{x}{t} = \frac{\partial \omega}{\partial k} \tag{12.68}$$

This differs from the so-called phase velocity v of a perfect monochromatic wave

$$v = \frac{\omega}{k} \tag{12.69}$$

These two speeds g and v will be equal only if

$$\frac{\omega}{k} = \frac{\partial \omega}{\partial k}$$

which corresponds to ω being proportional to k, and the medium being non-dispersive.

A simpler and somewhat more transparent discussion of eqn (12.68) is given in Section 14.4

12.12.4.2 Optics

We first examine the diffraction of light of wavelength λ by an object with structure in the x direction, and which is illuminated so that there is no phase variation of light across it. We assume that from each position x on the object, a secondary wavelet of amplitude $s(x)$ is produced; the way s varies as a function of x depends on the structure of the object (if it is a diffraction grating, it would consist of a large number of regions where $s = s_0$, alternating with ones where $s = 0$).

Now consider the amplitude of light diffracted at an angle θ with respect to the "straight-through" direction. The phase of wavelets originating from a position x (as compared with those from $x = 0$) is

$$\phi = \frac{2\pi x \sin \theta}{\lambda} = kx \sin \theta \tag{12.70}$$

(See fig. 12.21.) Then the resulting amplitude at angle θ from the whole object is

$$a(\theta) = \int_{-\infty}^{+\infty} s(x) e^{ikx \sin \theta} dx \tag{12.71}$$

From the Fourier transform relations (12.55) and (12.56), it then follows that for small θ such that $\sin \theta \approx \theta$

$$s(x) = \frac{k}{2\pi} \int_{-\infty}^{+\infty} a(\theta) e^{-ikx \sin \theta} d\theta \tag{12.72}$$

Thus the spatial structure of the object and its diffraction pattern in θ are

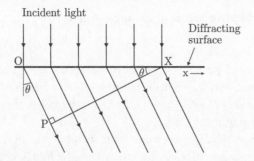

Fig. 12.21. Parallel monochromatic light falls on a diffracting surface along the x axis, and produces secondary wavelets with strength $s(x)$ which varies with x. For light diffracted at an angle θ, there will be an extra path OP of length $x \sin \theta$ for the light coming from the origin as compared with that from position X, and hence a phase difference of $2\pi x \sin \theta / \lambda = kx \sin \theta$. Then the total amplitude at angle θ is determined by adding the contributions from all the infinitesimal elements along the object, to give

$$\int_{-\infty}^{+\infty} s(x) e^{ikx \sin \theta} dx$$

related to each other as Fourier transforms. In particular, the structure of the object can in principle be determined by performing the required integration over the diffraction pattern.

There are two problems with doing this in practice. First, the integration specified in (12.72) is over an infinite range, while in reality we are restricted to $|\theta| < \pi/2$. This is not a serious difficulty, as diffraction patterns are usually confined to small angles, and so any contribution to the integral from the unphysical region in θ is likely to be unimportant.

Much more serious is the fact that $a(\theta)$ can be complex, whereas all that we observe in a usual diffraction pattern is $|a(\theta)|^2$. Thus skill and experience are required in making reasonable assumptions about the phases of the pattern, in order to deduce the structure of the object.

The above generalises in a fairly natural way to the problem of determining the structure of a crystal in three dimensions, from X-ray diffraction patterns obtained with the crystal in different orientations. Here a three-dimensional Fourier transform in x, y and z is required, rather than a one-dimensional transform in just x.

We now turn to a second example from the field of optics, the microscope. The fundamental point here is that, if the object being examined has harmonic structure with repeat distance s_0, the light entering the microscope objective lens is at an angle θ given by

$$s_0 \sin \theta = \lambda \qquad (12.73)$$

Hence if the microscope is capable of accepting light only up to some maximum value $\sin \theta_{\text{max}}$, it will be impossible to resolve features in the object on a scale below $s_{\text{min}} = \lambda / \sin \theta_{\text{max}}$, where λ is the wavelength of the light illuminating the object.

To derive eqn (12.73), we assume that the object is illuminated in such a way that the light entering the microscope from adjacent parts of the object is phase coherent. If we add two waves travelling at angles of $+\theta$ and $-\theta$ with respect to the y axis, we obtain

$$e^{i(\omega t - k\{y \cos \theta + x \sin \theta\})} + e^{i(\omega t - k\{y \cos \theta - x \sin \theta\})} = 2 e^{i(\omega t - ky \cos \theta)} \sin(kx \sin \theta)$$
$$(12.74)$$

That is, these two beams are equivalent to a sinusoidal variation in the x direction, with repeat distance

$$s_0 = \frac{2\pi}{k \sin \theta} = \frac{\lambda}{\sin \theta} \qquad (12.75)$$

Stated otherwise, a sinusoidal spatial variation of the amplitude of light coming from the object results in diffracted beams at angles $\pm\theta$ with respect to the axis.

Now consider a more realistic object whose spatial variation is not sinusoidal and does not even necessarily repeat. We can perform Fourier analysis on the spatial variation $l(x)$ of the light from the object, and then each Fourier component will produce its own pair of diffracted beams. Now the repeat distance s corresponding to higher Fourier components decreases, and hence the diffracted beams occur at larger and larger angles. Eventually they are outside the maximum acceptance of the microscope's lens, or even worse occur at unphysical angles corresponding to $\sin \theta$ being greater than unity (when $s < \lambda$ — see eqn (12.73)). To reproduce a perfectly faithful image of the source, the microscope would have to collect the diffracted beams from all the components of the Fourier transform of $l(x)$, and this is clearly impossible. The extent to which this causes serious distortions in the image depends on how significant are the Fourier components that are not collected by the microscope.

A simple example is again provided by a diffraction grating, consisting of alternate strips where $l(x)$ is either 0 or 1 (see fig. 12.22(a)). If this is infinite in extent, we need only the Fourier series, which apart from the constant term consists of an infinite series of odd harmonics. If

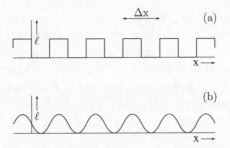

Fig. 12.22. With a diffraction grating as the object viewed in the microscope, the amplitude of the transmitted light $l(x)$ is as shown in (a). If the microscope is such that it accepts diffracted beams corresponding to the fundamental repeat distance $s = \Delta x$, but not to any higher harmonics, we would be unable to distinguish the object in (a) from a different one whose light output varied with x as shown in (b).

the microscope collects only the light diffracted from the first of these harmonics, we will be unable to distinguish whether the structure of the object is as shown in fig. 12.22(a) or (b).

Thus, even for a microscope with perfect lenses, the wavelength of the light used and the angular acceptance of the microscope set a fundamental limit on the size of structure that can be resolved in the object.

12.12.4.3 Quantum mechanics

In quantum mechanics, all that it is possible to know about, for example, an electron is contained in its wave function $\psi(x)$, which is a complex function of the spatial coordinate x. The momentum p of an electron whose wave function is

$$\psi = Ae^{ikx} \tag{12.76}$$

is

$$p = hk/2\pi \tag{12.77}$$

provided that the form (12.76) extends out to $\pm\infty$. Eqn (12.77) is consistent with the de Broglie hypothesis that an electron of momentum p is described by a wave of wavelength λ such that $\lambda = h/p$.

Actual electrons are usually confined in some limited spatial region (e.g. in a hydrogen atom; or as the free electron gas in a piece of metal of some specific size, etc.) and hence a more realistic wave function would be essentially zero outside the region of interest (see, for example, the envelope shown in fig. 12.18(a), although it must be remembered that

$\psi(x)$ is in fact complex). Then the momentum corresponding to such a wave function is no longer unique, since it does not extend in a harmonic way out to $\pm\infty$. To find the momentum distribution defined by such a wave function, we must perform a Fourier analysis to determine $a(k)$, the amplitude of each component as a function of its wave number, and this gives us the momentum distribution via eqn (12.77).

A fundamental property of Fourier transforms is the inverse relationship between (a) the length of the spatial region in x of a function such as $\psi(x)$, and (b) the width in k of its Fourier transform. This then results in the Heisenberg Uncertainty Principle

$$\Delta x \Delta p \geq h/4\pi$$

for the intrinsic uncertainty in the position x and the momentum p of an electron.

We conclude that Fourier transforms provide us with insights into a wide range of phenomena of physical relevance.

Problems

In all problems below where explicit Fourier coefficients are calculated, apply simple tests to check that they are reasonable.

12.1 Evaluate the integrals

$$\int_0^{2\pi} \sin mt \sin lt \, dt$$

$$\int_0^{2\pi} \cos nt \cos kt \, dt$$

$$\int_0^{2\pi} \sin mt \cos nt \, dt$$

for integral k, l, m, n. Consider separately the cases where the two integers in a given integral are: (a) different constants; (b) equal but non-zero; and (c) both zero.

Compare your answers with eqns (12.10) and (12.11).

12.2 Evaluate the coefficients of the Fourier series for the function shown in fig. 12.1(d). Verify that eqn (12.2) is correct.

12.3 The function $f(x)$ is defined to be equal to x^2 in the region $0 \leq x < \pi$. By a suitable extension of this function to the whole range of x, find the (half range) Fourier cosine and sine series

$$C(x) = \frac{a_0}{2} + \sum_{n=1}^{\infty} a_n \cos nx$$

$$S(x) = \sum_{n=1}^{\infty} b_n \sin nx$$

such that $C(x)$ and $S(x)$ both coincide with x^2 in the range $0 \leq x < \pi$. Sketch the functions C and S over the range $-2\pi < x < 2\pi$.

From these series, show that

$$\sum_{n=1}^{\infty} \frac{1}{n^2} = \frac{\pi^2}{6}$$

and that the corresponding sum over just the odd positive integers is $\pi^2/8$.

12.4 (i) The function $f(x)$ is an odd periodic function of x with period 2π and has an expansion as a Fourier sine series. Obtain expressions for the Fourier coefficients and evaluate them for the special case

$$f(x) = \cos x; \quad 0 < x < \pi$$
$$f(x) = -\cos x; \quad -\pi < x < 0$$

(ii) A new function $F(x)$ is defined by the integral

$$F(x) = \int_0^x f(x')dx'$$

where $f(x)$ is given in part (a) of this question. Sketch the function $F(x)$ for $-\pi < x < \pi$ and find its Fourier expansion.

12.5 A function $f(x)$ is defined as

$$f(x) = \begin{cases} x/d & \text{for} \quad 0 \leq x < d \\ 1 & \text{for} \quad d \leq x < \pi - d \\ (\pi - x)/d & \text{for} \quad \pi - d \leq x < \pi \end{cases}$$

It is expanded as a Fourier sine series, on the assumption that its repeat distance is 2π. Determine the Fourier coefficients b_n.

Determine the value of d such that the second non-vanishing coefficient is b_5. Draw a reasonably accurate diagram of $f(x)$ and the $b_1 \sin x$ term of the Fourier series, in order to convince yourself that it is reasonable that the coefficients b_2 to b_4 are zero in this

case. Compare this with the corresponding diagram for the choice
$d = \pi/4$.

12.6 (i) The temperature $T(x,t)$ along a bar of length π obeys the heat
conduction equation (12.36). The initial temperature distribution
$T(x,0)$ is given by $T_0 f(x)$, where T_0 is a constant and $f(x)$ is
defined as

$$f(x) = \begin{cases} x & \text{for} \quad 0 \le x < \pi/2 \\ \pi - x & \text{for} \quad \pi/2 \le x < \pi \end{cases}$$

The ends of the bar are maintained at a temperature $T = 0\,^\circ\text{C}$. Use
a suitable Fourier series to obtain an expression for the temperature
along the bar at a subsequent time t.

(ii) The initial displacement of a stretched string is given by $y_0 f(x)$,
where y_0 is a constant and $f(x)$ is as defined above. The ends of
the string are fixed, with zero displacement. Use a suitable Fourier
series to obtain an expression for the displacement along the string
at a subsequent time t.

(iii) If you have a programmable calculator or a computer, check
numerically that your solutions are consistent with figs. 12.11
and 12.10 respectively.

12.7 (i) A rod of length l is initially at 120 K. If one end is kept at
100 K and the other at 200 K, find the temperature of the rod as
a function of x and t.

(ii) Another rod of length L is initially at 400 K, with the end at
$x = L$ insulated. The end at $x = 0$ is, for $t > 0$, kept at 300 K.
Find the temperature of the rod as a function of x and t.

12.8 Laplace's Equation in two-dimensional polar coordinates is

$$\frac{1}{r}\frac{\partial}{\partial r}\left(r\frac{\partial V}{\partial r}\right) + \frac{1}{r^2}\frac{\partial^2 V}{\partial \theta^2} = 0$$

Find all possible solutions of the form

$$V = f(r)g(\theta)$$

where V tends to zero as r tends to infinity, and where $g(\theta) = g(\theta + 2\pi)$.

Find a solution of Laplace's Equation for $r > r_0$, which has the
following properties:

(a) V tends to zero as r tends to infinity, and

(b) for $r = r_0$

$$V = 2V_0\theta/\pi \qquad \text{for} \quad -\pi/2 < \theta \le \pi/2$$
$$V = 2V_0\left(1 - \theta/\pi\right) \quad \text{for} \quad \pi/2 < \theta \le 3\pi/2$$

12.9 This problem provides practice at using complex notation for Fourier series, as this is used for Fourier transforms in Section 12.12.
(i) Use eqns (12.12) and (12.12′) to determine the Fourier coefficients a_n and b_n for the function

$$f(t) = \begin{cases} t & \text{for} \quad 0 < t < \pi \\ 0 & \text{for} \quad -\pi < t < 0 \end{cases}$$

and which repeats with a period 2π in t.
(ii) Instead of using separate sine and cosine series, it is possible to use the complex notation of eqn (12.8). Show that this is equivalent to eqn (12.3), provided that

$$\text{Re}(C_n) = a_n$$

and

$$\text{Im}(C_n) = -b_n$$

and that the C_n are then given by

$$C_n = \frac{1}{\pi} \int_0^{2\pi} f(t)e^{-int}\,dt \qquad \text{(a)}$$

(iii) A more useful way to write the complex Fourier series is

$$f(t) = \frac{1}{2} \sum_{-\infty}^{+\infty} C_n e^{int} \qquad \text{(b)}$$

with the C_n as given in eqn (a). Show that this is equivalent to eqn (12.8), provided that

$$\text{Re}(C_{-n}) = \text{Re}(C_n)$$

and

$$\text{Im}(C_{-n}) = -\text{Im}(C_n)$$

(We are assuming that $f(t)$ is a real function of t.)
(iv) Use eqn (a) to determine the complex Fourier coefficients C_n for the function $f(t)$ defined in (i) above. Check that the Fourier series agrees with that obtained in (i) above.

12.10 (i) For the function (12.51), consisting of repeated square pulses, show that the coefficients of its Fourier series are given by eqn (12.52).

(ii) As the separation between the "on" parts of the function increases (as shown in fig. 12.17(b)), show that the Fourier coefficients are now given by eqn (12.54).

(iii) Determine the Fourier transform of the single square pulse of fig. 12.17(c).

12.11 Determine the Fourier transform of the Gaussian function

$$f(t) = \frac{\sqrt{\beta}}{\sqrt{2\pi}} e^{-\beta t^2/2}$$

(Notice that, apart from their normalisation, the Fourier transform of a Gaussian in t is a Gaussian in ω, and vice versa; and that the widths of the two Gaussians are reciprocals of each other.)

13

Normal modes

13.1 What is a normal mode?

A fairly common demonstration involves two identical pendula, which are close together with some small coupling between them. This can be achieved, for example, by hanging them from the same not too rigid supporting mechanism. One of the pendulum bobs is displaced slightly and then released; the other is left stationary in its equilibrium position.

Initially, the first pendulum oscillates in much the same way as a single isolated pendulum would. But gradually an interchange of energy takes place, with the first pendulum's oscillations being transferred to the second one, then back again to the first, and so on (see fig. 13.1). This is the type of phenomenon that we are going to describe in this chapter by the use of "normal modes".

The concept of normal modes arises in the following situations. We have a number N of oscillating systems (e.g. pendula, masses on springs, electric circuits with inductors and capacitors, electrons in atoms, etc.) which if independent are each characterised by their own natural frequency β_j.† Any damping is neglected. The essence of these phenomena is that the individual systems have some form of coupling to each other, just as the two pendula had in the introductory example. We then look for ways in which the separate components can all vibrate with the same frequency. i.e.

$$x_j = (x_0)_j \cos \omega t, \quad j = 1, 2, \ldots, N \tag{13.1}$$

where x_j is the displacement of the jth oscillator, $(x_0)_j$ is its (possibly

---

† We denote the natural (uncoupled) frequencies by β, in order to distinguish them from the frequencies ω of the normal modes. The notation used throughout this Chapter is summarised in Section 13.7.

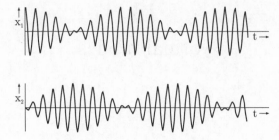

Fig. 13.1. The displacements x_1 and x_2 of two coupled pendula as a function of time. Initially the first one is displaced, while the second is in its equilibrium position $x_2 = 0$. As time goes by, the oscillation amplitude of the first decreases, while that of the second increases. Later still, the energy returns to the first pendulum. In Section 13.5.1, the two displacements are called x and y.

negative) amplitude† and ω is the common frequency of all the oscillators. It is such a set of motions that is called a normal mode.

What we shall find is that such modes of vibration for the system as a whole are possible, provided that the relative magnitudes and signs of the $(x_0)_j$ are suitably specified.‡ Then, for such an initial set of conditions, the individual oscillators will continue to vibrate at their common frequency, and at a time $2n\pi/\omega$ later (where n is an integer), all the displacements will be exactly the same as they were initially. This feature of a common frequency is what defines a normal mode, and applies even when the natural frequencies of the individual uncoupled oscillators differ from each other. It is the coupling which forces them to the same frequency in the normal mode.

In contrast, if the oscillators are given an initial set of displacements which do not correspond to the $(x_0)_j$ of a normal mode, then the individual oscillators will vibrate in a complicated, non-harmonic fashion, with the amplitude of motion varying as energy is exchanged via the coupling with other parts of the system. An example of this has already been provided by the case of the two coupled pendula, in which one was initially pulled to one side while the other remained in the equilibrium position; these displacements do not correspond to a normal mode.

---

† Because it is possible in a normal mode to have different parts of the system oscillating completely out of phase with other components, some of the $(x_0)_j$ can be negative.

‡ Arbitrary phases between different components can arise when two or more of the normal mode frequencies are degenerate (i.e. equal) — see, for example, Problem 13.4.

For a system with N components, there is not just one possible normal mode, but there are N, each specified by its own frequency ω_i, and by the required set of amplitudes† $(x_0)_{j,i}$. Here the subscript j specifies which of the N components we are considering, and i refers to the particular normal mode (of the N possible ones) of the system — see Section 13.7. Then even if we start from a set of oscillators all with the same natural frequency β, the effect of the coupling can be to produce a spectrum of N different normal mode frequencies for the system.

An example of this is provided by electrons in a solid. An isolated atom has a line spectrum, corresponding to the transition of an electron between two specific energy levels. This means that an excited atom will emit radiation of this particular frequency. If we take a large number of atoms in a gas, the separation of the atoms is large enough so that the electrons of one atom are to a good approximation unaffected by those in the others, and the atoms can be considered as independent. Thus the radiation from a gas has the same spectrum as that from an isolated atom, except that it is of course very much more intense. In contrast, in a solid the interactions are far from negligible. Indeed they are responsible for maintaining the solid structure. Their effect is to split a single line in the spectrum into N separate components. Since N can be very large, the net result is that solids exhibit band spectra, where the band in fact consists of the N components into which each individual line of the atomic spectrum is split, and which are close enough together to be considered continuous.

The general principles discussed here in this introduction will become clearer after we investigate a specific example.

13.2 A simple example: two identical coupled pendula

We now return to the case of two coupled pendula, with the coupling produced by a weak massless spring connecting the two bobs (see Fig. 13.2). This results in a force

$$F = k(y - x) \tag{13.2}$$

where x and y are the displacements of the pendulum bobs from the

---

† In fact there are N modes if each component has only one degree of freedom (e.g. it can move in only one direction). If instead each can move with m degrees of freedom (e.g. 3 for x, y and z motion), then there are Nm normal modes.

Normal modes

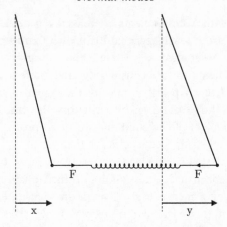

Fig. 13.2. Two identical pendula, whose bobs are coupled by a massless spring. The displacements of the pendula from their equilibrium positions are x and y. The force F exerted by the spring on the bobs is given by eqn (13.2).

undeflected positions,† and k is a constant related to the strength of the spring.

The equations of the two pendula are now

$$\left.\begin{array}{c} m\ddot{x} = -mgx/l + k(y - x) \\ \text{and} \quad m\ddot{y} = -mgy/l + k(x - y) \end{array}\right\} \quad (13.3)$$

where we have assumed that both pendula are of length l, and that their bobs each have mass m; and we have used the usual small oscillation approximation.

A useful check that these equations are sensible is obtained by adding them, which gives

$$m(\ddot{x} + \ddot{y}) = -\frac{mg}{l}(x + y) \quad (13.4)$$

This is a differential equation for $x + y$, and does not involve the spring constant k. This is as expected, since $(x + y)/2$ is the coordinate of the centre of mass of the pair of bobs; its motion cannot depend on the internal spring, which exerts equal but opposite forces on the two pendula.

If k is set to zero (i.e. the spring is effectively not there), then the

† This assumes that the spring is in its unstretched position when both bobs are undeflected. If this is not so, a suitable constant must be added to the right-hand side of eqn (13.2). The discussion of the remainder of this section carries through essentially unaltered, provided the displacements of the bobs are measured with respect to their equilibrium positions under the combined influence of gravity and the spring, rather than using their displacements from the vertical.

first equation of (13.3) involves only the variable x, and is independent of y, while the second contains only y. The equations are decoupled from each other (as are the pendula), and each equation can be solved independently of the other, to give simple harmonic solutions for x and y each at the frequency

$$\beta = \sqrt{g/l} \qquad (13.5)$$

13.2.1 Standard method

The essence of the current problem is that k is non-zero, and the equations are coupled. We look for solutions of the form

$$\left. \begin{array}{l} x = x_0 \cos \omega t \\ \text{and} \quad y = y_0 \cos \omega t \end{array} \right\} \qquad (13.6)$$

where the ωs in the trial solutions for x and y are the same. On substituting into the equations of motion (13.3), the second derivatives are simply replaced by $-\omega^2$, giving

$$\left. \begin{array}{l} (-mg/l - k)x_0 + ky_0 = -m\omega^2 x_0 \\ \text{and} \quad kx_0 + (-mg/l - k)y_0 = -m\omega^2 y_0 \end{array} \right\} \qquad (13.7)$$

This looks neater as

$$\left. \begin{array}{l} (A - \omega^2)x_0 + By_0 = 0 \\ \text{and} \quad Bx_0 + (A - \omega^2)y_0 = 0 \end{array} \right\} \qquad (13.7')$$

where

$$\left. \begin{array}{l} A = \dfrac{g}{l} + \dfrac{k}{m} \\ \text{and} \quad B = -\dfrac{k}{m} \end{array} \right\} \qquad (13.8)$$

The (13.7′) are two simultaneous equations for x_0 and y_0, but with the usual constant terms equal to zero (see Section 1.3). Thus if we want a solution other than the uninteresting case of $x_0 = y_0 = 0$ (i.e. both bobs are permanently stationary), ω must be chosen so that the two equations are consistent. It is this requirement which determines the allowed values of ω for the coupled pendula's normal modes.

The two equations of (13.7′) give respectively

$$\left. \begin{array}{l} \dfrac{x_0}{y_0} = -\dfrac{B}{A - \omega^2} \\ \text{and} \quad \dfrac{x_0}{y_0} = -\dfrac{A - \omega^2}{B} \end{array} \right\} \qquad (13.9)$$

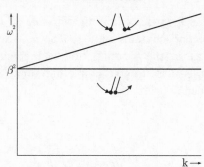

Fig. 13.3. The square of the angular frequency ω^2 as a function of the strength of the spring, for the two normal modes of the identical pendula of fig. 13.2. For $k = 0$, ω is equal to the natural frequency β. The schematic diagrams of the pendula indicate their relative motion in each of the normal modes.

so we need†

$$\frac{B}{A - \omega^2} = \frac{A - \omega^2}{B} \qquad (13.9')$$

This looks like a fourth order equation for ω which could be difficult to solve, but in fact it is only a quadratic in ω^2. It solutions are

$$\omega^2 = A + B \text{ or } A - B \qquad (13.10)$$

which, when the values of A and B from eqns (13.8) are substituted, become

$$\omega^2 = g/l \text{ or } g/l + 2k/m \qquad (13.10')$$

(see fig. 13.3).

It is important to realise what is the significance of these values of ω. They are the only angular frequencies at which the two bobs can oscillate with the same frequencies as each other, in perfectly harmonic motion. The reason that there are two of them is simply because there are two variables (x and y) in this problem. Soon we will understand why in this particular situation one of the solutions (13.10′) is the same as the original frequency of the uncoupled oscillators.

Having found ω, we now go back to eqns (13.9) to try to find x_0 and y_0. Although there are two equations, only one of them is useful, since ω has carefully been chosen to make the equations identical. The remaining one then determines x_0/y_0, but cannot be used to find x_0 and

† Alternatively, eqn (13.9′) can be derived directly from the fact that the determinant of coefficients in eqns (13.7′) must be zero — see Section 1.3.

y_0 separately. This corresponds to the fact that the coupled pendula in a normal mode of free oscillations have well-defined relative amplitudes, but the absolute size of their common oscillation is a free parameter, to be determined by the initial conditions of a given problem.

If we substitute the first solution for $\omega^2 = A + B$, we obtain

$$\frac{x_0}{y_0} = 1 \tag{13.11}$$

while for $\omega^2 = A - B$ (i.e. the higher frequency mode, since B is negative),

$$\frac{x_0}{y_0} = -1 \tag{13.11'}$$

Thus in the first normal mode, the pendula swing in phase with equal amplitudes, while for the second they again have equal amplitudes but are in antiphase.

It is now clear why, for our specific problem, the frequency of the first mode is independent of the spring constant (and hence is the same as for the uncoupled pendula). Since the pendulum bobs swing in phase with equal amplitudes, the spring is never extended, it exerts no force, and it thus makes no difference whether it is there or not. In contrast, for the second mode, the spring exerts an extra restoring force as the pendula move apart, and hence helps gravity in returning the bobs towards their equilibrium positions. It is thus no surprise that, for this mode, the frequency is larger than that for the free oscillations.

If the pendula are released from rest with initial displacements such that

$$x(t = 0) = y(t = 0) \tag{13.12}$$

they will continue to oscillate with angular frequency $\omega = \sqrt{g/l}$. Similarly, if initially

$$x(t = 0) = -y(t = 0) \tag{13.12'}$$

again the pendula will oscillate harmonically, but now at $\omega = (g/l + 2k/m)^{1/2}$ (see fig. 13.4). In either case these oscillations will continue as shown for ever, since damping has been omitted in the formulation of the problem.

In Section 13.5, we will return to the question of the subsequent motion of the pendula for other types of initial displacements.

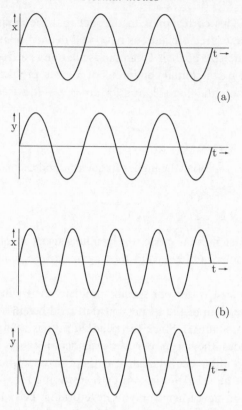

Fig. 13.4. The displacements x and y of the two pendula of fig. 13.2 as functions of time for the two normal modes. In (a), the separation of the pendula $(x - y)$ remains constant, and the normal mode frequency is independent of the spring constant; in (b), the relative motion is such as to increase the frequency as compared with that of the uncoupled pendula (see fig. 13.3).

13.2.2 *Spotting the solution*

For some simple problems, it may be possible to obtain the normal modes almost by inspection. Thus (13.4) already is an equation for the variable $x + y$, and is independent of any other combination of x and y separately. Then $x + y$ performs simple harmonic motion with $\omega^2 = g/l$.

Similarly, by subtracting the equations in (13.3), we obtain

$$m(\ddot{x} - \ddot{y}) = -mg(x - y)/l + 2k(y - x)$$
$$= -(mg/l + 2k)(x - y)$$

Now we see that the variable $x - y$ performs simple harmonic motion with $\omega^2 = mg/l + 2k$. This is the mode in which $x - y$ changes, and the bobs oscillate in antiphase.

To find the relative amplitudes x_0/y_0 for the first mode, we must make sure that the second mode is absent, i.e.

$$(x_0 - y_0) = 0$$

This agrees with eqn (13.11) for the mode with $\omega^2 = g/l$. Similarly, for the second mode, the absence of the first one requires

$$(x_0 + y_0) = 0$$

(Compare eqn (13.11') for the higher frequency mode.)

It is to be noted that, in this method, the relative amplitudes of the two bobs in a particular mode are determined by the linear combination of variables occurring for the other mode. Thus from an equation like (13.4), we deduce the frequency of one mode and the relative amplitudes for the other.

This technique depends on our ability to spot specific linear combinations of the original variables x and y, such that the differential equations for these combinations separate (i.e. each differential equation contains only one of the linear combinations). In more complicated problems, this may not be so trivial as here.

13.3 Two different coupled pendula

We now consider a generalisation of the previous section by having two pendula of different lengths l_1 and l_2, and hence of different natural frequencies. A simple arrangement would be to have their suspensions at suitably different heights so that the bobs are in the same horizontal plane. They can then be coupled in a similar way to that for the identical pendula.

In analogy with (13.3), the equations of motion are

$$\left.\begin{aligned} m\ddot{x} &= -mgx/l_1 + k(y - x) \\ \text{and} \quad m\ddot{y} &= -mgy/l_2 + k(x - y) \end{aligned}\right\} \tag{13.13}$$

Following through the argument about normal mode frequencies ω, we obtain instead of eqns (13.7')

$$\left.\begin{aligned} (A - \omega^2)x_0 + By_0 &= 0 \\ \text{and} \quad Bx_0 + (C - \omega^2)y_0 &= 0 \end{aligned}\right\} \tag{13.14}$$

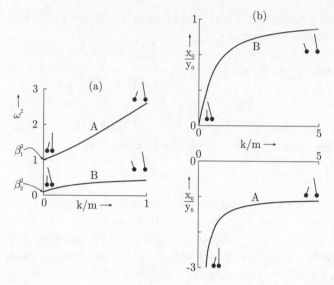

Fig. 13.5. Normal modes for two coupled pendula of unequal length, the longer one having displacement y. The natural frequencies of the uncoupled pendula are taken as $\beta_1 = 1.0$ and $\beta_2 = 0.3$. The frequencies of the two modes are shown in (a), and the amplitude ratios in (b), as functions of the coupling strength k. The curves labelled A correspond to the positive choice for the $\pm$ signs in eqn (13.16) for ω^2, and in (13.17) for x_0/y_0; the curves labelled B are for the negative choice. The schematic diagrams of the pendula indicate their relative motion in each of the normal modes.

where

$$\left.\begin{aligned} A &= \beta_1^2 + k/m \\ C &= \beta_2^2 + k/m \\ \text{and} \quad B &= -k/m \end{aligned}\right\} \tag{13.15}$$

Here β_1 and β_2 are the natural frequencies of the two uncoupled pendula (as opposed to the two values of ω, which are for the coupled normal mode oscillations).

The values of ω which are required to provide interesting solutions of eqns (13.14) for x_0 and y_0 are given by

$$\omega^2 = \frac{1}{2}\left[\beta_1^2 + \beta_2^2 + 2k/m \pm \sqrt{(\beta_1^2 - \beta_2^2)^2 + (2k/m)^2} \right] \tag{13.16}$$

The two values of ω are plotted in fig. 13.5 as functions of k/m, for a particular choice of the natural frequencies.

Two special cases of eqn (13.16) are worth checking. The first is the limit k tends to zero, when ω^2 reduces to β_1^2 or β_2^2. This is as it ought to be, since in this situation the spring becomes irrelevant and the two pendula are uncoupled.

The other limit is when β_1 and β_2 become equal, which is the situation considered in the previous section. Eqn (13.16) reduces to

$$\omega^2 = \beta_1^2 + k/m \pm k/m \qquad (13.16')$$

which is equivalent to eqn (13.10') (as it should be).

Again in analogy with the previous section, we next find the ratio x_0/y_0 of the amplitudes of the bobs for the normal modes. From eqns (13.14) and (13.16) we obtain

$$\frac{x_0}{y_0} = -\left[(\beta_1^2 - \beta_2^2) \pm \sqrt{(\beta_1^2 - \beta_2^2)^2 + (2k/m)^2}\right]/(2k/m) \qquad (13.17)$$

As required, this agrees with eqns (13.11) and (13.11') when $\beta_1 = \beta_2$.

Eqn (13.17) simplifies when k is either very small or very large. In the former case, x_0/y_0 tends to $-(\beta_1^2 - \beta_2^2)m/k$ or to $k/m(\beta_1^2 - \beta_2^2)$. These two modes thus correspond to y_0 being very much larger than x_0, or vice versa, respectively. Since the spring is very weak, the pendula are almost decoupled, and the modes correspond to one or other of the pendula oscillating at its own natural frequency, with the other almost stationary. In this situation, eqn (13.16) indeed gives $\omega \approx \beta_1$ or β_2.

On the other hand for very large k, x_0/y_0 tends to -1 or 1. Now the spring is so strong that in the first of these modes we can forget about the effect of gravity, and the motion of the two pendula with equal amplitudes in antiphase is dominated by the spring; this is the mode with $\omega^2 \approx 2k/m$. The other mode has the bobs in phase with nearly equal amplitude and so, even though the spring is very strong, it is almost unstretched, and the pendula oscillate such that ω^2 is the average of the squares of the natural uncoupled frequencies β_1^2 and β_2^2.

As is shown in fig. 13.5, not only do the frequencies of the normal modes change as the spring strength increases but so too does the nature of the normal modes. Thus for the curve labelled B in fig. 13.5(a), the normal mode at very small k consists of the longer pendulum oscillating and the shorter one being almost at rest; this pattern is reversed for the normal mode on the A curve at this value of k. As k increases, the normal modes gradually involve a larger amplitude for the previously stationary pendulum in each case, and when $k/m \gg \beta_1^2 - \beta_2^2$, the two modes involve equal amplitudes of the two pendula, being in phase for B and in antiphase for A. Thus the simplest types of motion of the system are different for different values of the coupling.

This is an elementary example of a phenomenon which recurs many times in physics. It involves using different basic descriptions of a system in different circumstances, with one basic set (for example, $\mathbf{v}_1''$ and $\mathbf{v}_2''$ of eqn (13.20)) being a linear combination of the other, e.g. $\mathbf{v}_1$ and $\mathbf{v}_2$ below. In the pendulum example, the simplest description at low k is in terms of the one pendulum moving, or the other. These two modes can thus be represented by the vectors

$$
\left.
\begin{aligned}
\mathbf{v}_1 &= \begin{pmatrix} 1 \\ 0 \end{pmatrix} \\
\text{and} \quad \mathbf{v}_2 &= \begin{pmatrix} 0 \\ 1 \end{pmatrix}
\end{aligned}
\right\}
\tag{13.18}
$$

where the two components denote x_0 and y_0, the amplitudes of the oscillations of the two bobs for this particular normal mode. Similarly, for very large spring strength, the modes are represented by

$$
\left.
\begin{aligned}
\mathbf{v}_1' &= \begin{pmatrix} 1/\sqrt{2} \\ -1/\sqrt{2} \end{pmatrix} \\
\text{and} \quad \mathbf{v}_2' &= \begin{pmatrix} 1/\sqrt{2} \\ 1/\sqrt{2} \end{pmatrix}
\end{aligned}
\right\}
\tag{13.19}
$$

where the vectors have been normalised to unity (which we are free to do since it is only the ratio x_0/y_0 which is defined for a normal mode).

For intermediate values of the coupling, the modes are†

$$
\left.
\begin{aligned}
\mathbf{v}_1'' &= C \begin{pmatrix} 1 \\ r \end{pmatrix} \\
\text{and} \quad \mathbf{v}_2'' &= C \begin{pmatrix} -r \\ 1 \end{pmatrix}
\end{aligned}
\right\}
\tag{13.20}
$$

where $1/r$ is given by the right-hand side of eqn (13.17) (with the positive choice for the $\pm$ sign, which makes r negative), and $C = 1/\sqrt{1 + r^2}$ is the normalisation factor.

It is almost a trivial remark that the pair of vectors for any of these three different sets of normal modes can be written as linear combinations of the pair for any other mode. In fact with our choice of normalised vectors for the modes, one set can be obtained by the relevant "rotation" from any other. The word "rotation" is written within quotation marks,

---

† $\mathbf{v}_2''$ can more obviously be written as $C'\begin{pmatrix} 1 \\ r' \end{pmatrix}$, where $1/r'$ is the right-hand side of eqn (13.17), with the negative choice of the $\pm$ sign. The equivalent form in (13.20) is used to emphasise the orthogonality of $\mathbf{v}_1''$ and $\mathbf{v}_2''$.

since it refers to a rotation, not in the ordinary space of the real three-dimensional world, but in the two-dimensional (x_0, y_0) space in which we have chosen to describe our normal modes.† Thus the "rotation" angle θ between $\mathbf{v}$ and $\mathbf{v}'$ is 45°, and that between $\mathbf{v}$ and $\mathbf{v}''$ is given by

$$\tan \theta = r \tag{13.21}$$

(See fig. 13.6.) Another point worth noting is that for each pair of modes, the vectors are orthogonal to each other, i.e.

$$\mathbf{v}_1 \cdot \mathbf{v}_2 = \mathbf{v}_1' \cdot \mathbf{v}_2' = \mathbf{v}_1'' \cdot \mathbf{v}_2'' = 0$$

Other examples of systems whose simplest descriptions change with different circumstances are:

(1) Polarised light: We can use either vertical and horizontal polarisation states, or alternatively left and right circularly polarised states, in order to describe any arbitrary state. As with all the examples here, one set is simply a specific linear combination of the others.

(2) Neutral K mesons: When these are produced in a typical nuclear collision, they are in states of well-defined strangeness, labelled K^0 and $\bar{K}^0$, but they decay as states known as K_1^0 and K_2^0.

(3) Quarks: The quark states d and s (known as down and strange) which participate in strong nuclear interactions are linear combinations of those (d' and s') which take part in the weak interactions, such as are responsible for β decay. The two different sets of quarks are related to each other by what is known as the Cabibbo rotation angle.

The point of the above examples was not so much to explain the phenomena involved, but to demonstrate a feature that is very common over a wide range of physics: what are effectively the normal modes of the system change as the physical circumstances vary. When you later come across such effects, you should remind yourself of the very simple analogy of the two coupled pendula. This should help you achieve a deeper insight into the new phenomena.

---

† Had our system consisted of ten coupled pendula, the normal modes would have needed 10-dimensional vectors to describe them.

(a)

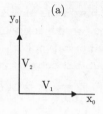

(b)

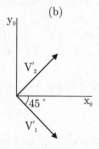

(c)

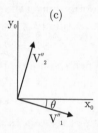

Fig. 13.6. Unit vectors, giving the ratio of the amplitudes x_0/y_0 for the normal modes of two pendula of different lengths: (a) No coupling between the pendula (see eqn (13.18)). (b) Strong coupling (see eqn (13.19)). (c) Intermediate coupling of strength k. The angle θ is given by eqn (13.21), and corresponds to a clockwise rotation of between zero and $\pi/4$ (see eqn (13.20)).

13.4 Matrix formulation

So far we have dealt with the case where our system consists of two components, and we found the two normal modes, each with its own frequency for the oscillations. In Sections 13.2 and 13.3, we solved these problems by writing explicitly the two equations of motion for the individual bobs, and then treating them as simultaneous equations. This was the most sensible way of approaching such a problem, but now we are going to repeat the analysis using matrix methods. This is in fact most useful for more complicated problems involving larger numbers of coupled components. It also provides additional insights concerning matrices and their eigenvalues. If you are not yet familiar with these topics, you should return to this section after working through Chapters 15 and 16.

Initially we consider the problem of the two identical pendula, and then go on to the case of the pendula of unequal length of Section 13.3.

Our starting point is to rewrite eqns (13.3) in matrix form

$$\mathbf{Mx} = -\ddot{\mathbf{x}} \tag{13.3'}$$

where the vector

$$\mathbf{x} = \begin{pmatrix} x \\ y \end{pmatrix} \tag{13.22}$$

and the matrix

$$\mathbf{M} = \begin{pmatrix} A & B \\ B & A \end{pmatrix} \tag{13.23}$$

with A and B as given in eqns (13.8).

With our usual assumption that the normal mode solution must have both components of $\mathbf{x}$ varying harmonically at the same frequency, the second derivative in eqn (13.3') is replaced by a factor of $-\omega^2$, so that the equation becomes

$$\mathbf{Mv} = \omega^2 \mathbf{v} \tag{13.3''}$$

Here $\mathbf{v}$ is simply the vector of the amplitudes of the oscillations:

$$\mathbf{v} = \begin{pmatrix} x_0 \\ y_0 \end{pmatrix} \tag{13.22'}$$

Now eqn (13.3'') is a standard eigenvalue problem. The eigenvalues are the values of ω^2 for each of the normal modes, while each of the corresponding eigenvectors gives the ratios of the amplitudes of the bobs in that mode. If the problem had involved N components, the matrix $\mathbf{M}$ would have been of size $N \times N$, and there would have been N eigenvectors, one for each of the N normal modes. Each of these eigenvectors would have had N components, one for the amplitude of each component in that particular mode.

The essence of the physics of a normal mode problem is thus contained in the matrix $\mathbf{M}$ that is obtained from the equations of motion. In our problem the matrix (13.23) has eigenvalues $A + B$ and $A - B$ (compare eqn (13.10)), and the corresponding eigenvectors are

$$\mathbf{v} = \begin{pmatrix} 1 \\ 1 \end{pmatrix} \text{ or } \begin{pmatrix} 1 \\ -1 \end{pmatrix} \tag{13.24}$$

This is equivalent to the previous eqns (13.11) and (13.11').

We have thus solved our problem. Not only does the matrix method provide a neat notation for dealing with situations with more than just a few components, but it also enables problems to be solved by standard matrix methods, without too much effort having to be expended concerning the details of the procedures for the normal mode analysis.

For the coupled pendula of different lengths, eqn (13.3″) still applies, but now

$$\mathbf{M} = \left(\begin{array}{cc} A & B \\ B & C \end{array} \right) \tag{13.25}$$

with the elements given by eqn (13.15). The eigenvalues of this matrix give the frequencies of the normal modes as in eqn (13.16); the eigenvectors (13.22′) are such that x_0/y_0 is as given in eqn (13.17). As with all eigenvalue and eigenvector problems, the situation is simplest when the eigenvalues differ from each other. Because the matrix $\mathbf{M}$ is symmetric, its eigenvectors are orthogonal (see Section 16.3.2.1); this explains why the normal mode vectors $\mathbf{v}_1$ and $\mathbf{v}_2$ of eqns (13.18–13.20) form orthogonal pairs.

When eigenvalues are degenerate (i.e. equal), any linear combination of the corresponding eigenvectors is also an eigenvector with the same eigenvalue. Similarly, if two or more normal modes have the same frequency, then any linear combination is still a normal mode at that frequency.

A rather uninteresting example of this is given by the two identical pendula without any coupling.† Then

$$\left(\begin{array}{c} x_0 \\ y_0 \end{array} \right) = \left(\begin{array}{c} 1 \\ 1 \end{array} \right) \quad \text{or} \quad \left(\begin{array}{c} 1 \\ -1 \end{array} \right) \tag{13.24'}$$

both correspond to normal modes with the same frequency. But so is any arbitrary $\mathbf{v}$ (even with complex components), since in the absence of coupling

$$\left. \begin{array}{l} x = \mathrm{Re}[x_0 e^{i\sqrt{g/l}\,t}] \\ \text{and} \quad y = \mathrm{Re}[y_0 e^{i\sqrt{g/l}\,t}] \end{array} \right\} \tag{13.26}$$

is a perfectly satisfactory solution.

A more interesting example of equal frequency normal modes is provided by one of the problems at the end of this chapter.

---

† The condition of no coupling is imposed merely to produce an example of equal frequencies for the two modes. The fact that any linear combination of such modes is also a mode depends on the equality of the frequencies, and not specifically on the lack of coupling.

13.5 Non-normal mode behaviour

So much emphasis has been placed on the normal modes of a coupled system that it is possible to forget that there are other types of motion. These are now examined, to see how they differ from the way a system oscillates in a normal mode.

13.5.1 Two identical pendula

We return to our example of two equal length coupled pendula, where initially one is displaced and then released. We actually know what the subsequent behaviour of the system looks like; this was displayed in fig. 13.1. What is now required is a mathematical derivation of this.

The initial conditions for the pendula are:

$$\left.\begin{array}{r} x(0) = u \\ y(0) = 0 \\ \dot{x}(0) = 0 \\ \text{and} \quad \dot{y}(0) = 0 \end{array}\right\} \tag{13.27}$$

Now this does *not* correspond to either of the two normal modes as given, for example, by eqns (13.11) or (13.11′), since for them the deflections are either equal or equal but opposite. However, there are always just the correct number of normal modes for any initial set of conditions to be expressed as a linear combination of the normal modes. The subsequent motion of the system is then determined by the fact that each normal mode has all of the components oscillating harmonically at the relevant frequency.

Thus for the two coupled pendula, it is possible to write

$$\begin{pmatrix} x \\ y \end{pmatrix} = \text{Re} \left\{ \alpha \begin{pmatrix} 1 \\ 1 \end{pmatrix} e^{i\omega_1 t} + \beta \begin{pmatrix} 1 \\ -1 \end{pmatrix} e^{i\omega_2 t} \right\} \tag{13.28}$$

Here α and β are the arbitrary constants which give the amount of each normal mode. Their values are determined by the initial conditions (13.27). In general α and β could be complex, corresponding to phase differences for the various normal modes; in this case, because the pendulum bobs are initially at rest, both α and β are real. The vectors $\begin{pmatrix} 1 \\ 1 \end{pmatrix}$ and $\begin{pmatrix} 1 \\ -1 \end{pmatrix}$ in eqn (13.28) are simply the normal modes of the system whose linear combination is taken in order to solve the problem; ω_1 and ω_2 are their corresponding frequencies.

Of course, eqn (13.28) can be written in component form as

$$\left.\begin{aligned} x &= \text{Re}\left(\alpha e^{i\omega_1 t} + \beta e^{i\omega_2 t}\right) \\ \text{and} \quad y &= \text{Re}\left(\alpha e^{i\omega_1 t} - \beta e^{i\omega_2 t}\right) \end{aligned}\right\} \tag{13.28'}$$

Even more simply, since α and β will turn out to be real, these equations reduce to

$$\left.\begin{aligned} x &= \alpha \cos \omega_1 t + \beta \cos \omega_2 t \\ \text{and} \quad y &= \alpha \cos \omega_1 t - \beta \cos \omega_2 t \end{aligned}\right\} \tag{13.28''}$$

It is most important to be aware of a feature of eqns (13.28') or (13.28'') that may not be immediately obvious. In the equations for x and y, the parameter α is multiplied by $+1$ in each case, while β is multiplied by $+1$ in the x equation but by -1 in that for y. This is because the normal mode of frequency ω_1 corresponds to equal amplitudes for x_0 and y_0, while that for ω_2 has them equal in magnitude but of opposite sign. In a more general case (for example, pendula of unequal lengths), these coefficients are more interesting numbers and hence would be more conspicuous in the corresponding equations. It is essential not to forget them.

One final point is worth noting before we proceed to solve this problem. It has been mentioned a few times that in a normal mode, the *ratios* of amplitudes are defined, but not their absolute normalisation. This arbitrariness is unimportant here, in that instead of the vector $\binom{1}{1}$, we could have used $\binom{2}{2}$, $\binom{-1}{-1}$, $\binom{3.14}{3.14}$, etc. Of course this would mean that the value of α determined from the given initial conditions changes, but these two effects cancel out to give the identical solution to the problem. On the other hand, the different signs of the two terms involving β in eqns (13.28') or (13.28'') are essential, as this corresponds to the fact that for the mode with frequency ω_2, the two bobs swing in *anti*phase.

At last we are going to determine the coefficients α and β in eqns (13.28'') from the initial conditions (13.27). On substitution of the first two conditions, we find

$$\left.\begin{aligned} \alpha &= u/2 \\ \text{and} \quad \beta &= u/2 \end{aligned}\right\} \tag{13.29}$$

The last two conditions of eqn (13.27) are also satisfied, albeit in a somewhat trivial manner. This was ensured by our choice of α and β in eqn (13.28') as being real, or equivalently by there being no extra phase angles in eqns (13.28'').

Thus the solutions are

$$\left.\begin{array}{r}x = u/2(\cos\omega_1 t + \cos\omega_2 t) \\ \text{and} \quad y = u/2(\cos\omega_1 t - \cos\omega_2 t)\end{array}\right\}\tag{13.30}$$

and this is really the end of the problem.

It is, however, instructive to look in more detail at this solution in the case where the coupling is weak. In that case, we find from eqn (13.10′) that

$$\omega_1 = \sqrt{g/l}$$

and

$$\begin{aligned}\omega_2 &= \sqrt{g/l + 2k/m} \\ &\approx \sqrt{g/l}(1 + kl/mg) \\ &\approx \omega_1 + \Delta\omega\end{aligned}$$

Thus the frequency of the second mode is close to that of the first one (see also fig. 13.3).

Then eqns (13.30) can be rewritten as

$$\left.\begin{array}{r}x = u\cos(\Delta\omega t/2)\cos\bar{\omega}t \\ \text{and} \quad y = -u\sin(\Delta\omega t/2)\sin\bar{\omega}t\end{array}\right\}\tag{13.30′}$$

where we have written $(\omega_1 + \omega_2)/2$ as $\bar{\omega}$. The solution (13.30′) for x can be regarded as consisting of the factor $\cos\bar{\omega}t$, giving oscillations at close to the natural frequency of the pendula in the absence of coupling, multiplied by $u\cos(\Delta\omega t/2)$; because $\Delta\omega$ is very much smaller than $\bar{\omega}$, this produces a variation in t which is very much slower than that corresponding to the $\cos\bar{\omega}t$ term. Thus $u\cos(\Delta\omega t/2)$ behaves as a slowly varying amplitude for the $\cos\bar{\omega}t$ oscillations, and provides the envelope of the oscillations of x as shown in fig. 13.1.

Similarly the envelope of the y oscillations is given by $-u\sin(\Delta\omega t/2)$. Thus, as expected, these start at zero, and gradually build up to their full amplitude at the stage where the $\cos(\Delta\omega t/2)$ term makes the x oscillations go to zero. There is indeed the transfer of energy backwards and forwards between the two pendula, as described in Section 13.1

It is worth looking back at this solution in order to understand how the behaviour of the system differs from that in a normal mode. A normal mode is defined as requiring all components of the system to oscillate at the same single frequency. Here the solutions (13.30′) for x and y show

that they move in a manner which is not perfectly harmonic.† It is this
which makes the behaviour not that of a normal mode.

13.5.2 Two pendula of different lengths

In complete analogy, we can treat the case of the two coupled pendula
of different lengths, discussed earlier in Section 13.3. The equations of
motion are (13.13), with the frequencies of the normal modes (13.20)
being given by eqn (13.16). Again we take the initial conditions as in
eqns (13.27), with just the first pendulum displaced; this results in not a
single mode but rather a linear combination of the two modes $\mathbf{v}_1''$ and $\mathbf{v}_2''$.
Thus, at $t = 0$,

$$\begin{pmatrix} x \\ y \end{pmatrix} = \begin{pmatrix} u \\ 0 \end{pmatrix} = \frac{u}{\sqrt{1+r^2}}(\mathbf{v}_1'' - r\mathbf{v}_2'') \tag{13.31}$$

where r in eqns (13.20) describes the amount of "rotation" required to
go from the normal modes $\mathbf{v}$ of eqn (13.18) in the absence of coupling,
to $\mathbf{v}''$.

Eqn (13.31) implies that at a later time t the displacements are

$$\begin{pmatrix} x(t) \\ y(t) \end{pmatrix} = \frac{u}{\sqrt{1+r^2}}(\mathbf{v}_1'' \cos \omega_1 t - r\mathbf{v}_2'' \cos \omega_2 t) \tag{13.32}$$

Thus

$$\left. \begin{array}{rl} (1+r^2)x(t) &= u(\cos \omega_1 t + r^2 \cos \omega_2 t) \\ \text{and} \quad (1+r^2)y(t) &= ur(\cos \omega_1 t - \cos \omega_2 t) \end{array} \right\} \tag{13.33}$$

We thus find that each of $x(t)$ and $y(t)$ is composed of two harmonic
oscillations. If ω_1 and ω_2 are close in frequency, x and y will exhibit beats.
Then $x(t)$ oscillates‡ at the angular frequency $\omega_1 \approx \omega_2$, with an amplitude
which varies between u when $\omega_1 t - \omega_2 t = 2n\pi$, and $u(1 - r^2)/(1 + r^2)$
when $\omega_1 t - \omega_2 t = (2n + 1)\pi$ (see fig. 13.7(a)). The beat period is then
$2\pi/\Delta\omega$, where $\Delta\omega = \omega_1 - \omega_2$. Similarly, we can write $y(t)$ of eqn (13.33)
as

$$(1+r^2)y(t) = -2ur \sin \bar{\omega}t \sin(\Delta\omega t/2) \tag{13.34}$$

† Admittedly, it is not much more complicated in that x or y can be expressed as the sum
  of two harmonic terms (see eqn (13.30)), but this is a very simple problem with only two
  normal modes. With N modes, any behaviour can be expressed as a linear combination
  of the N harmonic solutions corresponding to the N normal modes and can look quite
  complicated for large N.
‡ The relevant algebra very much resembles that of eqn (14.86).

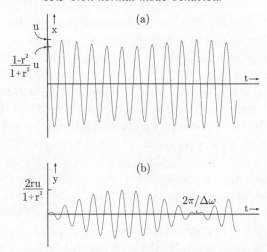

Fig. 13.7. Oscillations $x(t)$ and $y(t)$ of two coupled pendula of unequal but similar lengths. Initially one is displaced and then released from rest. An incomplete exchange of energy occurs between the two pendula. The maximum amplitude of the second pendulum's oscillation occurs at $t = \pi/\Delta\omega$, and is $2ur/(1 + r^2) =$ $u\sin 2\theta$, where θ is the "rotation" angle, related to r by eqn (13.21). The minimum of the $x(t)$ amplitude occurs at the same time, and is $u(1-r^2)/(1+r^2) = u\cos 2\theta$. The diagram is for $\Delta\omega = 0.1\omega$, and $r = 0.3$ ($\theta = 16.7°$).

where

$$\bar{\omega} = (\omega_1 + \omega_2)/2 \tag{13.35}$$

Hence $y(t)$ also exhibits beats, and has its maximum amplitude when $\sin(\Delta\omega t/2) = \pm 1$, i.e. when $\Delta\omega = (2n + 1)\pi$. Hence the maximum amplitude for the y oscillations occurs when the $x(t)$ amplitude is smallest (see fig. 13.7).

Just as in the case of the two equal pendula, there is an exchange of energy between the pendula, but here it is incomplete. Thus, for $|r| < 1$,† the envelope of the oscillations for x never decreases to zero; similarly that for y builds up to a maximum, but it reaches only $2ur/(1 + r^2)$, which again is less than u. In summary, the system starts out with a pure x oscillation, but later it decreases while that of y builds up. They vary in an oscillator manner, with the frequency of these envelope oscillations being determined by $\Delta\omega$, and with their amplitude depending on the parameter r and hence upon the coupling k.

† The variable r, as defined in eqn (13.21), reaches -1 only when the coupling k becomes infinitely strong.

This type of oscillations (i.e. the slow variation of the envelopes in fig. 13.7 rather than the fast oscillation of the pendula) is analogous to those found in other branches of physics. Thus neutrinos exist in three different varieties, v_e, v_μ and v_τ. If for simplicity we consider just the first two, the v_e and v_μ that are involved in elementary particle reactions are analogous to the $\mathbf{v}_1$ and $\mathbf{v}_2$ of eqn (13.18) for the uncoupled oscillations. However, the neutrinos that travel freely correspond more to the $\mathbf{v}_1''$ and $\mathbf{v}_2''$ of the normal modes in the coupled case. Thus a neutrino beam that initially contains only v_e (equivalent to only the x pendulum being displaced at $t = 0$) can after a while oscillate to contain some v_μ (equivalent to the y pendulum also moving).

Just as the time period for the energy transfer between the pendula depends on the difference between ω_1 and ω_2, and also requires the coupling to be non-zero, so in the neutrino case, oscillations between v_e and v_μ depend on them having different masses,† and on the "mixing angle" which describes their coupling being non-zero.

Again this is a situation where, when you later study neutrino oscillations in elementary particle physics, it will be very worth while to look again at the case of the humble pendulum.

13.5.3 Many components

In problems with more than two components, we proceed in a similar manner. In analogy with eqns (13.28), the solution for an N component problem is written as

$$\mathbf{x} = \mathrm{Re}\left\{ \sum_{i=1}^{N} \alpha_i \mathbf{v}_i e^{i\omega_i t} \right\} \tag{13.36}$$

Here $\mathbf{x}$ is the vector giving the displacements of each of the N components of the system; $\mathbf{v}_i$ is the vector‡ describing the system's ith normal mode, whose frequency is ω_i; and α_i is the arbitrary coefficient of the ith mode. The initial conditions of the problem determine the coefficients α_i, and the problem is solved. If all the initial velocities are zero, then the α_i are

---

† The three types of neutrinos are each consistent with having zero mass. It is possible, however, that they do have very small masses, below the limits of current experimental sensitivity. In that case, oscillations could be possible.

‡ The components of $\mathbf{v}_i$ will be real for the situation where all the normal mode frequencies ω_i are different from each other. This is the case we consider for the remainder of the paragraph (contrast Problem 13.4).

all real and the N initial displacements are just sufficient to determine the N real coefficients α_i. In that case eqn (13.36) reduces to a set of cosine terms, as in eqn (13.28''). In contrast, if all the initial displacements are zero but the velocities are not, the α_i are imaginary, and instead we have a sine series. For the general case when both displacements and velocities are non-zero at $t = 0$, the α_i are complex, and the $2N$ initial conditions (i.e. the position and velocity for each of the N components) are needed in order to determine the N complex coefficients α_i.

13.6 Resumé

The discussion in this chapter about normal modes has involved several digressions, and the logic has been explained at length. This may have given the false impression that normal mode problems are long and complicated. To put things in perspective, we give here a brief resumé of the steps required for solving a typical normal modes problem.

Step (i): Set up the equations of motion for the components of the system of the specific problem.

Step (ii): To find the normal modes, replace the second derivatives $\left(d^2/dt^2\right)$ by $-\omega^2$.

Step (iii): Rewrite the equations in matrix form. (For only two components, a reasonable alternative is to replace steps (iv) to (vi) by the equivalent non-matrix operations for solving two simultaneous equations.)

Step (iv): Find the eigenvalues of the matrix, and hence the frequencies of the normal modes.

Step (v): Obtain the eigenvectors of the matrix, and hence the ratios of the amplitudes of the motion of the various components in each mode separately.

Step (vi): For initial conditions which do not correspond to a single normal mode, express the motion as a linear combination of the modes, and determine the coefficients from the initial conditions.

Now that you are much more familiar with normal modes, it is worth while to return to Section 13.1, as you are in a much better position to appreciate the general description of normal mode behaviour to be found there.

13.7 Notation

We here summarise the notation used throughout this chapter.

j Label for the N coupled components

i Label for the N normal modes (also $\sqrt{-1}$)

β_j Natural frequencies of uncoupled components

ω_i Frequencies of normal modes

$\left.\begin{array}{l} x, y \\ x_j \\ \mathbf{x} \end{array}\right\}$ Displacements of components for general motion

$\left.\begin{array}{l} (x_0)_i, (y_0)_i \\ (x_0)_{j,i} \\ \mathbf{v}_i \end{array}\right\}$ Amplitudes of displacements of components in ith normal mode

$\left.\begin{array}{l} x_0, y_0 \\ (x_0)_j \end{array}\right\}$ Amplitudes of displacements in specific normal mode

$x(0), y(0)$ Initial displacements of components for general motion

$\left.\begin{array}{l} \alpha, \beta \\ \alpha_i \end{array}\right\}$ Coefficients of normal modes required for description of general motion of components

$\mathbf{M}$ Matrix of constants in equations of motion for components

l Length of pendulum

m Mass of pendulum bob

k Spring constant, responsible for coupling

Problems

13.1 Two identical pendula each of length l and each with bobs of mass m are free to oscillate in the same plane. The bobs are joined by a massless spring with a small spring constant k, such that the tension in the spring is k times its extension. The motion of the two bobs is governed by eqns (13.3).

Find the frequencies of the normal modes of the system, and also the relative displacements of the two bobs for each mode separately. Is the value of either of these frequencies obvious?

At $t = 0$, both bobs are at their equilibrium positions. One is stationary, but the other has a velocity v_0. If $k/m = 0.105g/l$, describe as fully as possible the subsequent velocities of the two bobs.

13.2 Two coupled simple pendula are of equal length l, but their bobs have different masses m_1 and m_2. Their equations of motion

are

$$\ddot{x} = -\frac{g}{l}x - \frac{k}{m_1}(x-y)$$

$$\text{and} \quad \ddot{y} = -\frac{g}{l}y + \frac{k}{m_2}(x-y)$$

(i) Use the method described in Section 13.2.1 to find the frequencies and the relative amplitudes of the bobs for the normal modes of the system.

(ii) By taking suitable linear combinations of the two equations of motion, obtain two uncoupled differential equations for the linear combinations (compare Section 13.2.2). Hence again find the normal mode frequencies and relative amplitudes.

(Hint: One of these linear combinations is fairly obvious. For the other, it is helpful to consider the centre of mass of the two bobs.)

13.3 Two massless springs with spring constants k_1 and k_2 are connected to masses m_1 and m_2 as shown below. Find the frequencies of the normal modes of this system for vertical displacements.

If $m_1 = m_2 = m$ and $k_1 = k_2 = \lambda m$, show that the angular frequencies of these modes are $(\sqrt{5} \pm 1)\sqrt{\lambda}/2$, and find the ratio of the amplitudes of the oscillations of the masses for each mode.

13.4 What is meant by a "normal mode" for a system of mutually coupled oscillating particles? How many normal modes does a system have? What effect does it have if the system is such that two of the normal modes have the same frequency?

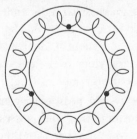

Three identical masses are confined to a circle around which are connected identical springs. The angular displacements with respect to the equilibrium positions (at 90°, 210° and 330°) are ϕ_1, ϕ_2 and ϕ_3 such that the equations of motion of the masses are

$$m\ddot{\phi}_i = c(\phi_j + \phi_k - 2\phi_i)$$

where i, j and k can equal 1, 2 or 3 and are all different from each other.

Find the frequencies of all the normal modes, and describe the motion of the three masses in each mode. Can you find a mode in which the sizes of the oscillation amplitudes of the three masses are the same, but they are out of phase with each other? (That is, $\phi_i = a\cos(\omega t - \alpha_i)$, with the three α_i being different.)

13.5 A stretched massless spring has its ends at $x = 0$ and $x = 3l$ fixed, and has equal masses attached at $x = l$ and $x = 2l$. The equations of the transverse motion of the masses are approximately

$$m\ddot{y}_1 = \frac{T}{l}(y_2 - 2y_1)$$
$$\text{and} \quad m\ddot{y}_2 = \frac{T}{l}(y_1 - 2y_2)$$

where T is the tension in the spring. (Gravity does not appear in this equation because: (a) y_1 and y_2 refer to motion with respect to the equilibrium positions; or (b) the motion takes place on a horizontal frictionless table.) Convince yourself that, for small oscillations, it is reasonable to neglect the changes in tension caused by the variation in length of the three sections of the spring resulting from the transverse motion of the masses.

Find the frequencies and the ratio of amplitudes of the transverse oscillations for the normal modes of the two masses. Is the relative motion of the higher frequency mode reasonable?

13.6 An infinitely long massless string is under tension T; it has equal masses m attached at separations δ. The transverse displacements of the masses are y_n, where n labels the position of each mass and varies in integral steps over the range $\pm\infty$. The y_n obey the differential equations $m\ddot{y}_n = (T/\delta)(y_{n+1} - 2y_n + y_{n-1})$.

Show that this system possesses normal modes such that the amplitudes of the motion of the masses are given by $A\sin kn\delta$, provided that the wave number k and the normal mode's angular frequency ω are related by

$$\sin^2\left(\frac{k\delta}{2}\right) = \frac{m\omega^2\delta}{4T}$$

Draw a graph of ω against k. What is the lowest value of k for the maximum value of ω? What is the relative displacement of the masses for the mode with this k? Hence explain why higher values of k do not lead to larger values of ω.

Show that as δ and m both tend to zero, but m/δ tends to ρ (the linear density of masses along the string), the relationship between k and ω reduces to

$$\frac{\omega}{k} = \sqrt{\frac{T}{\rho}}$$

This demonstrates that, as the discrete masses tend to a continuous distribution, the normal modes become continuous sine waves, with ω/k having the dependence on the tension and density as expected for waves on a string. This brings out the analogy between normal modes for a system with discrete masses, and Fourier series for the continuous case.

In a very different limit we can look again at Problem 13.5 for a spring connected to just two masses. The two normal modes there can be regarded, with a little imagination, as resembling the two lowest frequency sine waves here (in the figure below, the solid line shows the position of the spring, while the dashed line is the equivalent sine wave)

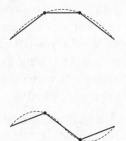

14
Waves

14.1 Waves are everywhere

It is useful to be completely at home with the properties of waves because they occur in so many different fields of physics. Thus quantum mechanics, optics, electromagnetism, stretched strings and membranes, seismology and sound are just some of the topics in which waves are very common. Even from the limited viewpoint of doing well in examinations, it pays to learn about waves because they are likely to occur in many different types of physics problems, as well as in mathematics papers.

For concreteness, most of the language we will use will be that for transverse waves on an elastic string; with suitable modification it can be applied to other types of examples. The string is taken as lying along the x axis when no wave is present, and in order to avoid end-effects will usually be assumed to extend to infinity in both directions. The transverse displacement is in the y direction. Thus we are interested in how y varies as we look at the wave; it will be a function of both the position x along the string and the instant t that we look at it.

Waves can usually be of almost any shape. For example, we could arrange that the initial displacement on the string (i.e. what would be seen in a photograph of the string) looks like a single square pulse, an infinite repetition of square pulses, five cycles of a sine wave, some arbitrary complicated shape, or an infinitely repeating sinusoidal wave (see fig. 14.1). Most of the time, we will consider this last possibility, for both concreteness and simplicity. Many of our comments, however, will apply to waves of arbitrary shapes.

230

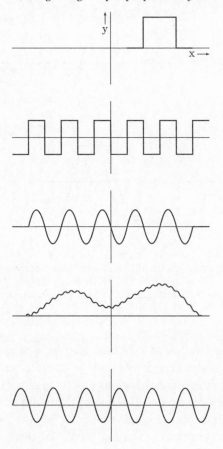

Fig. 14.1. Some possibilities for the displacement y of a piece of string, shown as functions of the position x along the string at $t = 0$.

14.2 Recognising simple properties of waves

There are two types of waves which are particularly common, at least in textbooks, even if the real world may be perhaps a little more complicated. The first of these is the travelling wave. This means that whatever pattern we could see for the displacement y at a given instant, at any later time t the displacement would be more or less the same shape, but shifted along by a distance that depends on the velocity of the wave and on t. Waves on the surface of the sea approximate to this.

The other basic type is the standing wave pattern. Here the wave does not move, in that the positions at which the vibration is a maximum

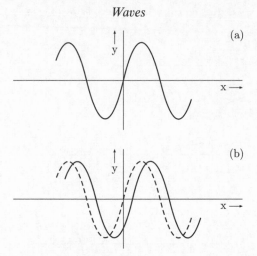

Fig. 14.2. Diagrams to illustrate which way the wave $y = A\sin[k(x - vt)]$ moves. In (a), it is drawn at $t = 0$. In (b), the solid curve corresponds to a slightly later time $t = \varepsilon$; as compared with the $t = 0$ curve (shown dashed in (b)), it is displaced to the right by a distance $v\varepsilon$. As time progresses, the initial curve thus moves smoothly to the right.

remain at the same values of x as time goes by, and similarly for those points which are nodes. It is simply that the displacements at every point on the string have their time variations exactly in phase (or in antiphase) with each other. A reasonable example of this is the motion of a violin string as the violinist plays a single note.

We start by writing down a moving wave, and then demonstrate that it has the required properties:

$$y = A\sin[k(x - vt)] \qquad (14.1)$$

First we note that, as expected, the formula for y contains both x and t as independent variables. Now usually when we draw a function, we have y being a function of only one variable, so if we want to see how eqn (14.1) behaves, we had better choose one of the variables as fixed in order to produce a simple diagram on two-dimensional paper. So let us keep t fixed, and the simplest value to choose for the time being is zero. Then

$$y(t = 0) = A\sin kx \qquad (14.2)$$

which we can easily draw (see fig. 14.2(a)). This is what we would obtain if we photographed our wave at $t = 0$.

Now we want to draw the displacement y at some later time. We choose $t = \varepsilon$, where ε is a small time (we shall see later how small ε has to be in order to be considered "small"). Then

$$y(t = \varepsilon) = A \sin[k(x - v\varepsilon)] \qquad (14.3)$$

This is of course sinusoidal in shape, but instead of passing through the origin, the curve is shifted over so that it "starts" at $x = +v\varepsilon$. This is shown as the solid curve of fig. 14.2(b); the dashed line is the curve for $t = 0$.

We next want to consider all the different curves we would obtain as we allow t to increase steadily. It should now be clear from eqn (14.1) that we will always obtain a sinusoidal-shaped curve, but that its "origin" will be found at $x = +vt$. That is, the origin moves steadily to the right with constant speed† v, and, of course, takes the whole curve along with it. Thus if we had drawn a very large number of these curves for y, each at a value of t slightly larger than the previous one, and then joined them together to make a movie picture, when we played back the film we would see a wave of sinusoidal shape moving smoothly to the right with constant speed.

Another way to convince ourselves that the wave is moving to the right is as follows. Instead of regarding eqn (14.1) as giving y as a function of x and t separately, we could instead think of it as specifying y as a function of a single variable $x - vt$. Then provided $x - vt$ keeps the same value, y will remain the same as well. For example, if we choose

$$x - vt = \pi/2$$

then y will be a maximum. Now if t changes by Δt, the necessary corresponding change in x in order to stay at the same maximum of the curve is given by

$$\Delta(x - vt) = \Delta x - v\Delta t = 0$$

Thus

$$\Delta x = +v\Delta t$$

and we must move in the positive direction to remain on the maximum.

The above wave was sinusoidal in shape, but we should now be able to guess that if someone were to draw us some arbitrary shape

$$y(t = 0) = f(x)$$

† In this chapter, the symbol c is used for the speed of light in a vacuum, and v for the speed of a general wave.

corresponding to the displacement of our string at time zero, then the way to make this into a wave of the same shape moving to the right with speed v is simply to write

$$y(x,t) = f(x - vt)$$

(Compare Section 11.4.) Thus, for example, if our initial displacement were

$$y(t = 0) = \begin{cases} +\dfrac{h}{L}(x + L) & \text{for } -L < x < 0 \\ -\dfrac{h}{L}(x - L) & \text{for } 0 < x < L \\ 0 & \text{for } |x| > L \end{cases} \tag{14.4}$$

then following our prescription above, we can convert it into a moving wave by writing the displacement at any time t as

$$y(x,t) = \begin{cases} +\dfrac{h}{L}(x - vt + L) & \text{for } -L < x - vt < 0 \\ -\dfrac{h}{L}(x - vt - L) & \text{for } 0 < x - vt < L \\ 0 & \text{for } |x - vt| > L \end{cases} \tag{14.5}$$

This generates for us the patterns shown in fig. 14.3.

For simplicity we now return to our sinusoidal wave (14.1), which moves to the right. Sometimes, of course, we may need to write down a wave moving towards the left. It does not need a great deal of imagination to realise that this is given by

$$y = A \sin k(x + vt) \tag{14.6}$$

It is worth noting that the $+$ sign in front of the vt implies a wave moving in an $-x$ direction, while the $-vt$ of eqn (14.1) corresponds to movement in the $+x$ sense.

Some other simple properties of a wave are its wavelength λ and its period τ. Since the wavelength is the distance for the wave pattern to repeat itself at constant t, we have

$$\Delta[k(x - vt)] = 2\pi$$
$$= k\Delta x - kv\Delta t$$
$$= k\lambda \tag{14.7}$$

Thus $\lambda = 2\pi/k$. In a similar manner, the period is the time taken for the pattern to repeat at a fixed position (i.e. $\Delta x = 0$). We find

$$\tau = 2\pi/kv$$

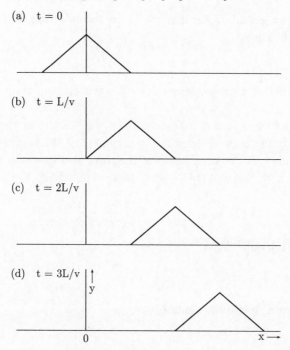

Fig. 14.3. A wave as specified by eqns (14.5), drawn at $t = 0$, and also at later times $t = L/v$, $2L/v$ and $3L/v$. As in fig. 14.2, this wave also moves to the right, because y is a function of $x - vt$.

If we want to, we can rewrite our wave (14.1) in terms of the period and wavelength as

$$y = A \sin \left[2\pi \left(\frac{x}{\lambda} - \frac{t}{\tau} \right) \right] \tag{14.8}$$

This is more symmetric in the variables x and t than the original equation.

Yet other possibilities arise from using the reciprocals of τ and λ. The former is the frequency v. The latter is the wave number and is sometimes denoted by k. Since we have already used k in eqn (14.1), we will write

$$\frac{1}{\lambda} = \bar{k} = \frac{k}{2\pi}$$

to obtain

$$y = A \sin \left[2\pi \left(\bar{k}x - vt \right) \right] \tag{14.9}$$

Finally, if we don't like the 2π appearing explicitly in the above equation, we can make use of the angular frequency ω (which is measured in radians per second, rather than cycles per second, and hence $\omega = 2\pi\nu$) and $k = 2\pi\bar{k}$. Then

$$y = A\sin(kx - \omega t) \tag{14.10}$$

These alternative representations of the wave are summarised in Table 14.1.

For any of these forms, we can read off the velocity of the wave by remembering the trick of keeping the argument of the sine constant, in order to remain at the same phase of the wave as it moves in space and time. Thus, if we use eqn (14.8), we obtain

$$\Delta\left[2\pi\left(\frac{x}{\lambda} - \frac{t}{\tau}\right)\right] = 2\pi\left(\frac{\Delta x}{\lambda} - \frac{\Delta t}{\tau}\right) = 0$$

to give the velocity v as

$$v = \frac{\Delta x}{\Delta t} = +\frac{\lambda}{\tau} \tag{14.11}$$

Similarly from (14.10), we could obtain

$$v = +\frac{\omega}{k} \tag{14.12}$$

Just to ensure we have not made an elementary algebraic mistake, we ought to check that our answers are dimensionally correct. Since the dimensions of λ and τ are length and time respectively, as are $1/k$ and $1/\omega$, we see that (14.11) and (14.12) both satisfactorily determine v as being length/time. In a similar way, we can verify that the arguments of all the sines and cosines are dimensionless.

Finally we can return to our choice of a "small" t in eqn (14.3). It is now clear that we ought to have $\varepsilon \ll \tau$, in order to be able to tell which way our wave has moved. Otherwise, because of the repetitive nature of the pattern, we could be confused about the correspondence of zeroes of the pattern at $t = 0$ and at finite ε.

Now we turn to stationary waves, and again adopt the procedure of writing down the answer, and then showing that it behaves reasonably. Our displacement is given by

$$y = A\sin kx \cos \omega t \tag{14.13}$$

In fact the sine and cosine in this equation have no deep significance, and could equally well have been replaced by two sines, or by two cosines, or by a cosine and a sine. The most general form would have a couple of phase angles, e.g.

$$y = A \sin(kx + \phi_x) \cos(\omega t + \phi_t)$$

It simply depends on how our wave starts off at $x = 0$ and at $t = 0$.

In order to investigate the properties of eqn (14.13), we again draw our y at various specific times. Thus at time zero, we have

$$y(t = 0) = A \sin kx \qquad (14.14)$$

This is drawn in fig. 14.4(a). A short time later, we have

$$y(t = \varepsilon) = A \sin kx \cos \varepsilon$$
$$= A' \sin kx \qquad (14.15)$$

where $A' = A \cos \varepsilon$, and is independent of x. Thus the shape of the curve is unchanged, and so is its position. All that has happened is that its amplitude A' is a little reduced compared with the amplitude A at $t = 0$. This is drawn in fig. 14.4(b).

Then, as time progresses, we will always be able to write

$$y(x, t) = [A \cos \omega t] \sin kx \qquad (14.13')$$

where the term in square brackets is the amplitude of the sine wave at that particular time; it varies between $+A$ and $-A$. The origin of the sine wave, however, is always at zero and never moves. This is why the wave is called a stationary wave.

An equivalent way of arriving at the same result is to write down the values of x at which the displacement is zero. From eqn (14.13), we have that

$$x(y = 0) = n\pi/k \qquad (14.16)$$

and this result is independent of t. Thus the nodes of the wave stay at fixed positions, and the whole curve is stationary.

The essential difference between the formulae (14.1) and (14.13) is that the latter factorises into independent functions of x and t. It is this feature which makes, for example, the nodes of the stationary wave remain stationary. On the other hand, (14.1) is a function of $x - vt$, and so any particular point on the wave moves along with a speed $+v$.

It is important to realise that the individual points on the string (i.e. fixed x) are in both cases performing simple harmonic oscillations in time. The difference between the two types of waves arises from the fact that in the moving wave case, different points oscillate with the same

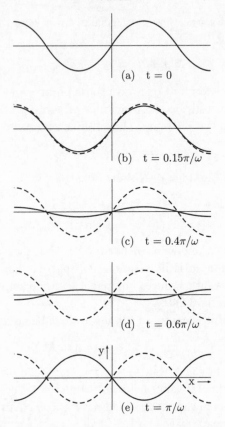

Fig. 14.4. The wave specified by eqn (14.13). In (a) it is shown at $t = 0$, and in (b) a short time $\varepsilon = 0.15\pi/\omega$ later. Subsequent pictures show the wave at $t = 0.4\pi/\omega$, $0.6\pi/\omega$ and π/ω. The positions where the wave has its nodes stay constant, and the wave is stationary. The dashed curves in (b)–(e) show the original ($t = 0$) displacement.

amplitude but are out of phase with respect to each other, while for the standing wave, they are all in phase but have different amplitudes.

As with the case of the moving wave, the stationary wave can be written in terms of variables other than the ω and k used in eqn (14.13). Other possibilities are shown in Table 14.1.

As a final illustration of the relationship between moving and stationary waves, we remember that one way of producing a stationary pattern is by sending two waves moving in opposite directions along the same piece of string. This can be achieved by waving one end of a piece of

Table 14.1. *Different ways of writing waves.*

The moving waves, which all travel in the $+x$ direction, are functions of $x - vt$, while the stationary waves are factorisable into separate functions of x and of t.

Form of moving wave	Velocity of moving wave	Wave-length	Period	Angular frequency	Form of stationary wave
$A\sin[k(x - vt)]$	v	$2\pi/k$	$2\pi/vk$	vk	$A\sin kx \sin vt$
$A\sin[2\pi(x/\lambda - t/\tau)]$	λ/τ	λ	τ	$2\pi/\tau$	$A\sin(2\pi x/\lambda)\times$ $\sin(2\pi t/\tau)$
$A\sin[2\pi(\tilde{k}x - vt)]$	$v/\tilde{k}$	$1/\tilde{k}$	$1/v$	$2\pi v$	$A\sin 2\pi\tilde{k}x \sin 2\pi vt$
$A\sin[kx - \omega t]$	ω/k	$2\pi/k$	$2\pi/\omega$	ω	$A\sin kx \sin \omega t$
$A\sin[2\pi(x - vt)/\lambda]$	v	λ	λ/v	$2\pi v/\lambda$	$A\sin(2\pi x/\lambda)\times$ $\sin(2\pi vt/\lambda)$

string, the other end of which is securely tied to a rigid support; the incident wave is reflected at this end and sets up a second wave moving in the opposite direction.

It is now a trivial exercise to show that our mathematical representation of waves reproduces this well-known fact. If the wave travelling to the right is $A\sin(kx - \omega t)$ and that to the left is $A\sin(kx + \omega t)$, then the net displacement is

$$y = A\sin(kx - \omega t) + A\sin(kx + \omega t)$$
$$= 2A\sin kx \cos \omega t \tag{14.17}$$

which is just a stationary wave.

It might be thought that we obtained a stationary wave because of our special choice of the moving waves. This is not so. Thus had either or both of them been cosines, or the second one $-A\sin(kx + \omega t)$ (as is, in fact, more appropriate for the reflection of the first wave — see Section 14.6), then we would still have obtained a stationary wave, although it would have differed in phase from that of eqn (14.17). Thus it can always be written as

$$2A\sin(kx + \phi_1)\cos(\omega t + \phi_2) \tag{14.18}$$

where ϕ_1 and ϕ_2 are constant phase angles. Indeed this is as we ought to have expected, since the relevant fact about our moving waves is that they are harmonic. The two phases in eqn (14.18) then are simply related to those of our moving waves $\sin(kx - \omega t + \phi_R)$ and $\sin(kx + \omega t + \phi_L)$, and all these phases simply depend on our fairly arbitrary choices of the space and time origins. Thus the effect is not to change the type of solution but only its phase.

If, however, the waves travelling in opposite directions are of different amplitude, then their sum will be a stationary wave plus a superimposed reduced-in-amplitude moving wave. Thus, for example,

$$A \sin(kx - \omega t) + B \sin(kx + \omega t)$$
$$= A[\sin(kx - \omega t) + \sin(kx + \omega t)] + (B - A) \sin(kx + \omega t)$$
$$= 2A \sin kx \cos \omega t + (B - A) \sin(kx + \omega t) \qquad (14.19)$$

This is the mathematical expression of the first sentence of this paragraph.†

14.3 The wave equation

We here obtain the differential equation

$$\frac{\partial^2 y}{\partial x^2} = \frac{1}{v^2} \frac{\partial^2 y}{\partial t^2} \qquad (14.20)$$

for small transverse displacements y on a string. Our derivation will serve both as a model for obtaining wave equations in other similar situations, and also to emphasise what are the approximations involved.

Fig. 14.5 shows the displacements y along a short segment δx of the string, and the forces acting on it. If we assume that the tension T is constant along the string,‡ then the vertical force on the segment is

$$V = T(\sin \theta_2 - \sin \theta_1)$$
$$\approx T(\tan \theta_2 - \tan \theta_1)$$
$$= T\left(\left(\frac{\partial y}{\partial x}\right)_2 - \left(\frac{\partial y}{\partial x}\right)_1\right) \qquad (14.21)$$

where we have assumed that the angles θ_1 and θ_2 are small,§ in order to replace the sines by tangents.

Now the gradients $\partial y/\partial x$ at the right- and left-hand ends of the

---

† It is perhaps not quite intuitively obvious that we can regard the sum as being either: (i) a stationary wave made from two moving waves of amplitude A, plus a wave of amplitude $(B - A)$ moving to the left (eqn (14.19) above); or alternatively (ii) a stationary wave composed from two moving waves of amplitude B, plus a wave of amplitude $(A - B)$ moving to the right.

‡ We examine this assumption shortly.

§ This tells us the sense in which the transverse displacements must be small. We require $y \ll \lambda$, in order to keep the angles small.

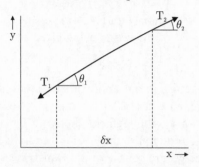

Fig. 14.5. The displacement y as a function of x for a short length δx of string which carries a transverse wave. The angles that the string makes with the horizontal are θ_1 and θ_2 at the two ends of the segment; these angles are assumed to be small. The tensions in the string at the ends of the segment are T_1 and T_2, although in the approximation to which we are working, they turn out to be equal.

segment of string are related by Taylor's Theorem. Thus

$$\left(\frac{\partial y}{\partial x}\right)_2 = \left(\frac{\partial y}{\partial x}\right)_1 + \frac{\partial}{\partial x}\left(\frac{\partial y}{\partial x}\right)\delta x \qquad (14.22)$$

and hence the vertical force becomes

$$V = T\frac{\partial^2 y}{\partial x^2}\delta x \qquad (14.21')$$

This exhibits a feature which is very common in problems of this type. From each end of the small region, the force (or heat flow, or diffusion rate, or whatever, depending on the nature of the specific problem) receives a contribution which is proportional to a derivative, in this case $\partial y/\partial x$. We are interested in the difference between the contributions from the two ends, and this results in a second derivative $\partial^2 y/\partial x^2$.

This vertical force is responsible for the vertical acceleration of the small segment of the string, and so by Newton's Second Law

$$V = \rho\delta x\frac{\partial^2 y}{\partial t^2} \qquad (14.23)$$

where ρ is the linear density of the string, and hence $\rho\delta x$ is its mass. By substituting for V from eqn (14.21'), we then obtain the wave equation

$$T\frac{\partial^2 y}{\partial x^2} = \rho\frac{\partial^2 y}{\partial t^2} \qquad (14.24)$$

where (again as is typical in these problems) the length of the segment

δx cancels from the equation. Eqn (14.24) is of the form (14.20) that we wanted, with the speed of the waves being given by

$$v = \sqrt{T/\rho} \qquad (14.25)$$

Now we want to return to our approximation that T is constant along the string. We obtained eqn (14.23) by applying Newton's Second Law to the vertical motion of the segment shown in fig. 14.5. If we instead resolve horizontally, and require that there is no horizontal displacement of the segment since we are considering a transverse wave, we obtain

$$T_1 \cos \theta_1 - T_2 \cos \theta_2 = 0 \qquad (14.26)$$

As we have already required our angles θ to be small, we can replace $\cos \theta$ by unity, whence

$$T_1 = T_2 \qquad (14.27)$$

i.e. the tension is, to this approximation, constant along the string.

14.4 Group velocity and phase velocity

Up till now we have been considering the straightforward case where the speed of the waves v was a constant. In many physical situations, however, the medium through which the waves travel is "dispersive". That is, different frequencies are transmitted at different speeds. Thus if we write our basic wave as

$$y = A \sin \omega \left(\frac{x}{v} - t\right) \qquad (14.28)$$

then $v = f(\omega)$

A well-known example of such behaviour is provided by light travelling through a glass prism. An incident beam of white light is split up by refraction into a series of outgoing beams with the various colours of the spectrum emerging at different angles. Since the refractive index μ is related to the speed v of light in the medium by

$$\mu = c/v$$

(where c is the speed of light in a vacuum), the appearance of the spectrum is evidence for dispersive effects in the propagation of light through glass.

In such situations, we encounter an interesting new phenomenon, in that the speed at which a signal is transmitted differs from the value we would first think of. From (14.28), we see that we have to be a bit more precise in specifying the nature of the signal, since the speed v of our wave depends on ω. So surely, if we have a monochromatic wave of frequency ω_0, its speed is $f(\omega_0)$, and isn't that the end of the problem?

The answer to this question is that indeed its speed is $f(\omega_0)$, but that is not the end of the problem. The reason is that the wave specified in (14.28) extends over the range $+\infty$ to $-\infty$ in x and in t, and this is not what is meant by a signal.

Our wave (14.28) corresponds, for example, to the light produced by a torch which was turned on in the dim and distant past, and which will remain on for ever. A signal, however, consists of some information. For example, you can arrange with a friend who lives nearby that, in the event of your home being invaded by Martians, you will signal to him with your torch, by turning it on for two lots of 3 seconds, separated by a 1 second interval. The essence of a signal is that it is necessary to *modulate* the infinite harmonic wave in some way, in order to transmit information. If your torch had been turned on all the time, and remains on for ever, there is no way that your friend will be able to deduce anything from this (except that the batteries are still working). In fact, he will probably rather quickly become oblivious of its existence.

We thus have to consider the speed at which modulations propagate. Some examples of such signals are shown in fig. 14.6. Again we might think that, if our signal is of frequency ω_0, with a suitable modulation on top of it, it would travel at a speed $f(\omega_0)$, just as for the unmodulated wave.

The reason that this is not so is rather subtle, and is related to what is meant by a monochromatic wave. This must be of the form

$$y = A \sin k(x \pm vt) \tag{14.29}$$

where k has a unique value, and the variable $x \pm vt$ extends over the full infinite range. Our modulated signal is not of this form, and hence is not monochromatic. We can imagine building up our modulated signal by combining a large number of waves of the form (14.29), each with a different value of k, such that they enhance each other where the modulated amplitude is large, but interfere destructively where it is small.

We can obtain a feeling for how this works by adding together just two waves in order to produce a modulated resultant. Thus with

$$\left. \begin{array}{l} y_1 = A \sin[(k + \delta k)x - (\omega + \delta\omega)t] \\ \text{and} \quad y_2 = A \sin[(k - \delta k)x - (\omega - \delta\omega)t] \end{array} \right\} \tag{14.30}$$

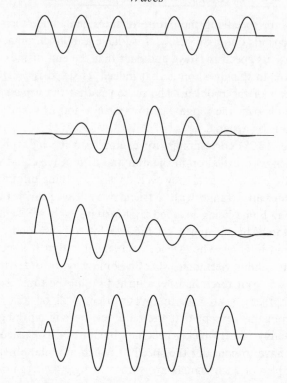

Fig. 14.6. Some examples of modulated sine waves, which carry more information than an unmodulated wave which extends to infinity in both directions.

we obtain

$$y = y_1 + y_2 = 2A \sin(kx - \omega t) \cos(\delta kx - \delta \omega t) \qquad (14.31)$$

Here y_1 and y_2 correspond to waves of frequencies just above and just below our original frequency ω, and each of them has the corresponding wave number $k \pm \delta k$.

If we consider the resultant wave (14.31) at $t = 0$, we obtain

$$y = 2A \cos \delta kx \sin kx \qquad (14.32)$$

That is, it has the well-known form $\sin kx$ of our usual sine wave, and its wave number k is just the average of those of the two components in (14.30). It is multiplied, however, by $2A \cos \delta kx$; since δk is small compared with k, it requires large changes in x in order for this factor to alter significantly. In other words, $2A \cos \delta kx$ can be thought of as a slowly changing amplitude for our sine wave.

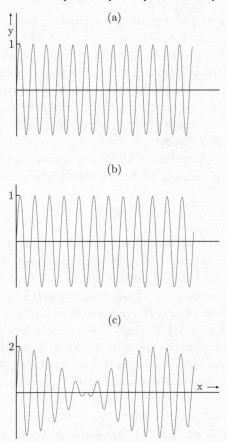

Fig. 14.7. An example of building a modulated wave from two infinitely long sine waves. The two components shown in (a) and (b) are the waves y_1 and y_2 of eqn (14.30) at $t = 0$, with $\delta k/k = 0.05$. The resultant y is drawn in (c), and is as given by eqn (14.32).

The components and the resultant at $t = 0$ are plotted in fig. 14.7. Indeed we have produced a modulated sine wave, even though the modulation is rather boring in that it extends for ever, whereas most signals are zero outside a certain region. Perhaps we should not be too surprised that our modulation is very simple in that we have added together only two components, rather that the large number we suggested were required for a more interesting modulation. Almost the only information our wave carries is the number of individual waves within each "sausage", and the sinusoidal shape of the "sausages".

As already mentioned, a realistic wave packet has an envelope which is zero outside a finite range. Thus it could look like any of those shown in fig. 14.6. In order to build up such a pattern, we in fact integrate a series of pure harmonic waves over a continuous range of values of k, each with an amplitude that depends on k. This is Fourier analysis (see Section 12.12). Here we merely note that the shorter the wave packet is, the larger will be the range of k values for which the Fourier components have significant amplitudes.

Our trivial two-component example also shows this sort of behaviour. The length of each sausage of the modulated amplitude is

$$\Delta x = \frac{\pi}{\delta k} \tag{14.33}$$

i.e. the spatial extent of the modulation is inversely proportional to the range of the wave numbers that were used in constructing the modulation.

We now consider how our pattern moves with time. Since the disturbance as a function of x and t is as given in eqn (14.31), the information can be extracted from there. The factor $\sin(kx - \omega t)$ describes the individual wavelets of fig. 14.7(c). We have learnt how to look at an expression like this, and to extract the speed of the waves; the answer is ω/k. The other factor — $\cos(\delta kx - \delta \omega t)$ — is the one which describes the envelope which modulates the previous sine factor. Again we can read off the speed directly as $\delta \omega / \delta k$, and this tells us how fast the envelope moves. In a dispersive medium, ω/k is a function of ω, and hence $\delta \omega / \delta k$ is not equal to ω/k. This implies that the individual waves and the envelope do not travel at the same speed; as the envelope moves, the individual waves move through it (either backwards or forwards relative to the envelope, depending on which speed is larger). It is the envelope that determines when the signal arrives, and hence it is $\delta \omega / \delta k$ that is relevant for determining the speed at which the signal travels.

Our analysis was performed for a wave packet made out of two harmonic components. The general result for more complicated wave packets is that the envelope travels with speed $d\omega/dk$, the rate of change of ω with k for waves in the dispersive medium.

The speed $d\omega/dk$ is known as the "group velocity", as it tells us how fast a group of waves of nearby frequencies will travel. We shall denote this by the symbol g. The expression ω/k in contrast is the "phase velocity" v, and is relevant for the idealistic case of an infinite wave train of frequency ω.

Because it is the group velocity that determines how fast a signal propagates, it is restricted by the Theory of Relativity to be not greater than c, the speed of light in a vacuum. However, no such restriction applies to phase velocities, which can be larger than c without contradicting Relativity.

The simplest example of a group velocity is for light in a non-dispersive medium of constant refractive index μ. Then

$$v = \frac{\omega}{k} = \frac{c}{\mu}$$

and

$$g = \frac{d\omega}{dk} = \frac{c}{\mu}$$

Thus in this case, because the medium is non-dispersive, the group and phase velocities are the same, and the individual waves maintain their position within the wave packet as it moves.

Next we consider the case where the refractive index is such that the phase velocity varies linearly with k, i.e.

$$v = \frac{\omega}{k} = a + bk \tag{14.34}$$

Thus

$$\omega = ak + bk^2$$

and

$$g = \frac{d\omega}{dk} = a + 2bk \tag{14.35}$$

Hence, for non-zero b, the group and phase velocities are different. This situation is illustrated in fig. 14.8.

It is important to realise that the above result that $g \neq v$ is independent of the details of the envelope of the wave packet, provided that it is not too short (see below). Thus even a long wave packet containing a very large number N of individual wavelets, which we might have guessed would not have been too different from an infinitely long harmonic wave, travels with speed g independent of N and different from v.

The group velocity itself will usually be a function of ω, or equivalently of k. This is so for the example we have just considered (see eqn (14.35)). Thus wave packets of different carrier wave frequencies will in such circumstances travel at different speeds. Now if our wave packet is too small, it will require a set of harmonic waves with a large range of wavelengths in order to construct it (see eqn (14.33)). If g varies with ω, which is the appropriate g to describe the way such a packet will travel?

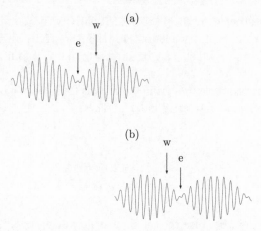

Fig. 14.8. Illustration of a situation in which the group and phase velocities differ. Part of the modulated wave of fig. 14.7(c) is shown in (a) at $t = 0$, and in (b) at a later time. The arrows marked e show the positions of a particular point on the envelope at the two times; those marked w show how an individual wavelet has progressed in this time interval. For the situation shown here, the envelope moves faster than the individual wavelets, and the group velocity is larger than the phase velocity (corresponding, for example, to positive b in eqns (14.34) and (14.35)).

In fact there is not a unique answer to this question. Because of dispersion, the different individual harmonic waves travel at significantly different speeds, and hence when they arrive at the receiver, they will have become so out of phase with each other that their sum results in an envelope which is not of the same shape as the initial envelope (see fig. 14.9). The "speed of the envelope" is then an ambiguous concept since we would obtain different results depending on the arbitrary nature of how we chose to define how far it had travelled in the given time (e.g. between the positions of the middle of the initial and final wave packets, between where the front edges rise to 10% of their maximum heights, between the corresponding heights for the trailing sides, etc.). The basic problem is that, in order to determine how far the wave packet has moved, we are hoping that the two wave packets can be related simply by a translation involving a single number, the distance travelled; this will be possible only if the initial and final waves are identical in shape. The fact that our definition of $g = d\omega/dk$ for a short packet yields a spread of values (depending on the available range in ω) simply reflects this genuine ambiguity in its meaning.

The longer our initial wave train is, the narrower will be the range

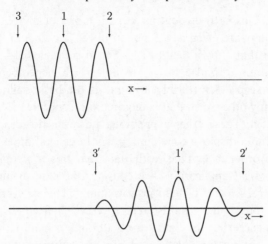

Fig. 14.9. A wave, with an envelope whose shape changes as it moves, gives rise to an ambiguity in the measurement of its group velocity. Thus differing answers are obtained if we define the distance moved in terms of the middle of the packet (1 and 1'), its front edge (2 and 2') or its trailing edge (3 and 3'). This effect is likely to be serious for a wave packet containing few wavelets, moving in a dispersive medium.

of frequencies that are significant in building it up, and the smaller the range of different $d\omega/dk$ values. Correspondingly, the smearing of the shape of the envelope will be less severe, and there will be less ambiguity in an experimental definition of its speed.

So far our only expression for the group velocity has been

$$g = \frac{d\omega}{dk} \qquad (14.36)$$

but because the phase velocity $v = \omega/k$, there are several alternative expressions. Thus

$$g = v + k\frac{dv}{dk} \qquad (14.36')$$

$$= v - \lambda\frac{dv}{d\lambda} \qquad (14.36'')$$

$$= \frac{c}{\mu}\left(1 + \frac{\lambda}{\mu}\frac{d\mu}{d\lambda}\right) \qquad (14.36''')$$

where we have used the relationships $k = 2\pi/\lambda$ and $\mu = c/v$. It is important to realise that the derivatives in these formulae refer to the way quantities vary with respect to k or λ, when k and λ are measured in the medium. This is to be contrasted with tables of the refractive index μ, which usually are given as a function of the wavelength in free space; this is longer than λ by a factor of μ. (See also Problem 14.1.)

The last three formulae for g are not worth remembering as such. It is best to keep (14.36) firmly in mind as the basic definition, and to realise that any other form that is required for a specific problem is easily derived from it. These other expressions, however, do bring out clearly the fact that in a dispersive medium, g will in general differ from v.

Finally, in order to emphasise the fact that this is a topic which has very real physical applications, we consider the case of an experiment to measure the velocity of light in a vacuum.† The experiment involves timing how long a light pulse takes to travel a fixed distance. If the apparatus is such that the light travels through a material (e.g. air), we must clearly correct for its effect on our measurement. Because we are using a pulse of light, rather than an infinite harmonic wave, what we measure is the group velocity g rather than the phase velocity v. Thus we see from eqn (14.36''') that it is not sufficient to multiply our determination of g by the refractive index μ, but we must also divide by $(1+(\lambda/\mu)d\mu/d\lambda)$. For the example given earlier of the phase velocity varying linearly with wave number (which for the case of $\mu \approx 1$ is approximately the same as μ varying linearly with λ), the complete correction is twice as big as that arising simply from μ.

For the remainder of this chapter, we return to the case of a non-dispersive medium.

14.5 Energy in a wave

Waves possess energy. For example, for our usual case of transverse waves on a string, there is kinetic energy associated with the transverse motion of the string. Not to be forgotten is the fact that the string also possesses potential energy, due to the fact that the string is stretched by the transverse displacement. We will calculate both these contributions, first for a moving wave given by

$$y = A \sin(kx - \omega t) \qquad (14.10)$$

The kinetic energy of a short length of string is given by

$$KE = \frac{1}{2}(\rho \delta x)\left(\frac{\partial y}{\partial t}\right)^2 \qquad (14.37)$$

† We assume that this experiment was performed before 1983, in which year the value of the velocity of light was fixed by definition.

since $\rho\delta x$ is the mass of the segment, and $\partial y/\partial t$ is the speed of its transverse motion. We then substitute for $\partial y/\partial t$, making use of (14.10), to obtain

$$KE = \frac{1}{2}\rho A^2\omega^2 \cos^2(kx - \omega t)\delta x \qquad (14.38)$$

Next we consider the kinetic energy, at a fixed time t, of a somewhat longer length l of the string that is an integral number of wavelengths long. Then

$$KE = \frac{1}{2}\rho A^2\omega^2 \int_x^{x+l} \cos^2(kx - \omega t)dx \qquad (14.39)$$

so that

$$KE/l = \frac{1}{4}\rho A^2\omega^2 \qquad (14.40)$$

where in the last line we have made use of the fact that the average value of $\cos^2\theta$ over a whole number of cycles is $\frac{1}{2}$.

In (14.40), we have derived the important result that the energy per unit length is proportional to A^2, the square of the amplitude of the wave. Of course, this is only kinetic energy, but we are just about to discover that the average potential energy is equal to the average kinetic energy.

As already mentioned, the string's potential energy derives from the fact that it is stretched more in the presence of the wave than in its absence; the potential energy is simply equal to the work performed in stretching the string. Again consider the short section of string of length δx in the absence of the wave, which becomes δl because of the transverse displacement y (see fig. 14.10). For small values of y, the tension T does not change appreciably. Thus for our small section

$$PE = T(\delta l - \delta x)$$

$$= T\delta x \left[\left(1 + \left(\frac{\delta y}{\delta x}\right)^2\right)^{\frac{1}{2}} - 1\right]$$

$$\approx \frac{1}{2}T\delta x \left(\frac{\partial y}{\partial x}\right)^2 \qquad (14.41)$$

where we have made use of Pythagoras' Theorem, and also the binomial expansion (and have neglected higher powers of $\partial y/\partial x$, since we are as usual assuming that the wave is such that the derivative is always small).

In analogy with our kinetic energy calculation, we substitute for $\partial y/\partial x$ from (14.10), and then average at fixed t over a length containing a whole

252 *Waves*

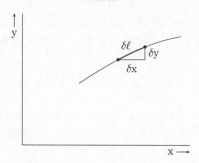

Fig. 14.10. A short segment of string on which there is a transverse wave. The original length of this segment in the absence of a wave was δx; the effect of the wave is to stretch it so that its length becomes $\delta l = (\delta x^2 + \delta y^2)^{1/2}$.

number of wavelengths, to obtain

$$PE/l = \frac{1}{4}TA^2k^2 \tag{14.42}$$

Finally we make use of the fact that the speed of waves on our string is given by

$$v = \omega/k = \sqrt{T/\rho}$$

so that we can substitute for Tk^2 in eqn (14.42) to obtain

$$PE/l = \frac{1}{4}\rho\omega^2A^2 = KE/l \tag{14.43}$$

This is our promised result that the mean potential and kinetic energies are equal.

In fact the brighter readers will have realised that there was no need to perform space averaging in order to deduce the equality of the two forms of energy for a small section of string carrying a moving wave, since each is equal to

$$\frac{1}{2}\rho A^2\omega^2 \cos^2(kx - \omega t)\delta x = \frac{1}{2}TA^2k^2 \cos^2(kx - \omega t)\delta x$$

However, for a stationary wave like

$$y = A \sin kx \cos \omega t \tag{14.13}$$

the kinetic energy is

$$KE = \frac{1}{2}\rho A^2\omega^2 \sin^2 kx \sin^2 \omega t\delta x \tag{14.44}$$

while the potential energy is

$$PE = \frac{1}{2} T A^2 k^2 \cos^2 kx \cos^2 \omega t \delta x \qquad (14.44')$$

Thus it is now necessary to average over a whole number of wavelengths in space, and over a whole number of periods in time, in order to obtain the result that the mean kinetic and potential energies are equal.†

A moment's thought shows that it is sensible to time and space average for the stationary wave (see fig. 14.11(a) and (b)). Thus the nodes N have plenty of potential energy in (a), since $\partial y / \partial x$ is large there, but much less at the later time depicted in (b) when the whole amplitude has decreased; in contrast the antinodes A have no potential energy. On the other hand, the antinodes have large kinetic energy in (b), when the string's deflection is changing faster than in (a); the nodes, however, don't move, and have no kinetic energy. Thus the time distributions of the two forms of energy differ, as also do their spatial dependences. It is only when we average over both space and time that there is the possibility of the potential and kinetic energies being equal.

For the moving wave, however, the maximum displacement is the position of zero potential energy (as the string's gradient is zero), and of zero kinetic energy (since at the maximum, the displacement does not change with time). Similarly at the positions of zero displacement, we have a large potential energy and also a large kinetic energy. Thus, even instantaneously, there is no obvious inconsistency in these two forms of energy being equal at each point on the string.

Our actual proof (that the potential and kinetic energies in a moving wave are equal at any x and any t) was only for a harmonic wave. It is, however, trivial to extend this to any moving wave. Thus for

$$y = f(x - vt)$$

the energies are given by

$$PE = \frac{1}{2} T \delta x \left(\frac{\partial y}{\partial x} \right)^2 = \frac{1}{2} T \delta x (f')^2$$

† If we average over both space and time, we find that the mean kinetic (or potential) energy per unit length is $\frac{1}{8}\rho A^2 w^2$, in contrast to $\frac{1}{4}\rho A^2 w^2$ for the moving wave. The difference arises from the conventions about amplitudes in (14.10) and (14.13). The stationary wave can be constructed as the sum of two moving waves, but then becomes twice as big as in (14.13). Since energy is proportional to the amplitude squared, this makes the standing wave's kinetic energy $\frac{1}{2}\rho A^2 w^2$. This is twice as big as the energy in either of the two moving waves from which it was constructed, which is sensible.

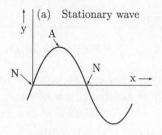

(a) Stationary wave

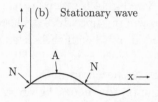

(b) Stationary wave

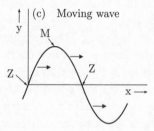

(c) Moving wave

Fig. 14.11. A stationary wave, at its maximum deflection in (a), and close to zero displacement about 1/4 of a period later in (b). The string is stretched near the node N in (a), but much less so in (b); at the antinode A the string is never stretched. Thus there is no contribution to the potential energy from the region near A, while at N there is a time-dependent contribution. In contrast the string at N does not move and so has no kinetic energy, but while A is momentarily stationary in (a), it is moving downwards at near its maximum speed in (b).

For the moving wave shown in (c), the string is stretched at the positions labelled Z, and is moving vertically; it thus has both potential and kinetic energy. However, at M, the string is unstretched and is not moving at the given instant (even though the wave is); it thus has neither form of energy.

and

$$KE = \frac{1}{2}\rho\delta x \left(\frac{\partial y}{\partial t}\right)^2 = \frac{1}{2}\rho\delta x v^2 (f')^2$$

Because $v^2 = T/\rho$, these are equal for all values of x and t.

We can see this explicitly for the sum of two harmonic waves:

$$y_1 = a \sin(k_1 x - \omega_1 t) \qquad (14.45)$$

and

$$y_2 = b \sin(k_2 x - \omega_2 t) \qquad (14.45')$$

travelling in the same direction. Then

$$\frac{\partial y}{\partial x} = \frac{\partial y_1}{\partial x} + \frac{\partial y_2}{\partial x} = ak_1 \cos(k_1 x - \omega_1 t) + bk_2 \cos(k_2 x - \omega_2 t)$$

The potential energy is given by

$$\frac{PE}{\delta x} = \frac{1}{2} T \left(\frac{\partial y}{\partial x} \right)^2$$

$$= \frac{1}{2} T (a^2 k_1^2 \cos_1^2 + b^2 k_2^2 \cos_2^2 + 2abk_1 k_2 \cos_1 \cos_2) \qquad (14.46)$$

where the subscripts 1 and 2 for the cosines are used to denote the arguments $k_1 x - \omega_1 t$ and $k_2 x - \omega_2 t$, respectively. Similarly for the kinetic energy we have

$$\frac{KE}{\delta x} = \frac{1}{2} \rho \left(\frac{\partial y}{\partial t} \right)^2$$

$$= \frac{1}{2} \rho (a^2 \omega_1^2 \cos_1^2 + b^2 \omega_2^2 \cos_2^2 + 2ab\omega_1 \omega_2 \cos_1 \cos_2)$$

$$= \frac{1}{2} T (a^2 k_1^2 \cos_1^2 + b^2 k_2^2 \cos_2^2 + 2abk_1 k_2 \cos_1 \cos_2) \qquad (14.47)$$

where we have used the relationships

$$v = \sqrt{\frac{T}{\rho}} = \frac{\omega_1}{k_1} = \frac{\omega_2}{k_2}$$

We thus see from (14.46) and (14.47) that the kinetic and potential energies of the moving wave form, given by the sum of the two harmonic waves, are equal at any specific values† of x and t.

Why is this not true for the stationary wave? This we can produce by adding

$$y_3 = a \sin(k_1 x + \omega_1 t) \qquad (14.48)$$

---

† However, we see that the energy of the sum of the waves is equal to the sum of the energies of the separate waves only after we average over a whole number of "sausages" of the envelope, so that the final term involving $\cos_1 \cos_2$ in (14.45) or (14.46) averages to zero.

to the wave of eqn (14.45). In this case

$$\frac{\partial y}{\partial x} = ak(\cos_+ + \cos_-)$$

$$\text{and} \quad \frac{\partial y}{\partial t} = a\omega(\cos_+ - \cos_-)$$

where the subscripts $+$ and $-$ represent the arguments $k_1 x + \omega_1 t$ and $k_1 x - \omega_1 t$ respectively. When we calculate $\frac{1}{2}T(\partial y/\partial x)^2$ and $\frac{1}{2}\rho(\partial y/\partial t)^2$ to find the potential and kinetic energies, we obtain

$$PE/\delta x = \frac{1}{2}Ta^2k^2(\cos_+^2 + \cos_-^2 + 2\cos_+ \cos_-) \tag{14.49}$$

and

$$KE/\delta x = \frac{1}{2}\rho a^2\omega^2(\cos_+^2 + \cos_-^2 - 2\cos_+ \cos_-) \tag{14.50}$$

The factors in front of the brackets are equal, but now it is the presence of the $\pm 2\cos_+ \cos_-$ terms which produces the difference. It is only when we average over space and time that this disappears, and the mean kinetic and potential energies become equal.†

Finally we consider the *energy flow* in a moving wave. It is simplest to consider a wave packet of length L such that it contains a whole number of waves‡ (see fig. 14.12). The energy of the wave is

$$E = KE + PE = \frac{1}{2}\rho\omega^2 A^2 L$$

and is associated with that region of space where the envelope of the wave is non-zero. As time passes, the wave moves, and hence so does the region which contains the energy. In a time interval δt, this region moves a distance $v\delta t$, and so the energy which has passed a given point is $\frac{1}{2}\rho\omega^2 A^2 v\delta t$. This implies a flux U of energy (i.e. the amount of energy passing in unit time) of

$$U = \frac{1}{2}\rho\omega^2 A^2 v = \frac{1}{2}\rho\omega^3 A^2/k$$

$$= \frac{1}{2}Tk^2 A^2 v = \frac{1}{2}T\omega k A^2 \tag{14.51}$$

These are equivalent expressions for the energy flow in a moving wave. We shall make use of the last version of (14.51) later.

---

† Again this can be proved for a general stationary wave $y = f(x)g(t)$. (See Problem 14.4(ii).)

‡ In practice, we are likely to have $L \gg \lambda$, and then it will not be crucial whether L/λ is a whole number, since the end effect due to a fraction of a wave would be relatively unimportant.

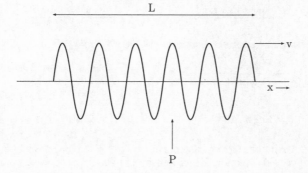

Fig. 14.12. The energy density E/L in the wave packet of length L is $\frac{1}{2}\rho\omega^2 A^2$. In time δt, the packet moves a distance $v\delta t$. The energy flux U is the rate at which this energy passes a point P in unit time, and so is $(E/L) \times v\delta t/\delta t = \frac{1}{2}\rho\omega^2 A^2 v$.

14.6 Reflection at a boundary

Let us imagine that we are in our living room at night. It is dark outside and we have the lights on. Suddenly we hear a strange noise outside, and we draw the curtains open to see what is happening. But as we try to look through the window, all we see is a reflection of ourselves in the glass. Of course, this reflection doesn't happen only at night. It is simply that it is more obvious when we are looking at the reflection of a bright object as compared with a dark background, rather than vice versa.

The reflection, which is about 4% as bright as the original object, arises because the light travelling from us towards the window suddenly encounters a new medium (glass) in which the speed of light differs from that of the air through which it was initially travelling. We are not going to analyse this problem in electromagnetism, but instead return to our favourite example of waves on a string. In the optics case, the different speeds in the two media were produced by the fact that glass has a refractive index of 1.5 relative to air. For the string, the wave's speed is

$$v = \sqrt{T/\rho}$$

Thus we shall consider a situation of a long string, with the region with x smaller than zero having density ρ_1, while at positive x there is a different density ρ_2 (see fig. 14.13). Since as usual the tension is constant throughout the string, the speeds v_1 and v_2 will differ in the two regions.

We now assume that someone has arranged for there to be an incident wave in the left half of the string, and it is moving towards the right. We

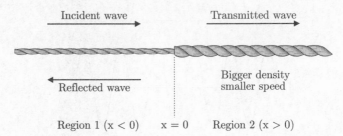

Fig. 14.13. An infinite string, the region at $x < 0$ having a small density ρ_1 and that at $x > 0$ having a larger density ρ_2. An incident wave coming from negative x towards the boundary at $x = 0$ will give rise to a reflected wave and a transmitted wave.

shall write this as

$$A\sin(-k_1 x + \omega_1 t), \qquad x \le 0 \tag{14.52}$$

where $\omega_1/k_1 = v_1$. We need opposite signs for $k_1 x$ and $\omega_1 t$ within the argument of the sine in order to ensure that the wave travels to the right. We have changed our convention† with respect to eqn (14.10) in order to ensure that all the waves in the problem have the same time dependence (i.e. $+\omega t$ within the sine).

Now part (or perhaps even all) of this wave will be transmitted beyond the discontinuity at $x = 0$, and so for the right-hand region, we write a transmitted wave as‡

$$T\sin(-k_2 x + \omega_2 t), \qquad x \ge 0 \tag{14.53}$$

We have written the amplitude as T, because this could well be different from A. Again we have a negative sign in front of $k_2 x$, as the wave again travels to the right.§ The speed is here v_2, so $\omega_2/k_2 = v_2$.

Now comes the reflected wave

$$R\sin(k_1 x + \omega_1 t), \qquad x \le 0 \tag{14.54}$$

† This change is not absolutely necessary. We could alternatively have kept all the kx terms positive (as in (14.10)) throughout this section, and where necessary compensated for this by changes in the signs of A and ωt. It would have given us $k_2 - k_1$ in the numerator of eqn (14.63) for R, which would then have been different from that in most other books.

‡ Beware. The symbol T is used here both for the tension and for the amplitude of the transmitted wave.

§ We may well wonder whether we ought to include an extra constant ϕ within the argument of the sine in eqn (14.53), to allow for the possibility that the phases of the incident and outgoing waves differ. However, it turns out that such phases are always either zero (so that we can forget about it altogether) or 180° (in which case it is equivalent to changing the sign of the amplitude of the wave, so again it is not necessary).

If the problem is such that there is no reflection, then we shall presumably find that $R = 0$, so that including the term unnecessarily should cause no problem. This reflected wave is in the left-hand region, and so the relevant k and ω have subscripts 1. As the wave travels to the left, there is a plus sign before the $k_1 x$.

The fourth possible wave is

$$N \sin(k_2 x + \omega_2 t) \qquad x \geq 0$$

This would be in the right-hand region travelling to the left. We do *not* include such a wave for our problem, since, with the incident wave as described, there is no way such an extra wave can be produced.

Of course, we could have imagined a different situation in which the incident wave comes in from the right. Alternatively, with another boundary with a third medium at $x = a$, we would have had such a wave in the region $0 \leq x \leq a$ (as well as another right-moving wave for $x \geq a$). But these are different problems, and we do not require such a wave for our case.

We have spent a fair amount of time describing how we write down these waves, as it is important to understand fully what we are doing. However, with a little practice, this stage becomes second nature, and it takes only a few seconds to do this for any specific problem. In fact we are soon to discover that $\omega_1 = \omega_2$, and so we rewrite the waves as:

$$\left.\begin{array}{lll} \text{Incident} & y_A = A\sin(-k_1 x + \omega t) & x \leq 0 \\ \text{Transmitted} & y_T = T\sin(-k_2 x + \omega t) & x \geq 0 \\ \text{Reflected} & y_R = R\sin(k_1 x + \omega t) & x \leq 0 \end{array}\right\} \qquad (14.55)$$

Then the waves in the two different spatial regions are

$$\left.\begin{array}{ll} y_1 = y_A + y_R & x \leq 0 \\ y_2 = y_T & x \geq 0 \end{array}\right\} \qquad (14.56)$$

We are now going to discover how the amplitudes T and R of the transmitted and reflected waves compare with that of the incident wave A. These are determined through consideration of the boundary conditions at $x = 0$, where the two pieces of string join together.

So far we have written down the separate waves in the two regions independently. If we are not careful, we could have an unphysical situation where the magnitude of the deflection at $x = 0$ as determined from the waves in the left-hand region differs from that as given in the right-hand region (see fig. 14.14(a)). In order to avoid this, we must insist that

$$(y_1)_{x=0} = (y_2)_{x=0} \qquad \text{at all } t \qquad (14.57)$$

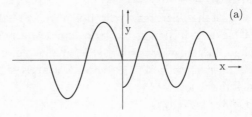

(a)

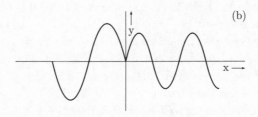

(b)

Fig. 14.14. Unphysical boundary conditions at the junction of the two different strings joined together at $x = 0$. In (a) there is a discontinuity in the displacement y. In (b), the gradient $\partial y/\partial x$ exhibits a discontinuity, which would give rise to an infinite acceleration of the small segment of string at $x = 0$.

If we substitute in our waves (14.52)–(14.54), this reduces to

$$(A + R)\sin \omega_1 t = T \sin \omega_2 t \qquad (14.58)$$

Thus, unless $R = -A$ and $T = 0$ (in which case there is no transmitted wave, and the value of ω_2 becomes irrelevant), we must have

$$\omega_1 = \omega_2 \qquad (14.59)$$

in order to ensure that the boundary condition (14.57) is satisfied at *all* values† of t.

This is an important result, which is worth emphasising. When a wave enters a new medium in which its speed is different, the frequency of the wave remains constant; it is the wave number (or equivalently the wavelength) which changes. This is our justification for writing the waves as in (14.55), without any subscripts for the ωs.

Our second boundary condition is that the gradients of the waves on

---

† If this were not so, we could solve eqn (14.58) to find the values of t at which it was satisfied. However, this is not enough, because the boundary condition (14.57) needs to be true for all possible values of t. We thus do require the frequencies to be identical.

the two sides of the boundary must match, i.e.

$$\left(\frac{\partial y_1}{\partial x}\right)_{x=0} = \left(\frac{\partial y_2}{\partial x}\right)_{x=0} \tag{14.60}$$

This can be understood by considering the very small section of string of length δx at $x = 0$. The forces on this segment arise from the tension in the string. The net upward force is

$$F = T\left(\frac{\partial y_2}{\partial x}\right)_{x \sim 0} - T\left(\frac{\partial y_1}{\partial x}\right)_{x \sim 0} \tag{14.61}$$

where the tension T is the same in the two different pieces of string. This force would result in an acceleration of the small segment of string, i.e.

$$F = T\left[\left(\frac{\partial y_2}{\partial x}\right)_{x \sim 0} - \left(\frac{\partial y_1}{\partial x}\right)_{x \sim 0}\right]$$

$$= \tfrac{1}{2}(\rho_1 + \rho_2)\delta x \left(\frac{\partial^2 y}{\partial t^2}\right)_{x=0} \tag{14.62}$$

Thus if the boundary condition (14.60) is not satisfied, F will be non-zero. Because the mass of this segment of string is infinitesimal, the resulting acceleration of the boundary point would be infinite,† and this again is unphysical. To avoid this, we require eqn (14.60) to be satisfied (see fig. 14.14(b)).

Thus the two boundary conditions (14.57) and (14.60) that we impose ensure that there is no discontinuity in either the displacement or its gradient across the boundary.‡

Now we are ready to substitute our waves into our boundary conditions. From (14.57), we obtain

$$A + R = T \tag{14.57'}$$

where we have used the fact that the boundary condition is for $x = 0$, and the factor $\sin \omega t$ is common to each term and can be cancelled.

† This argument would not apply if we had a finite mass M placed at the junction of the two strings. In that case, our second boundary condition would not have been (14.60) but instead

$$T\left[\left(\frac{\partial y_2}{\partial x}\right)_{x=0} - \left(\frac{\partial y_1}{\partial x}\right)_{x=0}\right] = M\left(\frac{\partial^2 y}{\partial x^2}\right)_{x=0}$$

and this would be entirely satisfactory.

‡ Of course, these requirements apply at every point along the string. However, there is no need to impose them explicitly elsewhere. This is because they will be automatically satisfied, as we have written down waves in a given region as smooth functions. It is only at the boundary between regions, where we have different functional forms for our solution, that problems could arise, and hence the boundary conditions must be imposed explicitly.

The second boundary condition (14.60) gives us

$$(R - A)k_1 = -Tk_2 \qquad (14.60')$$

We now have two equations (14.57') and (14.60') for our unknowns. At first sight it looks as if we have a difficulty because R, A and T seem to constitute three unknowns. However, physically it is reasonable that A, which is the amplitude of the incident wave, can be chosen arbitrarily. Hence we do not expect our equations to specify its value. Instead we would want to find only the ratios T/A and R/A. The structure of our two simultaneous equations is such that this happens automatically. We find

$$\left. \begin{array}{c} R = \dfrac{k_1 - k_2}{k_1 + k_2} A \\ \text{and} \quad T = \dfrac{2k_1}{k_1 + k_2} A \end{array} \right\} \qquad (14.63)$$

It is important to realise what we have achieved in our boundary problem, and what input was required. The only assumption was that the string behaves in a physically reasonable manner at the boundary. This we expressed via the two boundary conditions (14.57) and (14.60) on the string's displacement and on its gradient. From this, we discovered that the frequencies of the various waves had to be identical, and that the reflected and transmitted amplitudes were determined (see eqns (14.63)).

We see from eqn (14.63) that R/A is positive if $k_1 > k_2$, but negative for $k_2 > k_1$ (e.g. $\rho_2 > \rho_1$). A negative value of an amplitude is equivalent to a positive amplitude but with a phase change of 180°, i.e.

$$- R\sin(k_1 x + \omega t) \equiv R\sin(k_1 x + \omega t + \pi) \qquad (14.64)$$

This is the well-known change of phase for reflections at a "rare–dense" boundary.

We also see that, for the special case of $k_1 = k_2$, $R = 0$ and $T = 1$. In this case, the two media are identical, and the boundary between them becomes physically meaningless. It is no wonder that the "transmitted" wave is equal to the incident wave, and that there is no reflection.

The other simple example is when the right-hand string becomes very heavy. Our intuition tells us that as ρ_2 tends to infinity, this part of the string will not move and all the energy will be reflected. Indeed our eqns (14.63) in this case give us $R = -A$ and $T = 0$, since k_2 tends to infinity.

We now turn to what appears to be a paradox. We discovered in

Section 14.5 that the energy of a wave is proportional to its amplitude squared. Thus the energies of the three waves are proportional to A^2, R^2 and T^2, respectively. We would expect the sum of the outgoing reflected and transmitted energies to be equal to the incident energy. However, from eqn (14.63)

$$R^2 + T^2 = \left[1 - \frac{4k_1(k_1 - k_2)}{(k_1 + k_2)^2}\right] A^2 \qquad (14.65)$$

Now this will in general not be equal to A^2, and hence it looks as if energy is not conserved! This would be quite serious. The reader is urged to resolve this paradox before the answer is revealed later.

An alternative way of seeing that we have a paradox is to note that the boundary condition (14.57′) is

$$A = T - R$$

Hence

$$A^2 = T^2 + R^2 - 2TR \qquad (14.66)$$

and this is consistent with $A^2 = T^2 + R^2$ only if either T or R is zero. This agrees with what we can deduce from (14.65), i.e. $R^2 + T^2 = A^2$ if k_2 is infinite, or if $k_1 = k_2$. These correspond to $T = 0$ and to $R = 0$ respectively. The fact that we satisfy $R^2 + T^2 = A^2$ in the absence of a boundary is not surprising. That it is also true when we have complete reflection may perhaps give us a clue as to what has gone wrong with our attempt to demonstrate that energy is conserved.

If you still have not solved the paradox, then you should re-read Section 14.5 concerning the energy of a wave, and only then return to the next paragraph.

* * * * * * * * * * * * * * * * * * * *

The relevant points from Section 14.5 are that:

(1) the energy in a wave depends on factors other than just the square of the amplitude, and these differ in the two regions; and
(2) energy conservation requires that it is the energy *flows* in and out of the region near the boundary that must be equal.†

---

† If we take an aerial photograph of a section of road, which starts off well-surfaced and then becomes very bumpy, we will see well-separated faster cars on the good section, followed by lots of close, slowly-moving cars on the uneven part. A few drivers may even be discouraged by the bumpy surface, and turn round. Conservation of cars does not imply that the densities in the two regions are equal, but rather the fluxes (i.e. density × speed) towards and away from the boundary are.

Thus for each wave, we ought to calculate the energy flow U given by eqn (14.51). The last version is most useful, as the tension T is the same in the two parts of the string, and so is ω. Then the energy flux is proportional to kA^2.

Energy conservation thus implies

$$U_A = U_R + U_T \tag{14.67}$$

Substituting the relevant kA^2 for each term, we obtain

$$
\begin{aligned}
k_1 A^2 &= k_1 \left(\frac{k_1 - k_2}{k_1 + k_2} \right)^2 A^2 + k_2 \left(\frac{2k_1}{k_1 + k_2} \right)^2 A^2 \\
&= k_1 A^2 \frac{k_1^2 - 2k_1 k_2 + k_2^2 + 4k_1 k_2}{(k_1 + k_2)^2} \\
&= k_1 A^2
\end{aligned}
\tag{14.68}
$$

This confirms our energy conservation expectation.

It is worth noting that if we wish to calculate the fraction of energy flow reflected, it is given from eqn (14.63) simply by $(R/A)^2$, since the reflected and incident waves travel in the same medium. In contrast, the fraction of energy flow transmitted is $k_2 T^2 / k_1 A^2$.

We finally return to the subject of the reflection from a glass window. This of course differs from the problem we have been considering, but is turns out that the reflected amplitude is still given by eqn (14.63), where k_1 and k_2 are the wave numbers of the light in air and in glass respectively. Now glass has a refractive index of about 1.5, which is the ratio of the speed of light in air as compared with that in glass. Since the frequencies in the two media are the same, this is also equal to k_2/k_1. Thus

$$R = \frac{1 - 1.5}{1 + 1.5} A = -\frac{A}{5}$$

and the reflected intensity is $1/25$ of the incident energy. This thus explains the weak but non-negligible reflection from the glass.

The next time we try to look out of a window at night, we should thus remember that the reflection we see is simply due to the effect of the boundary conditions on the waves in the two media.

14.7 Polarisation

Up till now we have considered a string stretched along the x axis, and have called its displacement y. Now the real world has three dimensions, and so we could have had our wave represented not only as

$$y = A \sin(kx - \omega t + \phi_1) \tag{14.69}$$

but alternatively as

$$z = B \sin(kx - \omega t + \phi_2) \tag{14.69'}$$

These correspond respectively to the string vibrating in the y or in the z directions.†

These two separate possibilities are two independent linear polarisation states. This means that in either mode, the string is vibrating in only one direction perpendicular to the string's direction.

Of course, it is possible to have A and B in eqns (14.10) and (14.69) both non-zero. Provided $\phi_1 = \phi_2$ (i.e. the two independent vibrations are in phase with each other), this produces nothing fundamentally new since we can define rotated coordinates in the y–z plane:

$$\left. \begin{array}{l} y' = y \cos \alpha + z \sin \alpha \\ z' = -y \sin \alpha + z \cos \alpha \end{array} \right\}$$

Then with the choice of rotation angle given by $\tan \alpha = B/A$,

$$\left. \begin{array}{l} y' = \sqrt{A^2 + B^2} \sin(kx - \omega t + \phi) \\ z' = 0 \end{array} \right\} \tag{14.70}$$

This is identical in form to (14.10). Thus we again have a state of linear polarisation, but with the plane of vibration being rotated by an angle α with respect to the original y axis.

More interesting is the case where the phase difference $\phi_2 - \phi_1$ between the y and z vibrations is $\pm \pi/2$. Then, for example,

$$\begin{aligned} z &= B \sin(kx - \omega t + \phi_1 + \pi/2) \\ &= B \cos(kx - \omega t + \phi_1) \end{aligned} \tag{14.71}$$

Thus we have simultaneously both the y vibration of eqn (14.69) and this z displacement. We now consider the special case $A = B$, and define r and θ as the polar coordinates of the net transverse displacement in the y–z plane (see fig. 14.15). Thus r defines its magnitude, and θ determines in which direction in the y–z plane it is pointing. Then

$$\tan \theta = \frac{y}{z} = \tan(kx - \omega t + \phi_1)$$

† Although there are three space dimensions, we have only two independent vibrations of the transverse type. Vibrations in the x direction correspond to longitudinal waves (see Section 14.8).

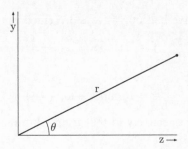

Fig. 14.15. Polar coordinates in the y–z plane, for describing the transverse displacement due to the presence of the two waves (14.69) and (14.69′) travelling along the $+x$ direction. The y and z axes are as shown, because we imagine we are looking along the direction the wave is moving (i.e. it is travelling into the page). With the x axis pointing into the page, a right-handed system of axes then defines the relative orientation of y and z. The angle θ, as defined by $\tan^{-1}(y/z)$, is as shown.

so that, with a suitable resolution of the $2n\pi$ ambiguity in the definition of a phase,

$$\theta = kx - \omega t + \phi_1 \qquad (14.72)$$

Furthermore

$$r^2 = y^2 + z^2 = A^2 \qquad (14.73)$$

The interpretation of eqns (14.72) and (14.73) is as follows. If we concentrate on the nature of the displacement (in the y–z plane, of course), then eqn (14.73) shows that its magnitude is constant, while from (14.72) we deduce that at a given position its polar angle θ varies linearly with time t. In simpler language, the displacement vector is just travelling round a circle in the y–z plane. Since θ decreases with t, the circle is described in a clockwise sense in the y–z plane (see fig. 14.15). On the other hand, at fixed t and with increasing x, the phase increases.

If we follow the way a displacement at a fixed polar direction moves, from eqn (14.72) we find that it does so in a way that keeps $kx - \omega t$ constant. Not surprisingly, this particular point on the wave moves in the positive x direction with a speed ω/k.

The conclusion from the last two paragraphs is that the displacement is of the form given by a long left-handed screw that is being moved along its length. This state is one of circular polarisation.

In an analogous manner, we could have chosen

$$y = A \sin(kx - \omega t + \phi_1) \qquad (14.69)$$

Table 14.2. *State of polarisation resulting from two linearly polarised waves.*

With two displacements defined by eqns (14.69) and (14.69′), the resulting polarisation state depends on the amplitudes A and B, and the phase difference $\phi_2 - \phi_1$.

A	B	$\phi_2 - \phi_1$	Resulting polarisation state
1	0	—	linear ↑
0	1	—	linear →
1	$\neq 0$	0	linear ↗
1	$\neq 0$	π	linear ↘
1	1	$\pi/2$	circular (left-handed)
1	1	$-\pi/2$	circular (right-handed)

and

$$z = A\sin(kx - \omega t + \phi_1 - \pi/2) \qquad (14.74)$$

which would have again given us a circular state of polarisation, but this time in the sense of a right-hand screw.

Table 14.2 summarises the different polarisation possibilities discussed above.

The above effects in the y–z plane can be displayed on a cathode ray oscilloscope by applying harmonically varying voltages of identical frequencies to the x and y plates.† The relevant phase difference between the voltages can be achieved using a circuit containing a resistor and a capacitor in series.

The most realistic example of polarisation effects is not with waves on strings, but with light waves. Even more amusing effects can occur in crystals which have a structure such that the different states of circular polarisation propagate with slightly different speeds. We will now consider what happens to a beam of linearly polarised light passing through such a medium.

Just as a state of circular polarisation can be thought of as a combination of two independent linear polarisations, so can our state of linear polarisation be considered as a combination of right- and left-handed circularly polarised light. If the speed of propagation were the same for the two circularly polarised beams, when they emerged at the far end of the material, they would recombine to give the original linearly polarised

---

† These are the conventional names for the oscilloscope's plates. Unfortunately, it differs from our choice of labels for the axes.

state. But since they travel through our special crystal at different speeds, they gradually get out of phase with each other, and when they leave the medium, their special phase relationship is lost, and they need no longer recombine to produce the original wave. Indeed, with a suitable length of crystal, they can even produce an outgoing beam which is linearly polarised in the direction orthogonal to the incident one.

The understanding of this effect required us to think of some states of polarisation alternatively in terms of pairs of orthogonal linearly polarised beams, or of pairs of oppositely circularly polarised ones. This type of description, in terms of different sorts of basic states, with the one being more suitable depending on the circumstances, is common in many different branches of physics. Look out for them later in your course.

14.8 Longitudinal waves

Up till now, we have been concerned mainly with a length of string and its transverse vibrations. If the string is elastic enough, however, we could imagine displacing it along its length, and then watching how this disturbance travelled along the string. These are called longitudinal waves. The wave properties in such a situation are very similar to those discussed previously, with one important exception: since there is only one direction in space parallel to the direction of propagation of the wave, we have only one mode of longitudinal waves, and there is no possibility of polarisation phenomena such as we discussed with transverse waves (see Section 14.7).

We can derive a wave equation for longitudinal waves on a string in direct analogy with our approach for a transverse wave. Once we have mastered the earlier problem, the most serious difficulty here is in drawing and understanding the diagram, since the displacement is now in the direction of propagation (see fig. 14.16, as compared with fig. 14.5 for the transverse wave).

We assume that the tension for an extension ε of a section of string of length l is given by

$$T = \lambda \varepsilon / l \qquad (14.75)$$

where λ is a constant, characteristic of the string. Initially the string is stretched uniformly with a constant tension T_0. The string lies beside a ruler along the x axis, and we imagine marks x' painted on the stretched but stationary string, so that x and x' coincide in the absence of longitudinal waves on the string.

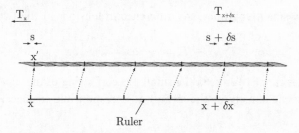

Fig. 14.16. A segment of string on which there is a longitudinal displacement s, which varies with position and time. x can be thought of as marked off on a ruler alongside the string, while x' is the new coordinate of the point on the dispaced string which was originally at x. Then the new displacement is given by $x' - x$. Because of the wave, the displacements s and $s + \delta s$ at the two end-positions x and $x + \delta x$ are different, and hence so, in general, are the tensions T_x and $T_{x+\delta x}$ (as seen from eqn (14.78)). It is this difference that provides the driving force for the longitudinal wave.

Now consider a wave on the string such that different parts of it are moved longitudinally (in the x direction) by different amounts. Then the distance s by which the string at x is displaced is

$$s(x) = x'(x) - x \tag{14.76}$$

where $x'(x)$ is the x coordinate of the point on the displaced string, which was at x when there was no wave present (see fig. 14.16).

If $s(x)$ is constant along the string, the tension remains at T_0 everywhere; there is thus no longitudinal motion, and no wave. But varying s results in differing tensions along the string, because its extension varies with x. It is thus the changing tension† which produces the force that makes the string move longitudinally, and hence drives the wave.

More specifically, if we consider a minute portion of string between x and $x + \mu$ (where $\mu \ll \delta x$ of fig. 14.16), the extension

$$\varepsilon(x) = s(x + \mu) - s(x) = \frac{\partial s}{\partial x}\mu$$

and hence from Hooke's Law (14.75), the tension is

$$T = T_0 + \lambda \left(\frac{\partial s}{\partial x}\mu \right) / \mu \tag{14.77}$$

† This contrasts with the transverse wave, where the tension was essentially constant along the string.

Waves

Then for a small length δx, the difference in tensions at the two ends is

$$T_{x+\delta x} - T_x = \frac{\partial T}{\partial x}\delta x = \lambda\frac{\partial^2 s}{\partial x^2}\delta x \qquad (14.78)$$

Since this is the net force on the small piece of string of mass $\rho\delta x$ (where ρ is the linear density of the string), it is responsible for accelerating it and so

$$\lambda\frac{\partial^2 s}{\partial x^2} = \rho\frac{\partial^2 s}{\partial t^2} \qquad (14.79)$$

Thus eqn (14.79) is the wave equation for the longitudinal vibrations of the string, and it yields solutions in complete analogy with those of eqn (14.24) for the transverse motion. In the current case, the speed of the waves is

$$v = \sqrt{\lambda/\rho} \qquad (14.80)$$

Now that we have derived the wave equation in two different situations, we should be capable of going on to do so for any other. The common features of such derivations are:

(1) consideration of a small region of the medium supporting the wave;
(2) applying to it Newton's Second Law, in the form:

net force = mass × second time derivative of the relevant coordinate w;

(3) expressing the force at one end of the region as proportional to a spatial derivative of w; and
(4) writing the net force as the difference between the forces at the two ends, which involves the spatial derivative of the force, and hence the second spatial derivative of w.

The net result is an equation of the form

$$f\frac{\partial^2 w}{\partial x^2} = \rho\frac{\partial^2 w}{\partial t^2} \qquad (14.81)$$

where f is a constant relating to the driving force that is responsible for the displacement w. The speed can immediately be read off as

$$v = \sqrt{f/\rho}$$

Sound provides another example of longitudinal waves. Here the density waves in the gas provide the pressure imbalance which is in turn responsible for the non-uniform density. In fact, at least in one sense, this is a better example of a longitudinal wave than the stretched string we have just considered, because in the sound case, we cannot have transverse waves — the gas is unable to provide the shear force that would be necessary to return a gas molecule with a transverse displacement towards its equilibrium position. As we have already seen, the elastic string can have transverse as well as longitudinal waves.

The derivation of the wave equation follows the above standard procedure. If the gas in which the sound is travelling were at constant temperature, so that its pressure p and volume V were related by

$$pV = \text{constant} \tag{14.82}$$

then we would obtain the speed v of the sound waves as

$$v = \sqrt{p/\rho} \tag{14.83}$$

This, however, is irrelevant. It turns out that any normal gas is incapable of conducting away the heat produced by the compressions in the sound wave, in a time which is small compared with the period of the sound waves. Thus the temperature at a compression is higher than that at a position of lower pressure, and the gas is not isothermal.

In fact it is a good approximation that almost no heat is lost from the compression region by conduction during one period of the wave, and the gas behaves in an almost adiabatic manner. Then eqn (14.82) is replaced by

$$pV^\gamma = \text{constant} \tag{14.82$'$}$$

where γ is the ratio of the specific heats of a gas, and is a number between 1 and 1.6. Finally the speed of sound becomes

$$v = \sqrt{\gamma p/\rho} \tag{14.83$'$}$$

It is amusing to note that v is within a factor of 2 of the RMS speed of the gas molecules due to their random thermal motion. When we stand in a field and see someone in the distance firing a gun, we can obtain a good idea of the speed of sound from the time lag between when we hear the event as compared with when we see it. We should be aware that this is also a reasonable measure of how fast the air molecules are travelling. Perhaps it should not be surprising that, since it is the motions of the molecules which are responsible for transmitting the sound through the air, these two speeds are similar.

Yet another example of longitudinal waves is provided by geophysics. The next time you are caught in an earthquake, you should see if, before the tremors die away, you can derive the wave equation and hence, by substituting in the relevant values of the parameters, calculate how long it takes the earthquake to travel from one side of the earth to the other.

14.9 Interference

Two or more waves which occupy the same region of space and time can interfere. This means that their combined effect is different from the sum of the separate effects of the individual waves. This can arise because the intensity of an observed effect is proportional to the square of the amplitude of the wave. When several waves are present, it is their amplitudes† which are added. Thus interference occurs because the square of a sum can differ from the sum of squares.

For example, when we look at an illuminated screen, we do not see the brightness changing at 10^{17} Hz, the frequency of the light. Instead we observe the average intensity of the light, which depends on its amplitude squared. Similarly, in quantum mechanics the probability of observing a particle at a given position is proportional to $|\psi|^2$ where ψ is its wave function there. Yet again, the energy of a wave on a string depends on the square of its displacement.

Interference effects can be observed in a wide variety of physical situations. Thus, if a screen is illuminated simultaneously by two light beams, each of which on its own produces a uniform distribution on the screen, the resultant pattern can consist of a series of light and dark bands.

In the rest of Section 14.9, we examine some of these interference situations in more detail.

14.9.1 Beats

Two waves w_1 and w_2 of slightly different frequencies exhibit beats. Thus for‡

$$w_1 = A\cos(\omega_0 + \delta\omega)t \left.\right\}$$
$$\text{and} \quad w_2 = A\cos(\omega_0 - \delta\omega)t \left.\right\}$$

---

† The reader is warned that in this section the term "amplitude" is used in the senses of both w and A of waves such as that of eqn (14.87).

‡ The argument of the cosines should also contain $(k \pm \delta k)x$ terms. However, we are assuming that the waves are observed at some constant position, so that these are simply constant phase factors, which do not affect the essence of the discussion.

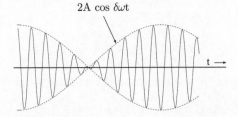

2A cos $\delta\omega$t

Fig. 14.17. The beat pattern produced as the sum of the amplitudes of two waves with similar frequencies $(\omega_0 \pm \delta\omega)/2\pi$. The resulting wave oscillates as $\cos\omega_0 t$, but with an amplitude $2A \cos\delta\omega t$. The intensity, which depends on the square of the amplitude, thus varies at a frequency $\delta\omega/\pi$.

we obtain

$$w = w_1 + w_2 = 2A \cos\delta\omega t \cos\omega_0 t \tag{14.84}$$

That is, the resultant oscillates at ω_0, the mean of the values of the individual waves, but its amplitude is $2A \cos\delta\omega t$. For small $\delta\omega$, this is slowly varying, so that the intensity of the waves changes harmonically (see fig. 14.17); this is the phenomenon of beats. The minimum intensities are separated by times Δt such that

$$\delta\omega\Delta t = \pi \tag{14.85}$$

Thus the frequency of the beats is $1/\Delta t = 2\omega/2\pi$, which is equal to the difference in frequencies of the individual waves.† An example of beats is provided by two very similar but not quite identical tuning forks, providing notes at the same time.

If the two waves are of unequal amplitudes A and B, beats still occur, but the net intensity does not reduce to zero, i.e. for $w_2 = B \cos(\omega_0 - \delta\omega)t$,

$$\begin{aligned}
w &= w_1 + w_2 \\
&= A \cos(\omega_0 + \delta\omega)t + B \cos(\omega_0 + \delta\omega - 2\delta\omega)t \\
&= A \cos(\omega_0 + \delta\omega)t \\
&\quad + B[\cos(\omega_0 + \delta\omega)t \cos 2\delta\omega t + \sin(\omega_0 + \delta\omega)t \sin 2\delta\omega t] \\
&= (A + B \cos 2\delta\omega t)\cos(\omega_0 + \delta\omega)t + B \sin 2\delta\omega t \sin(\omega_0 + \delta\omega)t \\
&= C \cos[(\omega_0 + \delta\omega)t + \phi] \tag{14.86}
\end{aligned}$$

† Note that, although from eqn (14.84) the amplitude varies like $2A \cos\delta\omega t$, the angular frequency of the beats is $2\delta\omega$, and not $\delta\omega$.

where

$$C^2 = A^2 + B^2 + 2AB \cos 2\delta\omega t$$

and

$$\tan \phi = -B \sin 2\delta\omega t / (A + B \cos 2\delta\omega t)$$

Thus the amplitude C varies between $A + B$ and $A - B$ as $\cos 2\delta\omega t$ changes between ± 1, and as before the angular frequency of the beats is $2\delta\omega$, the corresponding difference between the two components (compare fig. 13.7(a)). A new feature here is that the phase ϕ of the resultant wave (14.86) changes slowly at the beat frequency.

14.9.2 Phase variation between two waves

Another type of interference arises between two waves of identical frequency, but where there is a relative phase variation between the two waves in the region of interest. This gives rise to a characteristic interference pattern.

A very simple example is provided by two waves propagating in opposite directions, i.e.

$$\left. \begin{array}{l} w_+ = A \cos(kx - \omega t) \\ \text{and} \quad w_- = A \cos(kx + \omega t) \end{array} \right\} \qquad (14.87)$$

As previously discussed in Section 14.2, the resultant is a stationary wave

$$w = w_+ + w_- = 2A \cos kx \cos \omega t \qquad (14.88)$$

At different positions along the x axis, the fast oscillation $\cos \omega t$ has amplitude $2A \cos kx$. That is, there are places where the amplitude is zero (when $kx = (n + \frac{1}{2})\pi$, where n is any integer), interleaved by positions of amplitude $\pm 2A$ where the intensity is a maximum. The separation of the maxima (or of the zeroes) is thus

$$\Delta x = \pi / k \qquad (14.89)$$

A way of appreciating how the interference pattern arises is to consider the phases of the two waves, as a function of position along the x axis (see fig. 14.18). At $x = x_0$, the two waves happen to be in phase. At a slightly larger x, the phases of the two waves have changed in opposite senses, and the resultant is reduced (fig. 14.18(b)). Even further along (fig. 14.18(c)), the phase of each wave has changed by almost 90°, and the resultant is nearly zero, i.e. we are close to a minimum of the intensity.

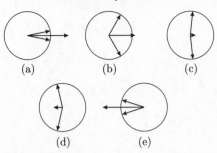

Fig. 14.18. The complex amplitudes of the two waves travelling in opposite directions, plotted on Argand diagrams (compare fig. 4.8). As we move through the interference pattern, the phase of one wave increases and that of the other decreases. Between (a) and (e), the phases of each of the waves change by $\approx \pm\pi$, and the resultant intensity goes from one maximum to the next.

Larger x results in the net amplitude increasing with the opposite phase, but the intensity again is near a maximum in (e).

If w_+ and w_- were light waves, the above set-up of waves travelling in antiparallel directions would not be a very good one for observing the interference pattern. This is because the pattern's repeat distance is only half a wavelength (see eqn (14.89)). A method of obtaining an increased repeat distance, which makes it easier to observe the pattern, is discussed in Section 14.9.3.

An interesting alternative is to consider the above as a beats problem. If we imagine travelling at speed v in the positive x direction, as a result of the Doppler effect, the frequency v of the waves of (14.87) is changed; that of the first is lowered and that of the second is raised (see fig. 14.19). To first order in v/c, where c is the speed of the waves, the magnitude δv of each of these changes is the same, and is given by

$$\frac{\delta v}{v} = \frac{v}{c}$$

Then the two waves, of slightly different observed frequencies, exhibit beats with a frequency

$$v_B = 2\delta v = 2vv/c$$

Thus the maxima of the beat pattern are observed at a time separation of

$$\Delta t_B = 1/v_B = c/2vv$$

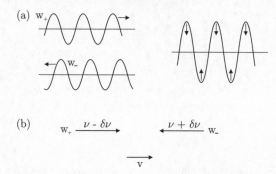

(b)

$w_+ \xrightarrow{\nu - \delta\nu}$ $\xleftarrow{\nu + \delta\nu} w_-$

$\xrightarrow{v}$

Fig. 14.19. (a) Two waves w_+ and w_- travelling in opposite directions produce a standing wave interference pattern. (b) Alternatively, if we imagine travelling along the $+x$ direction, the observed frequencies of the waves would be Doppler shifted in opposite directions, and a beat frequency would result. The distance separation of the observed beat maxima is identical to that for the maxima of the interference pattern of (a).

or, since we are travelling at speed v, at a distance apart of

$$\Delta x = v \Delta t_B = c/2v = \lambda/2 \tag{14.90}$$

This agrees with eqn (14.89).

14.9.3 Young's slits arrangement

In order to produce fringes which are more widely separated in x, the antiparallel beams w_+ and w_- of (14.87) are replaced by nearly parallel ones.

Thus we consider two waves

$$\left.\begin{array}{rl} & w_1 = A\sin(k_x x + k_y y - \omega t) \\ \text{and} & w_2 = A\sin(-k_x x + k_y y - \omega t) \end{array}\right\}$$

which are travelling at small angles $\theta \approx \pm k_x/k_y$ to the y axis. Their resultant is

$$w = w_1 + w_2 = 2A\sin(k_y y - \omega t)\cos k_x x \tag{14.91}$$

At fixed y but varying x, the wave intensity is thus modulated by the factor $\cos^2 k_x x$. With small k_x, the interference pattern's repeat distance is

$$\Delta x = \pi/k_x \approx \pi/k_y \theta \approx \pi/k\theta \tag{14.92}$$

For small θ, this is very much enlarged compared with the case of antiparallel waves (compare eqn (14.89)).

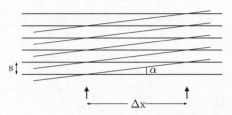

Fig. 14.20. Two transparent pages each have a set of equispaced parallel thin lines ruled on them. They are orientated so that the two sets of lines are almost but not quite parallel. The arrows show where the lines cross. The separation Δx of these crossing positions is given by $\alpha \approx s/\Delta x$, where s is the separation of the parallel lines, and α is the small angle between the two sets. Thus Δx is inversely proportional to α.

Compared with the optics case discussed in the text, the parallel lines correspond to the maxima of the individual waves, giving rise to maximum intensity in the interference pattern whenever they cross the other set. Thus s is equivalent to λ, and α to 2θ, the angle between the light beams.

This essentially corresponds to the situation shown in fig. 14.20: the separation Δx of the positions where the almost parallel lines overlap is inversely proportional to the small angle α between the sets of lines.

In the Young's slits arrangement, almost parallel light from two sources† is allowed to fall onto a screen (see fig. 14.21). As different positions on the screen are considered, the relative phase of the two waves varies, and so an interference pattern may be observed.

Since the phase of the wave at a given position on the screen depends on the distance d from its source, we need to calculate (see fig. 14.21)

$$\left.\begin{array}{c} d_1 = \sqrt{L^2 + (x-s)^2} \approx L\left[1 + \tfrac{1}{2}\left(\frac{x-s}{L}\right)^2\right] \\[2mm] \text{and} \quad d_2 \approx L\left[1 + \frac{1}{2}\left(\frac{x+s}{L}\right)^2\right] \end{array}\right\}$$

where we have made use of the fact that both x and the source separation $2s$ are very much smaller than L. Then

$$d_2 - d_1 \approx 2xs/L \tag{14.93}$$

The bright positions on the screen are determined by the path difference $d_2 - d_1$ corresponding to a whole number of wavelengths. If this occurs at some specific position x_1, then the next constructive interference will

† See Section 14.9.4 concerning the important subject of the coherence of the sources.

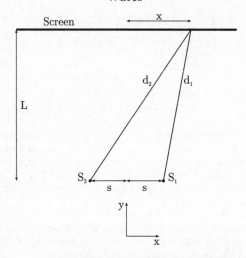

Fig. 14.21. Simplified version of Young's slits. The sources S_1 and S_2 are separated by $2s$. The screen is a large distance L away. At a position x on the screen, the resultant amplitude is determined by the phase difference between the waves from S_1 and S_2; this depends on $d_1 - d_2$.

In order to have coherent sources of light, S_1 and S_2 are in fact two slits, illuminated from a single light source.

be when $d_2 - d_1$ changes by one wavelength. They are thus separated by

$$\Delta x = \lambda L/2s \tag{14.94}$$

This separation is thus larger than λ by a factor of $L/2s$, the ratio of the screen distance to the source separation. It is left as an exercise for the reader to make sure that this discussion is consistent with that resulting in eqn (14.92).

Once again, this problem can alternatively be treated as the beats observed by someone travelling across the screen (see Problem 14.9).

14.9.4 Coherence

In order to observe the optical interference pattern on the screen, it is necessary for the sources S_1 and S_2 to be coherent. This means that if the source S_1 suddenly jumps in phase by $\Delta\phi$, then so does S_2. The reason for this is that a change in one source only would keep unaltered the separation of the bright bands of the interference pattern, but would move it bodily in x. If such changes happened rapidly and randomly, the superposition of the different interference patterns would give rise to a uniform illumination, which would be simply the non-interfering sum of the intensities of the two sources separately (see below).

The reason that this is relevant for typical light sources is that they emit bursts of radiation of length some tenths of a metre, which are uncorrelated in phase with each other. This results in a coherence time for the phase of the emitted radiation which is very small compared with the time of ≈ 0.1 s for the human eye to register a pattern. Thus two independent light sources would indeed give rise to a non-interference pattern.

This problem is overcome by ensuring that the sources are coherent. This can be achieved by illuminating two slits (S_1 and S_2) from a single light source behind them.

With sound waves, this problem does not arise. A tuning fork will emit a note for tens of seconds, which is larger than the time needed to appreciate a sound. Thus an interference pattern can be observed with two independent tuning forks as the sources S_1 and S_2.

We now give a slightly more mathematical justification of the statement at the beginning of this section, about sources of varying phase not giving rise to interference effects. We consider

$$\left. \begin{array}{l} w_1 = A \cos \omega t \\ \text{and} \quad w_2 = A \cos[\omega t + \phi_1(x) + \phi_2(t)] \end{array} \right\}$$

where $\phi_1(x)$ is a phase which varies with position, and hence is responsible for the interference pattern when $\phi_2(t)$ is constant, and $\phi_2(t)$ is the relative phase of the sources, and changes on a time scale much shorter than the observation time. Then the intensity is

$$I = (w_1 + w_2)^2 = 4A^2 \cos^2 \left(\frac{\phi_1 + \phi_2}{2} \right) \cos^2 \left(\omega t + \frac{\phi_1 + \phi_2}{2} \right) \quad (14.95)$$

In this equation, two different time scales are involved. The first is the extremely short one of $2\pi/\omega$, associated with the oscillations of the light wave; the second is that characterising the jumps in $\phi_2(t)$. With $\phi_2(t)$ constant, the observed intensity is

$$I' = 2A^2 \cos^2 \left(\frac{\phi_1 + \phi_2}{2} \right) \quad (14.96)$$

because the last factor in (14.95) averages to $1/2$ over the extremely short time scale. Eqn (14.96) corresponds to an interference pattern, because $\phi_1(x)$ results in I' having a varying spatial pattern.

In contrast, when $\phi_2(t)$ takes on a whole series of random values in the range 0–2π, we also need to integrate over ϕ_2, to find the resultant observed intensity.† The effect of the $\phi_2(t)$ in eqn (14.96) is that, when the integration is performed, the average is A^2, and the x dependence which arose from $\phi_1(x)$ disappears. That is, there is now no interference pattern in the phase-averaged intensity. Furthermore, the resultant average intensity A^2 is just the sum of the average intensities $A^2/2$ of each of the two individual waves.

Problem 14.8 shows that, when there are two very different scales of time variation, it is indeed possible to average first over the very short time scale.

14.10 Other types of waves

In this final section, we briefly discuss a few more, somewhat unrelated, aspects of wave motion.

14.10.1 Electromagnetic waves

Electromagnetism provides a famous example of waves. The equations of Maxwell lead to the prediction of the existence of such waves, which were subsequently discovered by Hertz. Electromagnetic waves now include such a range of phenomena as television transmissions, radio-astronomy, heat radiation, light, X-rays, nuclear gamma rays, etc.

They are transverse in character, but differ from the waves on a string discussed earlier in two important respects. First, they do not require a medium to transport them. Thus light from distant galaxies reaches us across the almost perfect vacuum of outer space. Once it used to be thought that even the vacuum was permeated by something called the aether, which had been invented simply as an artificial medium which could support these waves. This hypothesis was made unlikely by direct experiments, and then ruled out by the Theory of Relativity. Nowadays it is simply accepted that electromagnetic waves can exist in a vacuum, and do not require any medium to be displaced.

The other difference is that, as is apparent from their name, electromagnetic waves have transverse electric and magnetic fields, $\mathbf{E}$ and $\mathbf{H}$ respectively. The $\mathbf{E}$ and $\mathbf{H}$ vectors and the direction of motion of the wave are mutually perpendicular; thus for a wave travelling in the x direction, if $\mathbf{E}$ were in the y direction, the $\mathbf{H}$ would be along z. Thus this

---

† This assumes that our integration time is long enough for ϕ_2 to have a large number of different values, which approximate to a uniform distribution.

wave could be written as

$$\left.\begin{array}{l} E_y\mathbf{j} = \mathbf{E}_0\sin(kx - \omega t) \\ H_z\mathbf{k} = \mathbf{H}_0\sin(kx - \omega t) \end{array}\right\} \qquad (14.97)$$

where $\mathbf{j}$ and $\mathbf{k}$ are unit vectors in the y and z directions respectively. The relative magnitudes of $\mathbf{E}_0$ and $\mathbf{H}_0$ are determined by the medium (or vacuum) through which the waves travel.

Of course another linear polarisation state could be chosen involving $E_z\mathbf{k}$ and $-H_y\mathbf{j}$. Again other linear polarisation states could be obtained by combining the previous two orthogonal linear states; or circular polarisation states could be obtained from them with a suitable phase difference (compare Section 14.7).

14.10.2 Decaying waves

So far our waves have maintained their amplitude as they travel. In many realistic situations, the intensity of a wave decreases because of dissipative effects. Thus a parallel laser beam could be scattered by dust in the air through which it is travelling; or a wave on a string may diminish because of frictional effects within the string; etc.

For a wave travelling to the right, we could approximate these effects by writing

$$y = y_0 e^{-\gamma x}\sin(kx - \omega t) \qquad (14.98)$$

where the $e^{-\gamma x}$ factor gives an envelope, whose shape is a decaying exponential, for the moving sine waves. It is γ which controls the rate at which the amplitude decays; it falls off by a factor of e when x increases by a distance of $1/\gamma$.

For a wave in the opposite direction, we need to write

$$y = y_0 e^{\gamma x}\sin(kx + \omega t) \qquad (14.99)$$

where the $e^{\gamma x}$ factor is required in order to give a decaying amplitude as the wave progresses in the $-x$ direction.

A very similar phenomenon of harmonic waves with decreasing amplitude arises in the transmission of thermal waves in a medium (see eqn (11.50')).

Another situation where waves decrease in amplitude is mentioned at the end of Section 14.10.3.

14.10.3 More dimensions

Because the real world has three space dimensions, we can have more general waves than the ones considered up till now, which merely travelled in the positive or negative x direction.

The simplest generalisation is to consider plane waves travelling in arbitrary directions. These have a constant amplitude across any plane which is perpendicular to the direction of motion. Thus the pressure in such a sound wave could be written as

$$p = p_0 \sin(k_1 x + k_2 y + k_3 z - \omega t)$$
$$= p_0 \sin(\mathbf{k} \cdot \mathbf{r} - \omega t) \qquad (14.100)$$

where $\mathbf{k}$ is a vector with components $(k_1 \ k_2 \ k_3)$ and is parallel to the direction in which the wave is moving, $\mathbf{r}$ is the vector of spatial coordinates $(x \ y \ z)$, and p_0 is the amplitude of the pressure variations. In analogy with the wave along the x axis, $2\pi/|\mathbf{k}|$ is the wavelength of the wave, and its speed is $\omega/|\mathbf{k}|$.

For an electromagnetic wave travelling in the direction specified by $\mathbf{k}$ (here not the unit vector along the z direction!), the electric field is given not by eqn (14.97), but by

$$\mathbf{E} = \mathbf{E}_0 \sin(\mathbf{k} \cdot \mathbf{r} - \omega t) \qquad (14.101)$$

where $\mathbf{E}_0$ is a constant vector perpendicular to $\mathbf{k}$.

To demonstrate that eqn (14.100) (or (14.101)) represents plane waves, it is only necessary to note that p (or $\mathbf{E}$) has a fixed value provided that the $\sin(\mathbf{k} \cdot \mathbf{r} - \omega t)$ factor is constant. Thus at any particular time, $\mathbf{k} \cdot \mathbf{r}$ must be constant, which is the vector equation of a plane.

For such disturbances travelling in a general direction, the wave equation is no longer of the form (14.81) since the displacement-type variable w depends not only on x but also on y and z. When the other spatial coordinates are taken into account, it becomes

$$\rho \frac{\partial^2 w}{\partial t^2} = f \left(\frac{\partial^2 w}{\partial x^2} + \frac{\partial^2 w}{\partial y^2} + \frac{\partial^2 w}{\partial z^2} \right)$$

By a suitable redefinition of our choice of spatial axes, all the above plane waves could be made to travel along the new x' axis. More interesting is the situation where the wave spreads out from a source, decreasing in amplitude as the distance increases. This fall-off is like $1/r^2$ for a wave in three-dimensional space. The exact form of these solutions is beyond the scope of this book.

14.10.4 Quantum mechanics

In quantum mechanics, the way in which we interpret the concept of a particle changes. Rather than being able to describe exactly where it is at any given moment, we instead make do with a wave function $\psi(x)$, which is such that $|\psi(x)|^2$ gives the probability of finding the particle at any specified position x. It turns out to be necessary to allow ψ to be complex, even though the probability is of course real. Thus a typical quantum mechanical wave for a particle of momentum p and energy E moving along the x axis is

$$\psi = \psi_0 e^{ikx - i\omega t} \tag{14.102}$$

where ψ_0 is the amplitude of the wave, and where

$$\left.\begin{array}{c} hk/2\pi = p \\ \text{and} \quad h\omega/2\pi = E \end{array}\right\} \tag{14.103}$$

h being Planck's constant. Eqn (14.102) is not meant to imply that ψ is given by the real part of the right-hand side; it is indeed the complex function as shown there.

Again it is possible to extend the above to the case where the quantum mechanical waves travel in three-dimensional space. These can be plane waves travelling in a particular direction; waves spreading out from the origin; etc.

In addition to the above travelling waves, standing wave solutions exist. For particular values of the frequency, these correspond to the eigenfunction solutions of Schrödinger's Equation (see eqn (11.83), and Section 16.1.2). These stationary wave solutions describe bound states of the particle, for example an electron bound in an atom. In contrast, the travelling waves correspond to the particle being free.

We conclude that, rather than the aether pervading all space, waves occur in almost every branch of physics. It is indeed very worthwhile to be familiar with their properties.

Problems

14.1 Use eqn (14.36″) to derive the relation

$$g = v\left(1 - \frac{1}{1 + \frac{v}{\lambda}\frac{d\lambda}{dv}}\right)$$

between the group velocity g and the phase velocity v, where here λ is the wavelength in free space.

14.2 In quantum mechanics, a particle with momentum p and energy E has associated with it a wave of wavelength λ and frequency v given by

$$\lambda = h/p \quad \text{and} \quad v = E/h$$

where h is Planck's constant. Find the phase and group velocities of these waves when the particle is: (a) non-relativistic, such that

$$p = m_0 v \quad \text{and} \quad E = \frac{1}{2} m_0 v^2$$

and (b) relativistic, in which case

$$p = m_0 v / \sqrt{1 - v^2/c^2} \quad \text{and} \quad E = m_0 c^2 / \sqrt{1 - v^2/c^2}$$

(The particle's rest mass is m_0, its speed is v and c is the speed of light.)

Comment on your answers.

14.3 An infinite string lies along the x axis, and is under tension T. It consists of a section at $0 < x < a$, of linear density ρ_1, and two semi-infinite pieces of density ρ_2. A moving wave $A \cos 2\pi(x/\lambda + vt)$ travels along the string at $x > a$ towards the short section.

Show that, if $a = n\lambda_1$ (where λ_1 is the wavelength on the short section, and n is an integer), the amplitude of the wave that emerges at $x < 0$ is A. What is the amplitude of the wave in the short section?

14.4 (i) The kinetic energy U and the potential energy V for a length $\lambda = 2\pi/k$ of a transverse wave on a string of density ρ and at tension T are given by

$$\left. \begin{aligned} U &= \int_0^\lambda \frac{1}{2} \rho \left(\frac{\partial y}{\partial t} \right)^2 dx \\ \text{and} \quad V &= \int_0^\lambda \frac{1}{2} T \left(\frac{\partial y}{\partial x} \right)^2 dx \end{aligned} \right\}$$

Evaluate these for the wave

$$y = A \cos(kx + \omega t + \phi)$$

and show that $U = V$.

(ii) Show that the potential and kinetic energies in a stationary wave $y = f(x)g(t)$ are equal, after averaging over both the repeat distance of the wave and its period.

(Hint: Remember (a) the formula for integrating by parts, and (b) the fact that y obeys the wave equation.)

14.5

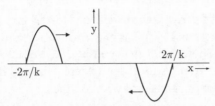

Two transverse waves are on the same piece of string. The first has displacement y non-zero only for $kx + \omega t$ between π and 2π, when it is equal to $A \sin(kx + \omega t)$. The second has $y = A \sin(kx - \omega t)$ for $kx - \omega t$ between -2π and $-\pi$, and is zero otherwise. When $t = 0$, the displacement is as shown in the figure. Calculate the energy of the two waves.

What is the displacement of the string at $t = 3\pi/2\omega$? Calculate the energy at this time.

14.6 Two semi-infinite strings are connected at $x = 0$, and stretched to a tension T. They have linear densities ρ_1 and ρ_2. A harmonic travelling wave, given in complex form as

$$Ae^{i\omega(t - x_1/v_1)}$$

travels along string 1 towards the boundary at $x = 0$. Determine the amplitudes of the reflected and the transmitted waves.

Check that these amplitudes are such that energy conservation in the region at $x \approx 0$ is obeyed.

14.7 An infinite massless string lies along the x axis. It is stretched to a tension T, and has masses m attached at every $x_n = nl$, where n is an integer. Show that the differential equations for the small deflections y_n of the masses are

$$m\ddot{y}_n = \frac{T}{l}(y_{n+1} + y_{n-1} - 2y_n)$$

Show that this string can have waves of the form

$$y_n = Ae^{i(kx_n - \omega t)}$$

provided that k and ω are suitably related. For the case where l and m both tend to zero, but $m/l = \rho$, demonstrate that this relationship between k and ω reduces to the standard one for a continuous string of linear density ρ.

14.8 A screen is illuminated by light, whose amplitude is

$$w(x,t) = A\cos[kx + \phi(t)]\cos\omega t$$

where ω is very large, corresponding to the very rapid oscillation of a light wave, and $\phi(t)$ is a phase which varies on a much slower time scale τ. Show that if the intensity $I(x,t)$ of the wave is averaged over the time from T to $T + \delta$ (where δ is very much larger than $2\pi/\omega$, but very much shorter than τ), then

$$\bar{I} \approx \frac{A^2}{2}\cos^2[kx + \phi(T)]$$

(This agrees with the intuitive idea that, in calculating the mean intensity of a wave, averaged over a shortish time scale, it is possible to integrate over the very fast time variation of w, while regarding the $\phi(t)$ factor as essentially constant — compare Section 14.9.4.)

14.9 For the Young's slits arrangement of fig. 14.21, ensure that the same interference pattern is obtained by considering the beats between the Doppler-shifted sources, as noticed by an observer travelling across the screen with velocity v.

   (For small v/c, the Doppler shift δv of a source of frequency v is given by $\delta v/v = v_L/c$, where v_L is the component of the observer's velocity in the direction of the source.)

15

Matrices

15.1 Linear transformations and matrices

We often have to deal with changes from one set of coordinates to another. Here we consider transformations between variables (x, y) and (x', y') which are of the type

$$\left. \begin{array}{l} x' = ax + by \\ y' = cx + dy \end{array} \right\} \tag{15.1}$$

where a, b, c and d are coefficients. That is, the transformations are linear in the variables, and contain no constant terms. They thus include reflections, rotations and stretchings, but not translations which would require constant terms. Also non-linear transformations are excluded, e.g.

$$\left. \begin{array}{l} x' = xy \\ y' = x/y \end{array} \right\} \tag{15.2}$$

or

$$\left. \begin{array}{l} x'^2 = x^2 + y^2 \\ y' = \tan^{-1}(y/x) \end{array} \right\} \tag{15.3}$$

Eqns (15.1) are for a two-dimensional situation in that the vectors involved $((x, y)$ and $(x', y'))$ each have two components. Almost everything discussed in this chapter applies in any number of dimensions n. In such cases, eqns (15.1) will be replaced by n equations relating the new variables $(x'_1, x'_2, x'_3, \ldots, x'_n)$ to the old ones. Instead of the four coefficients a, b, c and d of eqns (15.1), we now need n^2 constants. Thus

287

we write

$$\left.\begin{array}{l} x'_1 = a_{11}x_1 + a_{12}x_2 + \cdots + a_{1n}x_n \\ x'_2 = a_{21}x_1 + a_{22}x_2 + \cdots + a_{2n}x_n \\ \vdots \\ x'_n = a_{n1}x_1 + a_{n2}x_2 + \cdots + a_{nn}x_n \end{array}\right\} \tag{15.4}$$

where a_{ij} are the n^2 constants.

Eqns (15.4) can be written more compactly as

$$x'_i = a_{i1}x_1 + a_{i2}x_2 + \ldots a_{in}x_n, \quad (i = 1, \ldots, n) \tag{15.5}$$

or in an even more condensed way as

$$x'_i = \sum_j a_{ij}x_j \quad (i = 1, \ldots, n) \tag{15.6}$$

Now we introduce the idea of matrices. If we return to eqns (15.1), the essence of the transformation is contained in the four constants a, b, c and d. We write them in the same format as they appear in the equations, as

$$\begin{pmatrix} a & b \\ c & d \end{pmatrix} \tag{15.7}$$

where the brackets show that these coefficients are to be considered together as a single entity; this is our matrix.

We are now ready to rewrite our transformation equations (15.1) in a way that incorporates our matrix (15.7). This is

$$\begin{pmatrix} x' \\ y' \end{pmatrix} = \begin{pmatrix} a & b \\ c & d \end{pmatrix} \begin{pmatrix} x \\ y \end{pmatrix} \tag{15.8}$$

$$\underset{\substack{\uparrow \\ \text{New} \\ \text{vector}}}{} \qquad \underset{\substack{\uparrow \\ \text{Transformation} \\ \text{matrix}}}{} \qquad \underset{\substack{\uparrow \\ \text{Old} \\ \text{vector}}}{}$$

The fact that we have written x' and y' one underneath the other looks sensible, as they have that structure in eqns (15.1); the brackets surrounding them again show that they are the two components of a single vector $\mathbf{r}'$. That x and y are written similarly is a little surprising; we might rather have expected them to appear as

$$(x \ y)$$

The logic behind the convention used for these vectors will become clearer when we consider matrix multiplication in the next section. Till then, however, we will simply remember to write these vectors vertically.

When we write out the matrix and the vectors completely as in eqn (15.8), we do not save all that much as compared with the original eqn (15.1). The real improvement comes when we denote the transformation matrix (15.7) by $\mathbf{M}$, and the vectors by $\mathbf{r}'$ and $\mathbf{r}$.† Then eqn (15.8) can be written as

$$\mathbf{r}' = \mathbf{M}\mathbf{r} \tag{15.9}$$

This means that the matrix $\mathbf{M}$ multiplying (or operating on) the vector $\mathbf{r}$ gives a new vector $\mathbf{r}'$.

The larger set of eqns (15.4) can similarly be written as

$$\begin{pmatrix} x_1' \\ x_2' \\ \vdots \\ x_n' \end{pmatrix} \begin{pmatrix} a_{11} & a_{12} & \cdots & a_{1n} \\ a_{21} & a_{22} & \cdots & a_{2n} \\ \vdots & \vdots & \ddots & \vdots \\ a_{n1} & a_{n2} & \cdots & a_{nn} \end{pmatrix} \begin{pmatrix} x_1 \\ x_2 \\ \vdots \\ x_n \end{pmatrix} \tag{15.10}$$

$$\uparrow \qquad\qquad \uparrow \qquad\qquad \uparrow$$

Vector $\mathbf{x}'$ Matrix $\mathbf{A}$ Vector $\mathbf{x}$

or even more neatly as

$$\mathbf{x}' = \mathbf{A}\mathbf{x} \tag{15.11}$$

That this is identical in structure to (15.9) simply expresses the fact that in both cases, the new variables are given as linear transformations of the old ones; and this is true independent of the dimensionality of $\mathbf{x}$ or $\mathbf{r}$.

Matrices always have rectangular shapes with a certain number of rows and columns (r and c, respectively). I find it easy to forget which way the rows go, but I do remember that the columns are vertical (just like columns on a building). Thus the first column of the matrix $\mathbf{A}$ in (15.10) contains the elements

$$\left.\begin{matrix} a_{11} \\ a_{21} \\ \vdots \\ a_{n1} \end{matrix}\right\} \tag{15.12}$$

The rows, which go in the other direction, are thus horizontal; for example, the second row is

$$a_{21}\ a_{22}\ \cdots\ a_{2n} \tag{15.13}$$

† Throughout this chapter and the next one, boldface capital letters are used to denote matrices, and boldface lower case for vectors.

The matrices **M** and **A** are square, i.e. they have $r = c$. Vectors can be thought of as matrices with a single column ($c = 1$).

One of the problems that worried me when I first came across matrices was "What is the matrix equal to?" Thus I knew that if we had a square matrix, its determinant was given by sums and differences of products of the elements. But if we don't take the determinant, or if the matrix is not square, what is its numerical value?

The answer is that the question is not a sensible one. A matrix can be equal only to another matrix,† but not to a number. Thus a matrix does not have a value, but just sits there looking pretty.

This is not quite the whole story. A vector is a special sort of matrix, and by now we have a strong intuitive feeling of what a vector is. So cannot we have a similar insight into what a matrix represents? Hopefully you will achieve this by the end of this chapter.

15.2 Pairs of transformations, and matrix multiplication

Eqns (15.1) gave a transformation from $\begin{pmatrix} x \\ y \end{pmatrix}$ to a new vector $\begin{pmatrix} x' \\ y' \end{pmatrix}$.

Now we perform a further transformation that converts $\begin{pmatrix} x' \\ y' \end{pmatrix}$ into $\begin{pmatrix} x'' \\ y'' \end{pmatrix}$, e.g.

$$\left. \begin{aligned} x'' &= ex' + fy' \\ y'' &= gx' + hy' \end{aligned} \right\} \tag{15.14}$$

We can, of course, use eqns (15.1) to replace x' and y' by x and y, and hence write x'' and y'' in terms of the original variables x and y as

$$\left. \begin{aligned} x'' &= (ea + fc)x + (eb + fd)y \\ y'' &= (ga + hc)x + (gb + hd)y \end{aligned} \right\} \tag{15.15}$$

Alternatively, we could have written the transformations in matrix form. Thus (15.14) becomes

$$\mathbf{r}'' = \mathbf{L}\mathbf{r}' \tag{15.14'}$$

where

$$\mathbf{L} = \begin{pmatrix} e & f \\ g & h \end{pmatrix} \tag{15.16}$$

---

† This would mean that each element of the first matrix is equal to the corresponding element of the second. It follows that the two matrices must have the same number of rows as each other, and similarly for columns.

Then using equation (15.9), we have

$$\mathbf{r}'' = \mathbf{LMr} \tag{15.17}$$

This means that $\mathbf{r}''$ is obtained by first transforming $\mathbf{r}$ according to the matrix $\mathbf{M}$, and then transforming the result ($\mathbf{Mr}$) by the matrix $\mathbf{L}$. This, of course, is just what we said in the first couple of sentences of this section.

(It is important to realise that in eqn (15.17), the combination $\mathbf{LMr}$ is to be taken as meaning that the first transformation we apply to $\mathbf{r}$ is that corresponding to the matrix $\mathbf{M}$, and afterwards we apply that of $\mathbf{L}$. Because $\mathbf{L}$ is written in front of $\mathbf{M}$, it is easy to make the mistake of thinking that the $\mathbf{L}$ transformation is first.)

Instead of considering the two separate steps that took $\mathbf{r}$ into $\mathbf{r}''$, we can try to find one transformation that will do this directly. We could then write this as

$$\mathbf{r}'' = \mathbf{Nr} \tag{15.18}$$

where $\mathbf{N}$ is a matrix representing this single transformation we are trying to find. Comparison of eqns (15.17) and (15.18) leads us directly to†

$$\mathbf{LM} = \mathbf{N}$$

Thus $\mathbf{N}$ is said to be the matrix product of the two matrices $\mathbf{L}$ and $\mathbf{M}$.

From eqn (15.15), we can identify the matrix $\mathbf{N}$ as

$$\mathbf{N} = \begin{pmatrix} ea + fc & eb + fd \\ ga + hc & gb + hd \end{pmatrix} \tag{15.19}$$

Thus if we write out the product, we have

$$\begin{pmatrix} e & f \\ g & h \end{pmatrix} \begin{pmatrix} a & b \\ c & d \end{pmatrix} = \begin{pmatrix} ea + fc & eb + fd \\ ga + hc & gb + hd \end{pmatrix} \tag{15.20}$$

We can rewrite this in terms of the elements of the matrices as

$$\begin{pmatrix} L_{11} & L_{12} \\ L_{21} & L_{22} \end{pmatrix} \begin{pmatrix} M_{11} & M_{12} \\ M_{21} & M_{22} \end{pmatrix} = \begin{pmatrix} L_{11}M_{11} + L_{12}M_{21} & L_{11}M_{12} + L_{12}M_{22} \\ L_{21}M_{11} + L_{22}M_{21} & L_{21}M_{12} + L_{22}M_{22} \end{pmatrix}$$

$$= \begin{pmatrix} N_{11} & N_{12} \\ N_{21} & N_{22} \end{pmatrix} \tag{15.21}$$

$\uparrow$ $\uparrow$ $\uparrow$

$\mathbf{L}$ $\mathbf{M}$ $\mathbf{N}$

† In general, it is not a valid procedure to compare bits of a matrix equation and to cancel out the rest from both sides. (See Section 15.9.3.)

where L_{ij} means the element of the matrix $\mathbf{L}$ in its ith row and its jth column. We can thus summarise the result of the combined operation as producing a matrix $\mathbf{N}$ whose elements are given by the formula

$$N_{ij} = \sum_{k=1}^{2} L_{ik} M_{kj} \tag{15.22}$$

Because the answer involves combinations of products of the various elements of the matrices $\mathbf{L}$ and $\mathbf{M}$, the operation is known as matrix multiplication.

We have just discussed in detail the case of multiplying two 2×2 matrices. In general, for matrices of other shapes and sizes,† the product

$$\mathbf{A} = \mathbf{BC} \tag{15.23}$$

is such that the elements of $\mathbf{A}$ are given in analogy with eqn (15.22) by

$$A_{ij} = \sum_{k} B_{ik} C_{kj} \tag{15.24}$$

In order for such a multiplication to be possible, the numbers of elements denoted by B_{ik} and by C_{kj} (each with k varying) must be equal; one reason for this is that we have to be able to match up these elements in order to form the products as required by (15.24). Thus the number of columns of the matrix $\mathbf{B}$ must be the same as the number of rows of $\mathbf{C}$. For example, we can multiply matrices whose sizes are

$$
\left.
\begin{array}{lll}
(2 \times 2) & \text{and} & (2 \times 2) \\
(8 \times 2) & \text{and} & (2 \times 1) \\
(1 \times 2) & \text{and} & (2 \times 2) \\
(1 \times 5) & \text{and} & (5 \times 1) \\
(5 \times 1) & \text{and} & (1 \times 5),
\end{array}
\right\} \text{ allowed matrix multiplications} \tag{15.25}
$$

but for pairs like

$$
\left.
\begin{array}{lll}
(2 \times 2) & \text{and} & (1 \times 2) \\
(2 \times 1) & \text{and} & (2 \times 1) \\
(3 \times 3) & \text{and} & (2 \times 2)
\end{array}
\right\} \text{ forbidden matrix multiplications} \tag{15.26}
$$

matrix multiplication cannot be performed. When written out in this way, it is very easy to spot whether matrix multiplication is legal; we simply check that the middle two numbers (columns of left-hand matrix, and rows of the right-hand one) are identical.

---

† An explanation of how matrices can have the numbers of rows and columns different is given in Section 15.4.

If we now look back on Section 15.1, we can see that, given the rule for multiplying matrices, (15.8) is identical with eqns (15.1), provided we regard $\begin{pmatrix} x \\ y \end{pmatrix}$ as a matrix with two rows and one column. Similarly in n dimensions, a vector can be represented as an $(n \times 1)$ matrix.

One feature of matrix multiplication is at first sight surprising. If instead of considering the product **BC**, we had instead calculated

$$\mathbf{D} = \mathbf{CB}, \tag{15.27}$$

then in analogy with (15.24) we obtain

$$D_{ij} = \sum_k C_{ik} B_{kj} \tag{15.28}$$

which, in general, will not be identical with (15.24). Thus usually

$$\mathbf{BC} \neq \mathbf{CB} \tag{15.29}$$

For ordinary numbers, we are used to the property that

$$7 \times 3 = 3 \times 7$$

This is described by saying that multiplication of real (or complex) numbers is commutative. But for matrices this need not be so. Perhaps some name other than multiplication should have been chosen for the process of eqns (15.23) and (15.24), but the convention is by now firmly established. (A similar situation concerning non-commuting entities occurred with vector products in Section 3.3.3.)

Not only can the product **BC** differ from **CB**, but sometimes one may contain a legal combination of numbers of rows and columns, while the other is forbidden for matrix multiplication. An example of this can be found in the lists (15.25) and (15.26).

Perhaps a way to obtain insight into the fact that the order of a matrix product is important is as follows. Matrices often correspond to operations (e.g. rotation, reflection, etc.), and hence a matrix product represents a pair of consecutive operations. It is a well-known fact from everyday life that the order in which operations are performed can have a crucial influence on the result. Examples of this are:

(i) Wash your hands
     Dry your hands
(ii) Get dressed
     Go for a walk in the morning

(iii) Revise for an examination
       Take the examination
(iv)   Add 3
       Multiply by 5
 (v)   Multiply a function by x
       Differentiate with respect to x

In Section 15.7.1 which deals with rotations in three dimensions, we will consider an example which illustrates the importance of the order of performing the two consecutive operations.

The rule (15.24) for multiplying matrices is easiest to apply by using the following mechanical procedure. To calculate an element of the matrix **A** of eqn (15.23), you move your *left hand horizontally to the right* along a row of **B**, while at the same time moving your *right hand vertically downwards* along a column of **C**. While doing this, you calculate the sum of the products of pairs of elements that your fingers are pointing to at the same time. If you do this for the ith row of **B** (left hand) and the jth column of **C** (right hand), the answer is the element A_{ij} of the product. If one hand runs out of matrix elements before the other, then this is a forbidden product.

It is worth including this movement of

$$\left.\begin{array}{l}\text{left hand moves horizontally to the right}\\\text{right hand moves vertically downwards}\end{array}\right\} \qquad (15.30)$$

in your daily exercises, so that it becomes automatic. This will ensure that you can multiply matrices confidently and correctly.

At this stage it is well worth attempting a few of the simple matrix multiplication problems at the end of this chapter.

15.3 Matrix addition

Matrix addition is simpler than multiplication, but not nearly so interesting. For

$$\mathbf{A} = \mathbf{B} + \mathbf{C} \qquad (15.31)$$

the elements of **A** are given in terms of those of **B** and **C** by

$$A_{ij} = B_{ij} + C_{ij} \qquad (15.32)$$

Thus we can add matrices only if they are the same shape and size, i.e. the numbers of rows of B and of C must be equal, and similarly for their columns.

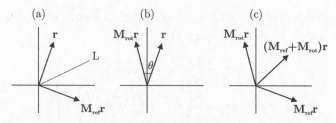

Fig. 15.1. (a) The matrix $\mathbf{M}_{\text{ref}}$ corresponds to reflecting a vector $\mathbf{r}$ in the line *L*. (b) The vector $\mathbf{M}_{\text{rot}}\mathbf{r}$ is the vector $\mathbf{r}$ rotated anticlockwise by an angle θ. (c) $(\mathbf{M}_{\text{ref}} + \mathbf{M}_{\text{rot}})\mathbf{r}$ is the vector sum of the reflection and the rotation of $\mathbf{r}$.

We would need matrix addition if, for example, we wanted to add the results of reflecting a vector and rotating it. Thus

$$\mathbf{r}' = \mathbf{M}_{\text{ref}}\mathbf{r} + \mathbf{M}_{\text{rot}}\mathbf{r}$$
$$= (\mathbf{M}_{\text{ref}} + \mathbf{M}_{\text{rot}})\mathbf{r} \tag{15.33}$$

where $\mathbf{M}_{\text{ref}}$ and $\mathbf{M}_{\text{rot}}$ are the matrices corresponding to reflection about a given plane, and rotation by a specified angle about a particular axis. (See fig. 15.1 for a two-dimensional version of this.)

An extension of the addition idea is that the multiplication of a matrix by a positive integer is equivalent to repeated addition. Thus

$$3 \begin{pmatrix} a & b & c \\ d & e & f \end{pmatrix} = \begin{pmatrix} 3a & 3b & 3c \\ 3d & 3e & 3f \end{pmatrix} \tag{15.34}$$

This generalises in an obvious manner to multiplication by any constant, or to division by such a number.

This type of multiplication by a number such as 3 is not to be confused with matrix multiplication by the 1×1 matrix whose sole element is 3. This would be forbidden for the matrix of eqn (15.34), which has two rows.

15.4 Other shaped matrices

Although we have mentioned the existence of matrices with arbitrary numbers of rows and columns, the actual examples we have met so far have been either square (e.g. (15.7)), or a single column (see eqn (15.10)). We now show how other shapes can arise.

For two separate functions $f_1(x, y, z)$ and $f_2(x, y, z)$, the small changes† δf_1 and δf_2 that result as a consequence of small changes δx, δy and δz in the coordinates are given by

$$\left.\begin{aligned} \delta f_1 &= \frac{\partial f_1}{\partial x}\delta x + \frac{\partial f_1}{\partial y}\delta y + \frac{\partial f_1}{\partial z}\delta z \\ \text{and}\quad \delta f_2 &= \frac{\partial f_2}{\partial x}\delta x + \frac{\partial f_2}{\partial y}\delta y + \frac{\partial f_2}{\partial z}\delta z \end{aligned}\right\} \tag{15.35}$$

These can be combined into a single matrix equation

$$\delta\mathbf{f} = \mathbf{D}\delta\mathbf{r} \tag{15.36}$$

where

$$\delta\mathbf{f} = \begin{pmatrix} \delta f_1 \\ \delta f_2 \end{pmatrix} \tag{15.37}$$

is the vector giving the changes of the functions;

$$\delta\mathbf{r} = \begin{pmatrix} \delta x \\ \delta y \\ \delta z \end{pmatrix} \tag{15.37'}$$

is the vector change in the coordinates; and

$$\mathbf{D} = \begin{pmatrix} \dfrac{\partial f_1}{\partial x} & \dfrac{\partial f_1}{\partial y} & \dfrac{\partial f_1}{\partial z} \\ \dfrac{\partial f_2}{\partial x} & \dfrac{\partial f_2}{\partial y} & \dfrac{\partial f_2}{\partial z} \end{pmatrix} \tag{15.37''}$$

is the derivative matrix, and consists of two rows and three columns. You should practice the matrix multiplication hand-moving exercise of (15.30) to see that these matrices do indeed reproduce eqns (15.35).

We can also imagine calculating some linear combination of δf_1 and δf_2. For example

$$g = c_1\,\delta f_1 + c_2\,\delta f_2 \tag{15.38}$$

where c_1 and c_2 are specified constants. This we could write in matrix notation as

$$g = \tilde{\mathbf{c}}\delta\mathbf{f} \tag{15.39}$$

---

† The reason for considering small changes (rather than just the functions themselves) is that matrices are useful for linear relationships. These we obtain via the first terms of the Taylor series for the increments (see eqn (15.35) and Chapter 7), even if the functions themselves are non-linear.

where $\delta\mathbf{f}$ is given by (15.37), and

$$\tilde{\mathbf{c}} = (c_1 \ c_2) \qquad (15.40)$$

We have written the two constants c_1 and c_2 together as a (1×2) matrix, rather than as a (2×1) column vector, in order that the matrix multiplication of $\tilde{\mathbf{c}}$ and $\delta\mathbf{f}$ is allowed according to the rules of Section 15.2. From there we can also see that the product

$$\begin{pmatrix} c_1 \\ c_2 \end{pmatrix} \begin{pmatrix} \delta f_1 \\ \delta f_2 \end{pmatrix}$$

is forbidden, while

$$(c_1 \ c_2) \begin{pmatrix} \delta f_1 \\ \delta f_2 \end{pmatrix}$$

is allowed, and that the result is a matrix of size (1×1), i.e. simply a constant. Since we represent vectors by column matrices, a row matrix like (15.40) is denoted† by $\tilde{\mathbf{c}}$.

We can now combine (15.39) and (15.36) to write

$$g = \tilde{\mathbf{c}}\mathbf{D}\delta\mathbf{r} \qquad (15.41)$$

which involves the product of three matrices of various shapes and sizes, but with numbers of rows and columns compatible for matrix multiplication.

15.5 Specific example of matrices

Having so far in this chapter considered some fairly general properties of matrices, we now look at examples of matrices $\mathbf{M}$ for specific transformations. For simplicity we will restrict ourselves for the time being to two-dimensional vectors $\mathbf{r} = \begin{pmatrix} x \\ y \end{pmatrix}$, which after transformation becomes $\mathbf{r}' = \begin{pmatrix} x' \\ y' \end{pmatrix}$. Thus

$$\mathbf{r}' = \mathbf{M}\mathbf{r} \qquad (15.42)$$

† The twiddle over any matrix (called a tilde) denotes its transpose, i.e. $\tilde{\mathbf{A}}$ is related to $\mathbf{A}$ by

$$\tilde{A}_{ij} = A_{ji}$$

Thus for a square matrix, $\tilde{\mathbf{A}}$ is the reflection of $\mathbf{A}$ about its diagonal. Since a vector $\mathbf{v}$ is a single column, $\tilde{\mathbf{v}}$ is a single row. (See also Section 15.9.1.)

15.5.1 Stretching and shrinking

A very simple transformation is to stretch the whole of the x–y plane by a factor k. Then

$$\left.\begin{array}{c} x' = kx \\ \text{and} \quad y' = ky \end{array}\right\} \tag{15.43}$$

The matrix for this transformation is

$$\mathbf{M}_1 = \left(\begin{array}{cc} k & 0 \\ 0 & k \end{array}\right) \tag{15.44}$$

The matrix eqn (15.42) then reproduces the required transformation equations (15.43).

Another possibility is to stretch the x coordinates by a factor k, but to leave the y coordinates unaffected, i.e.

$$\left.\begin{array}{c} x' = kx \\ \text{and} \quad y' = y \end{array}\right\} \tag{15.45}$$

The corresponding matrix is

$$\mathbf{M}_2 = \left(\begin{array}{cc} k & 0 \\ 0 & 1 \end{array}\right) \tag{15.46}$$

Similarly if the y coordinates are multiplied by a factor of l but the xs are left unaltered, the transformation is described by the matrix

$$\mathbf{M}_3 = \left(\begin{array}{cc} 1 & 0 \\ 0 & l \end{array}\right) \tag{15.47}$$

If we multiply the two matrices $\mathbf{M}_2$ and $\mathbf{M}_3$, we obtain

$$\mathbf{M}_4 = \mathbf{M}_2\mathbf{M}_3 = \left(\begin{array}{cc} k & 0 \\ 0 & l \end{array}\right) \tag{15.48}$$

This corresponds to first stretching the y coordinates by the factor l, and then the xs by k. In this case, we obtain the same answer if we calculate $\mathbf{M}_3\mathbf{M}_2$. Thus the order of performing these two transformations is unimportant, as is clear physically.

A more interesting example is provided by stretching by a factor of r along a line inclined at an angle $+\theta$ to the x axis, but leaving distances perpendicular to this line unchanged (see fig. 15.2(e)).

To derive the transformation equations, we see that, with respect to

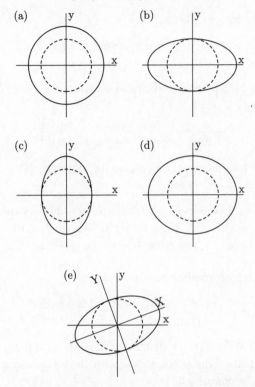

Fig. 15.2. Some simple transformations, corresponding to the matrices $\mathbf{M}_1$–$\mathbf{M}_5$ in the text. The solid curves show the effect of these transformations on the dashed circle: (a) uniform stretch of the plane (matrix $\mathbf{M}_1$); (b) stretch of the x coordinates only ($\mathbf{M}_2$); (c) stretch of the y coordinates only ($\mathbf{M}_3$); (d) combined effect of (b) and (c) ($\mathbf{M}_4$); and (e) stretch along a direction inclined at an angle θ with respect to the x axis ($\mathbf{M}_5$).

the rotated set of axes XY, the effect of the transformation is (in analogy with (15.45))

$$\left. \begin{array}{l} X' = rX \\ \text{and} \quad Y' = Y \end{array} \right\} \tag{15.49}$$

We now simply need to write these equations in terms of x', y', x and y, instead of X', Y', X and Y.

Since

$$\left. \begin{array}{l} X = x\cos\theta + y\sin\theta \\ \text{and} \quad Y = -x\sin\theta + y\cos\theta \end{array} \right\} \tag{15.50}$$

we can rewrite eqns (15.49) as

$$\left. \begin{array}{l} X' = r(x\cos\theta + y\sin\theta) \\ \text{and} \quad Y' = (-x\sin\theta + y\cos\theta) \end{array} \right\} \qquad (15.49')$$

Finally we need the inverse of the rotation relations (15.50) for the primed variables

$$\left. \begin{array}{l} x' = X'\cos\theta - Y'\sin\theta \\ \text{and} \quad y' = X'\sin\theta + Y'\cos\theta \end{array} \right\} \qquad (15.51)$$

which, together with eqns (15.49'), give the required relationship between $(x'$ and $y')$ and $(x$ and $y)$ as

$$\left. \begin{array}{l} x' = r(x\cos\theta + y\sin\theta)\cos\theta - (-x\sin\theta + y\cos\theta)\sin\theta \\ \quad = x[r\cos^2\theta + \sin^2\theta] + y[(r-1)\sin\theta\cos\theta] \\ \text{and} \quad y' = x[(r-1)\sin\theta\cos\theta + y[r\sin^2\theta + \cos^2\theta] \end{array} \right\}$$

$$(15.52)$$

The required transformation matrix is thus

$$\mathbf{M}_5 = \left(\begin{array}{cc} r\cos^2\theta + \sin^2\theta & (r-1)\sin\theta\cos\theta \\ (r-1)\sin\theta\cos\theta & r\sin^2\theta + \cos^2\theta \end{array} \right) \qquad (15.53)$$

The matrix $\mathbf{M}_5$ differs in structure from $\mathbf{M}_4$ in having non-zero off-diagonal elements. This is connected with the stretching axes not being aligned with the coordinate axes.

It is worth checking that $\mathbf{M}_5$ reproduces the simpler results obtained earlier for the relevant special cases (e.g. $\theta = 0°$ or $90°$). Also for $r = 1$, $\mathbf{M}_5$ is clearly correct.

15.5.2 Rotations

In the previous section, we have already made use of a rotation of axes. Here we consider a rotation of the point (x, y) with respect to fixed axes (see fig. 15.3). The relevant formulae are

$$\left. \begin{array}{l} x' = x\cos\theta - y\sin\theta \\ \text{and} \quad y' = x\sin\theta + y\cos\theta \end{array} \right\} \qquad (15.54)$$

(The signs of the sine terms are opposite from those for the rotation of the axes, as explained in Appendix A2.2.) The matrix for this rotation is thus

$$\mathbf{M}_6 = \left(\begin{array}{cc} \cos\theta & -\sin\theta \\ \sin\theta & \cos\theta \end{array} \right) \qquad (15.55)$$

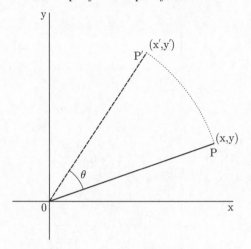

Fig. 15.3. A rotation anticlockwise by an angle θ moves the point P to P'. The corresponding matrix is $\mathbf{M_6}$.

If we perform a further rotation by an angle ϕ, the relevant matrix for the combined operation must be given by the product

$$\mathbf{M_7} = \left(\begin{array}{cc} \cos\phi & -\sin\phi \\ \sin\phi & \cos\phi \end{array} \right) \left(\begin{array}{cc} \cos\theta & -\sin\theta \\ \sin\theta & \cos\theta \end{array} \right) \qquad (15.56)$$

When we evaluate this matrix product, we find

$$\mathbf{M_7} = \left(\begin{array}{cc} \cos(\theta+\phi) & -\sin(\theta+\phi) \\ \sin(\theta+\phi) & \cos(\theta+\phi) \end{array} \right) \qquad (15.56')$$

This is of the same form as the matrix $\mathbf{M_6}$, but with θ replaced by $\theta + \phi$. It thus corresponds to a rotation by an angle of $\theta + \phi$. This is hardly surprising, but it is nevertheless satisfactory.

Had we performed the rotations in the opposite order, or correspondingly multiplied the matrices in (15.56) the other way round, we would have obtained the identical result, as is physically sensible. This is because the rotations are about the same axis. We will find in Section 15.7.1 when we consider rotations in three dimensions that the order is important if the axes of the two rotations do not coincide.

Rotation matrices can be used to deduce the matrix $\mathbf{M_5}$ of Section 15.5.1 somewhat more easily. The operations required there were:

$$\left. \begin{array}{l} \text{rotate axes anticlockwise by } \theta \\ \text{stretch the new } X \\ \text{rotate axes clockwise by } \theta \end{array} \right\} \qquad (15.57)$$

Matrices

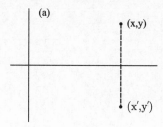

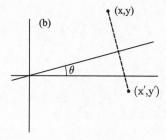

Fig. 15.4. Reflections in: (a) the x axis; and (b) an inclined line. The corresponding matrices are $\mathbf{M}_8$ and $\mathbf{M}_9$ respectively.

Thus

$$\mathbf{M}_5 = \begin{pmatrix} \cos\theta & -\sin\theta \\ \sin\theta & \cos\theta \end{pmatrix} \begin{pmatrix} r & 0 \\ 0 & 1 \end{pmatrix} \begin{pmatrix} \cos\theta & \sin\theta \\ -\sin\theta & \cos\theta \end{pmatrix}$$

which reproduces eqn (15.53) when the required multiplications are performed.

15.5.3 Reflections

A reflection in the x axis (see fig. 15.4(a)) results in

$$\left. \begin{array}{c} x' = x \\ \text{and} \quad y' = -y \end{array} \right\} \tag{15.58}$$

The matrix describing this transformation is thus

$$\mathbf{M}_8 = \begin{pmatrix} 1 & 0 \\ 0 & -1 \end{pmatrix} \tag{15.59}$$

We can determine the matrix for a reflection in a line inclined at an angle θ to the x axis by analogy with the technique used above, for the case of stretching along an inclined direction (see fig. 15.4(b)). We obtain

$$\mathbf{M}_9 = \begin{pmatrix} \cos 2\theta & \sin 2\theta \\ \sin 2\theta & -\cos 2\theta \end{pmatrix} \tag{15.60}$$

15.5.4 Identity matrix

The very simplest possibility is to leave the coordinates as they are. Thus

$$\left. \begin{aligned} x' &= x \\ y' &= y \end{aligned} \right\} \tag{15.61}$$

The corresponding matrix is known as the identity matrix. It is denoted by $\mathbf{I}$ and is

$$\mathbf{I} = \begin{pmatrix} 1 & 0 \\ 0 & 1 \end{pmatrix} \tag{15.62}$$

It has the property that, if we multiply any 2×2 matrix by $\mathbf{I}$, it is left unaltered. That is,

$$\mathbf{IA} = \mathbf{AI} = \mathbf{A} \tag{15.63}$$

This can be verified by explicit multiplication for a general matrix

$$\mathbf{A} = \begin{pmatrix} a_{11} & a_{12} \\ a_{21} & a_{22} \end{pmatrix} \tag{15.64}$$

Indeed the requirement $\mathbf{IA} = \mathbf{A}$ for any matrix $\mathbf{A}$ can be used as the definition of $\mathbf{I}$, from which we can deduce that $\mathbf{I}$ must be given by (15.62).

We see from eqn (15.63) that the matrix $\mathbf{I}$ plays a role corresponding to the number 1 in ordinary multiplication. That is why it is called the identity matrix.

The generalisation to a larger number of dimensions is straightforward. Thus in three dimensions,

$$\mathbf{I} = \begin{pmatrix} 1 & 0 & 0 \\ 0 & 1 & 0 \\ 0 & 0 & 1 \end{pmatrix} \tag{15.65}$$

15.6 Inverse

15.6.1 Definition

All the square matrices $\mathbf{M}_1$–$\mathbf{M}_9$ of Section 15.5, and most square matrices in practice, have an inverse. The inverse of a matrix $\mathbf{A}$ is denoted by $\mathbf{A}^{-1}$.

It is defined by

$$\mathbf{A}\mathbf{A}^{-1} = \mathbf{A}^{-1}\mathbf{A} = \mathbf{I} \qquad (15.66)$$

where $\mathbf{I}$ is the identity matrix of the previous section. Thus $\mathbf{A}^{-1}$ represents the transformation that, if it is performed either after or before that corresponding to $\mathbf{A}$, gets us back where we started.

For the 2×2 case (e.g. where $\mathbf{A}$ is given by eqn (15.64)), we can determine the elements of the inverse matrix $\mathbf{A}^{-1}$ simply by

(i) writing $\mathbf{A}^{-1} = \begin{pmatrix} b_{11} & b_{12} \\ b_{21} & b_{22} \end{pmatrix}$;

(ii) multiplying out $\mathbf{A}$ by $\mathbf{A}^{-1}$ explicitly, and equating the product to the 2×2 identity matrix $\mathbf{I}$; and

(iii) solving the resulting four simultaneous equations (which, in fact, turn out to be two sets of pairs of simultaneous equations) for the unknown elements b_{ij}.

We find that

$$\begin{aligned} \mathbf{A}^{-1} &= \begin{pmatrix} a_{22}/d & -a_{12}/d \\ -a_{21}/d & a_{11}/d \end{pmatrix} \\ &= \frac{1}{d}\begin{pmatrix} a_{22} & -a_{12} \\ -a_{21} & a_{11} \end{pmatrix} \end{aligned} \qquad (15.67)$$

where d is the determinant of the matrix $\mathbf{A}$, i.e.

$$d = a_{11}a_{22} - a_{21}a_{12} \qquad (15.68)$$

This procedure can be extended to calculate the inverse of larger square matrices, but becomes increasingly messy as the size of the matrix increases. The result is as follows. For a matrix $\mathbf{X}$ of size $n \times n$ whose elements are x_{ij}, the inverse $\mathbf{Y} = \mathbf{X}^{-1}$ has elements y_{kl} which are given by

$$y_{kl} = (-1)^{k+l} c_{lk}/d \qquad (15.69)$$

where

d is the determinant of the matrix $\mathbf{X}$;
$(-1)^{k+l}$ is a sign which starts off as positive for y_{11} and then alternates as we proceed along the elements of the inverse matrix; and
c_{lk} is known as the minor of the element x_{lk} of the original matrix $\mathbf{X}$. It is defined as the determinant of the $(n-1) \times (n-1)$ matrix, obtained by deleting the lth row and kth column from the matrix $\mathbf{X}$. The indices

$$\mathbf{X} = \begin{pmatrix} 1 & \text{-}3 & 0 \\ 4 & \text{-}1 & 1 \\ 2 & 2 & 3 \end{pmatrix} \qquad \text{Determinant} = 25$$

$$\begin{pmatrix} \cancel{1} & \cancel{\text{-}3} & \cancel{0} \\ 4 & \cancel{\text{-}1} & 1 \\ 2 & \cancel{2} & 3 \end{pmatrix} \longrightarrow \begin{vmatrix} 4 & 1 \\ 2 & 3 \end{vmatrix} = 10$$

$$(-1)^{1+2} = -1$$

$$\therefore \mathbf{Y} = \begin{pmatrix} \cdot & \cdot & \cdot \\ -\frac{10}{25} & \cdot & \cdot \\ \cdot & \cdot & \cdot \end{pmatrix}$$

Fig. 15.5. To determine the element y_{21} of the matrix $\mathbf{Y}$ which is the inverse of $\mathbf{X}$, it is first necessary to obtain the minor of the element $x_{12} = -3$. This is achieved by deleting the row and column of $\mathbf{X}$ containing x_{12}, and calculating the determinant of the remaining 2×2 matrix; this is equal to 10. The determinant of the original matrix is 25, and the factor $(-1)^{k+l}$ for this element is -1, so eqn (15.69) gives the required element of $\mathbf{Y}$ as $-10/25 = -2/5$.

deserve attention: the minor of x_{lk} becomes part of the element y_{kl} of the inverse.

This procedure is demonstrated in fig. 15.5.

The definition of the minors is such that the determinant of the matrix $\mathbf{X}$ is given by

$$d = \sum_m (-1)^{1+m} x_{1m} c_{1m}$$

$$= \sum_m (-1)^{r+m} x_{rm} c_{rm} \qquad (15.70)$$

Then from eqns (15.69) and (15.70),

$$1 = \sum_m x_{rm} y_{mr} = (\mathbf{XY})_{rr}$$

which is obviously correct because

$$\mathbf{XY} = \mathbf{XX}^{-1} = \mathbf{I}$$

This provides partial justification for the definition (15.69) of the inverse. The combination $(-1)^{k+l}c_{kl}$ is known as the cofactor of the element x_{kl}.

For a 2×2 matrix, the general recipe (15.69) successfully reproduces our previous specific result (15.67), as, of course, it ought to do.

15.6.2 Examples

In order to illustrate how the calculations work out, and what the significance of the inverse matrix is in practice, we will calculate the inverses of a few of the matrices from Section 15.5. Since these are all 2×2 matrices, we can use (15.67) to obtain the inverses.

For the uniform stretching matrix M_1 of eqn (15.44), we find

$$\mathbf{M}_1^{-1} = \begin{pmatrix} 1/k & 0 \\ 0 & 1/k \end{pmatrix} \tag{15.44'}$$

This has the same form as $\mathbf{M}_1$, except that the stretch factor is now $1/k$ instead of k. That is, $\mathbf{M}_1^{-1}$ corresponds to a uniform shrinking by a factor of k. This is as it should be, since the combined operations of stretching and shrinking by the same factor will result in any vector in the x–y plane being left unaltered, i.e.

$$\mathbf{M}_1^{-1}\mathbf{M}_1 = \mathbf{I}$$

as is implied by the definition of an inverse.

The matrix $\mathbf{M}_5$ (eqn (15.53)) corresponds to a stretch by a factor of r along a direction inclined at an angle θ to the x axis. To determine its inverse, we first calculate its determinant, which turns out to be r. Then

$$\mathbf{M}_5^{-1} = \begin{pmatrix} \sin^2\theta + \dfrac{1}{r}\cos^2\theta & -\left(1 - \dfrac{1}{r}\right)\sin\theta\cos\theta \\ -\left(1 - \dfrac{1}{r}\right)\sin\theta\cos\theta & \cos^2\theta + \dfrac{1}{r}\sin^2\theta \end{pmatrix} \tag{15.53'}$$

This has the same structure as the matrix $\mathbf{M}_5$, but with r replaced by $1/r$. Again as expected, $\mathbf{M}_5^{-1}$ thus corresponds to shrinking by a factor of r along the same direction θ.

If we next consider the rotation matrix $\mathbf{M}_6$, its determinant is unity, and

$$\mathbf{M}_6^{-1} = \begin{pmatrix} \cos\theta & \sin\theta \\ -\sin\theta & \cos\theta \end{pmatrix}$$

$$= \begin{pmatrix} \cos(-\theta) & -\sin(-\theta) \\ \sin(-\theta) & \cos(-\theta) \end{pmatrix} \tag{15.55'}$$

The second form of $\mathbf{M}_6^{-1}$ above shows that it corresponds to a rotation by an angle $-\theta$. This, of course, is exactly what is required to be combined with a rotation of $+\theta$ in order to leave any vector unaltered.

Finally for the reflection matrix (15.60), the determinant is -1, and

$$\mathbf{M}_9^{-1} = \begin{pmatrix} \cos 2\theta & \sin 2\theta \\ \sin 2\theta & -\cos 2\theta \end{pmatrix} \tag{15.60'}$$

This is identical with $\mathbf{M}_9$. Thus $\mathbf{M}_9$ is its own inverse. Again this is sensible, since a second reflection through a given line restores any vector to its original position.

15.6.3 Determinants and singular matrices

As we have seen from the general formula (15.69), or from (15.67), or from the specific examples above, the first stage in calculating an inverse is to evaluate the determinant d of the original matrix. Since we then divide each of the cofactors by d, our prescription will be useless if the determinant happens to be zero. In this case, the matrix is said to be "singular", and its inverse simply does not exist.

We can understand why this is so by considering a specific 2×2 example with $d = 0$, e.g.

$$\left. \begin{array}{l} x' = x - y \\ y' = -2x + 2y \end{array} \right\} \tag{15.71}$$

for which the transformation matrix

$$\mathbf{M}_{10} = \begin{pmatrix} 1 & -1 \\ -2 & 2 \end{pmatrix} \tag{15.72}$$

Now eqns (15.71) imply that $y' = -2x'$. This means that, whatever point in the x–y plane we start with, it will always be transformed onto the line $y' = -2x'$. The transformation is thus curious in that we effectively lose a dimension in going from (x, y) to (x', y').

A similar way of viewing the problem is to see that, for example, the (x, y) points $(1, 0), (2, 1), (3, 2)$ etc. all transform into the same (x', y') point $(1, -2)$. The inverse matrix is supposed to tell us how to transform from (x', y') to (x, y). In this case we cannot know whether the (x', y') point $(1, -2)$ corresponds to $(1, 0), (2, 1), (3, 2)$ or to something else (see fig. 15.6). On the other hand, there are no (x, y) points which transform to an (x', y') point like $(1, -1)$, which does not lie on the line $y' = -2x'$. These are the reasons that the matrix $\mathbf{M}_{10}$ does not have an inverse.

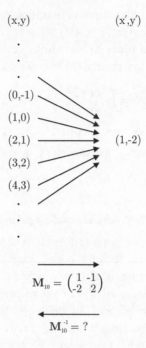

Fig. 15.6. The matrix $\mathbf{M}_{10}$ of eqn (15.72) transforms the shown (x, y) combinations into the same point $x' = 1$, $y' = -2$. The inverse transformation matrix $\mathbf{M}_{10}^{-1}$ is supposed to provide a prescription for converting (x', y') back to (x, y). Because there are an infinity of (x, y) values corresponding to this particular (x', y'), no such recipe is possible and the inverse of $\mathbf{M}_{10}$ does not exist.

The lack of an inverse simply expresses the fact that eqns (15.71) are such that we cannot invert them to determine x and y in terms of x' and y'.

We have so far examined only one specific example of a 2×2 matrix with zero determinant, but it is a general property that a singular $n \times n$ matrix is such that it transforms vectors $\mathbf{r}$ in n-dimensional space into new vectors $\mathbf{r}'$ which occupy a lower number of dimensions in the transformed space. For example, any point in three dimensions, when transformed by a singular 3×3 matrix, ends up on a specific plane, or on a line, or at the origin. It then is impossible to find a unique vector $\mathbf{r}$ which corresponds to a given $\mathbf{r}'$. For a specific $\mathbf{r}'$, there will either be an infinite number of possibilities for $\mathbf{r}$, or none at all. It is for this reason that there is no inverse.

It is clear from the above discussion that the determinant of a matrix

Table 15.1. *Simple transformation matrices, and their determinants.*

Transformation	Example	Determinant
One way stretch by r	$\mathbf{M}_5$	r
Two way stretch by k and l	$\mathbf{M}_4$	kl
Rotation by θ	$\mathbf{M}_6$	1
Reflection	$\mathbf{M}_9$	-1
Identity	$\mathbf{I}$	1
Singular	$\mathbf{M}_{10}$	0

is of significance. In Table 15.1 are listed the determinants of some of the specific 2×2 matrices we have already considered. Since the transformations are linear, they have the property that any region of area a in (x, y) is transformed into a region of area λa in (x', y'), where λ is a constant for the given transformation. This factor λ is simply the magnitude $|d|$ of the determinant of the corresponding matrix. A consequence of this is that the determinant of the inverse matrix will be $1/d$.

Thus the fact that the singular matrix $\mathbf{M}_{10}$ has determinant zero is an expression of the fact that any area in (x, y) is transformed into a line in (x', y'), which has zero area.

The reflection matrix has determinant -1 because it does not alter the area of a region but it does change its "parity" (see Appendix A8). Thus if we had drawn a right-handed glove in (x, y), the reflection converts it into a left-handed one. Thus we can regard the sign of a determinant as keeping track of the "parity" of the transformation, while its magnitude determines the ratio of the areas after and before the transformation.

The extension of these ideas to larger size matrices is straightforward.

15.6.4 Solving simultaneous equations

We now can use the inverse to provide a very neat solution for a set of simultaneous equations (compare Chapter 1). Thus if we are given n equations

$$\left.\begin{array}{l} a_{11}x_1 + a_{12}x_2 + a_{13}x_3 + \cdots = b_1 \\ a_{21}x_1 + a_{22}x_2 + a_{23}x_3 + \cdots = b_2 \\ a_{31}x_1 + a_{32}x_2 + a_{33}x_3 + \cdots = b_3 \\ \vdots \end{array}\right\} \qquad (15.73)$$

where a_{ij} and b_i are constants and x_i are the variables we want to determine, we rewrite our equations as

$$\mathbf{Ax} = \mathbf{b} \tag{15.74}$$

where $\mathbf{A}$ is the matrix of the coefficients a_{ij}, and $\mathbf{x}$ and $\mathbf{b}$ are column vectors of the unknown x_i and the constants b_i respectively. It is thus tempting to divide both sides of the equation by $\mathbf{A}$, but unfortunately this is forbidden since matrix division does not exist (for good reason). We thus use our standard trick (compare vectors, Fourier series, etc.) of multiplying both sides of eqn (15.74) by something clever so that we are left with whatever we are interested in (i.e. $\mathbf{x}$) on its own on one side of the equation. It does not need too much imagination to realise that the magic factor is $\mathbf{A}^{-1}$. This is because our equation then becomes

$$\mathbf{A}^{-1}\mathbf{Ax} = \mathbf{A}^{-1}\mathbf{b} \tag{15.75}$$

Then from the definition of $\mathbf{A}^{-1}$ (see eqn (15.66))

$$\mathbf{Ix} = \mathbf{A}^{-1}\mathbf{b} \tag{15.76}$$

and finally using eqn (15.63) for the identity matrix $\mathbf{I}$ we obtain

$$\mathbf{x} = \mathbf{A}^{-1}\mathbf{b} \tag{15.77}$$

This is our desired result, giving $\mathbf{x}$ in terms of $\mathbf{A}^{-1}$, which we can calculate from $\mathbf{A}$ (provided it is not singular), and $\mathbf{b}$.

It is important in the above procedure to start by multiplying eqn (15.74) by $\mathbf{A}^{-1}$ on the left so that $\mathbf{A}^{-1}$ and $\mathbf{A}$ appear next to each other, and so that they can be combined to give $\mathbf{I}$. In fact, multiplication on the right would have given $\mathbf{AxA}^{-1}$, which is a forbidden combination for multiplication since $\mathbf{x}$ and $\mathbf{A}^{-1}$ are of size $n \times 1$ and $n \times n$ respectively (see Section 15.2).

Eqn (15.77) is a very compact solution of our simultaneous equations. Of course, if we actually have to evaluate $\mathbf{A}^{-1}$ ourselves, and have to perform the matrix multiplication $\mathbf{A}^{-1}\mathbf{b}$, we will be involved in the same arithmetic as if we were to solve the simultaneous equations conventionally. What it does save is the need to think what we ought to be doing next, as the matrix procedure is exactly defined. It is also useful to have a neat notation, in much the same way as vector products and divs, grads and curls help very much with vector manipulations, as compared with writing out everything in components. And sometimes we will be satisfied with a matrix solution like (15.77), and not be particularly concerned about its numerical evaluation anyway.

If we do need to perform numerical matrix operations on a more than an occasional basis, then it is a very good idea to use computer routines for so doing. Any reasonable computer library contains packages for matrix calculations (e.g. inversion, multiplication, finding eigenvalues and eigenvectors — see Chapter 16). Since computers merrily manipulate matrices without becoming bored, they are more likely than human beings to produce the correct answers regularly, and somewhat faster too.

15.7 Rotations in three dimensions

15.7.1 Pairs of rotations

The matrix M_6 of eqn (15.55) refers to a rotation in two dimensions. We live in a three-dimensional world, so we want to extend the earlier discussion. An anticlockwise rotation of a point (x, y, z) by an angle θ about the z axis (with the axes left fixed) results in it moving to (x', y', z'), given by

$$\left.\begin{array}{l} x' = x \cos\theta - y \sin\theta \\ y' = x \sin\theta + y \cos\theta \\ z' = z \end{array}\right\} \tag{15.78}$$

The matrix describing this transformation is thus

$$M_{11} = \begin{pmatrix} \cos\theta & -\sin\theta & 0 \\ \sin\theta & \cos\theta & 0 \\ 0 & 0 & 1 \end{pmatrix} \tag{15.79}$$

As in Section 15.5.2, we find that if we consider M_{12} as the matrix corresponding to a rotation of ϕ about the z axis, then $M_{11}M_{12}$ or $M_{12}M_{11}$ both give a matrix describing a rotation by $\theta + \phi$. Thus M_{11} and M_{12} commute. This is not generally true for three-dimensional rotation matrices, as we are about to discover. The present example is special, in that the rotations are about the same axis.

Since M_{11} describes a rotation about the z axis, it does not require too much imagination to write down the matrix for a rotation by an angle λ about the x axis as

$$M_{13} = \begin{pmatrix} 1 & 0 & 0 \\ 0 & \cos\lambda & -\sin\lambda \\ 0 & \sin\lambda & \cos\lambda \end{pmatrix} \tag{15.80}$$

As before this corresponds to a rotation of the vector, rather than the axes, and is for an anticlockwise rotation.

Now we consider the result of the combined operation $M_{13}M_{11}$. This implies performing the rotation by θ about z first, and then a rotation of λ about the x axis after. Explicit multiplication of the matrices gives

$$\mathbf{M}_{14} = \mathbf{M}_{13}\mathbf{M}_{11} = \begin{pmatrix} \cos\theta & -\sin\theta & 0 \\ \cos\lambda\sin\theta & \cos\lambda\cos\theta & -\sin\lambda \\ \sin\lambda\sin\theta & \sin\lambda\cos\theta & \cos\lambda \end{pmatrix} \quad (15.81)$$

If on the other hand we perform the two rotations in the opposite order, the matrix describing the resulting combined operation is

$$\mathbf{M}_{15} = \mathbf{M}_{11}\mathbf{M}_{13} = \begin{pmatrix} \cos\theta & -\sin\theta\cos\lambda & \sin\theta\sin\lambda \\ \sin\theta & \cos\theta\cos\lambda & -\cos\theta\sin\lambda \\ 0 & \sin\lambda & \cos\lambda \end{pmatrix} \quad (15.82)$$

The matrices $\mathbf{M}_{14}$ and $\mathbf{M}_{15}$ are not equal, and hence correspond to different transformations. In fact they each correspond to a rotation, but about different axes. In the next chapter, we will find out how to determine the axis and the rotation angle for a matrix such as (15.81) or (15.82).

We can convince ourselves that the non-commutation of the matrices is sensible by physically performing the pair of operations $\mathbf{M}_{13}\mathbf{M}_{11}$ and $\mathbf{M}_{11}\mathbf{M}_{13}$ on two identical objects. For simplicity we take the specific case $\theta = \lambda = 90°$. Fig. 15.7 shows that these two operations do indeed produce different final orientations, and hence $\mathbf{M}_{13}\mathbf{M}_{11} \neq \mathbf{M}_{11}\mathbf{M}_{13}$.

We can also check that for this case our matrices give the correct transformation. Thus, when both rotations are 90°, $\mathbf{M}_{14}$ becomes

$$\begin{pmatrix} 0 & -1 & 0 \\ 0 & 0 & -1 \\ 1 & 0 & 0 \end{pmatrix}$$

which implies

$$\left.\begin{array}{l} x'' = -y \\ y'' = -z \\ \text{and} \quad z'' = x \end{array}\right\} \quad (15.83)$$

As usual these equations are to be interpreted in the following sense: for any point (x, y, z) on the original object, (x'', y'', z'') gives the coordinates with respect to the same axes after the object has undergone the two rotations.

Inspection of the upper part of fig. 15.7 does indeed show that points

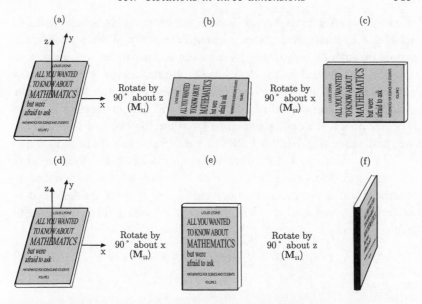

Fig. 15.7. The effect of two 90° rotations about two different axes, applied to an everyday object. The final orientation depends on the order of the two rotations. (Because we are using a transformation of points and not of axes, a second rotation about x (or about z) is with respect to the original unrotated axes.)

For the upper transformation, the matrix $\mathbf{M}_{14} = \mathbf{M}_{13}\mathbf{M}_{11}$ gives eqns (15.83) for the relationships between the new and old coordinates of any point. Thus the point on the object along the x axis with coordinates $(x, y, z) = (1, 0, 0)$ is expected after the rotations at $(x'', y'', z'') = (0, 0, 1)$. The diagram confirms that this is correct.

The lower transformation is given by $\mathbf{M}_{15} = \mathbf{M}_{11}\mathbf{M}_{13}$.

along

$$
\left.
\begin{array}{l}
\text{the old } x \text{ axis now lie along the } z \text{ direction} \\
\text{the old } y \text{ axis now lie along the } -x \text{ direction} \\
\text{and} \quad \text{the old } z \text{ axis now lie along the } -y \text{ direction}
\end{array}
\right\} \quad (15.83')
$$

in agreement with (15.83).

15.7.2 Rotation about a general axis

We now turn to the problem of deriving the matrix corresponding to a rotation by an angle θ about an axis in a general direction $\hat{\mathbf{n}}$. As usual, we interpret this as a rotation of a vector with respect to a fixed set of coordinate axes.

An example of the type of problem for which this could be required is as follows. An underground detector for cosmic rays is being used to look at radiation coming from a particular galaxy. When a cosmic ray passes through the apparatus, its direction is determined with respect to a set of axes defined by the detector. During the course of 24 hours, the earth performs a complete rotation and hence so does the detector. Thus if we want to know the direction of the cosmic ray in space (and hence whether or not it does indeed come from the direction of that galaxy), we must rotate our detector back to a set of axes that are fixed in space.

We now return to the mathematics. As before, we first need to write the transformation equations, in order to read off the corresponding matrix. These are most easily obtained in vector form, as described in Section 3.5.2 (see also fig. 3.9). The result obtained there (eqns (3.46) and (3.45)) was

$$\mathbf{r}' = (\mathbf{r} \cdot \hat{\mathbf{n}})\hat{\mathbf{n}} + [\mathbf{r} - (\mathbf{r} \cdot \hat{\mathbf{n}})\hat{\mathbf{n}}]\cos\theta + \hat{\mathbf{n}} \wedge \mathbf{r}\sin\theta \qquad (15.84)$$

which we rewrite as

$$\mathbf{r}' = \mathbf{r}\cos\theta + \hat{\mathbf{n}}(\mathbf{r} \cdot \hat{\mathbf{n}})(1 - \cos\theta) + \hat{\mathbf{n}} \wedge \mathbf{r}\sin\theta \qquad (15.84')$$

We consider in turn each term on the right-hand side of this equation, writing them out in component form and obtaining the matrix for each of them; the matrix we want is then just the sum of these three.

The first term gives a contribution to $\mathbf{r}'$ such that

$$\left.\begin{array}{r} x_1' = \cos\theta\, x \\ y_1' = \cos\theta\, y \\ \text{and} \quad z_1' = \cos\theta\, z \end{array}\right\} \qquad (15.85)$$

Its matrix is thus

$$\mathbf{R}_1 = \cos\theta \begin{pmatrix} 1 & 0 & 0 \\ 0 & 1 & 0 \\ 0 & 0 & 1 \end{pmatrix} \qquad (15.86)$$

The second term produces

$$\left.\begin{array}{r} x_2' = (1 - \cos\theta)(n_x x + n_y y + n_z z)n_x \\ y_2' = (1 - \cos\theta)(n_x x + n_y y + n_z z)n_y \\ \text{and} \quad z_2' = (1 - \cos\theta)(n_x x + n_y y + n_z z)n_z \end{array}\right\} \qquad (15.87)$$

and so

$$\mathbf{R}_2 = (1 - \cos\theta) \begin{pmatrix} n_x^2 & n_x n_y & n_x n_z \\ n_y n_x & n_y^2 & n_y n_z \\ n_z n_x & n_z n_y & n_z^2 \end{pmatrix} \qquad (15.88)$$

Finally from the third term

$$\left.\begin{array}{c} x_3' = \sin\theta(n_y z - n_z y) \\ y_3' = \sin\theta(n_z x - n_x z) \\ \text{and} \quad z_3' = \sin\theta(n_x y - n_y x) \end{array}\right\} \quad (15.89)$$

whence

$$\mathbf{R}_3 = \sin\theta \begin{pmatrix} 0 & -n_z & n_y \\ n_z & 0 & -n_x \\ -n_y & n_x & 0 \end{pmatrix} \quad (15.90)$$

Thus finally our transformation is given by

$$\mathbf{r}' = \mathbf{R}\mathbf{r}$$

where

$$\mathbf{R} = \mathbf{R}_1 + \mathbf{R}_2 + \mathbf{R}_3$$

This is a fairly messy result. It is worth checking that for the simple case of the axis being along the z axis (i.e. $n_x = n_y = 0$, $n_z = 1$), $\mathbf{R}$ does indeed reduce to the previous rotation matrix $\mathbf{M}_{11}$ of eqn (15.79).

In the next chapter, we will address the inverse problem of looking at a matrix $\mathbf{R}$, and deciding whether it is consistent with a rotation, and if so determining the axis and angle of rotation.

15.7.3 Euler angles

Finally on the subject of three-dimensional rotations, we describe an alternative to the last section for obtaining a matrix which can describe the rotated position of a solid body with respect to a fixed set of axes. A rotation requires three independent parameters to specify it. In Section 15.7.2, these were the rotation angle θ and two constants giving the orientation of the axis $\hat{n}$. Here instead we use three independent rotations defined as follows:

(i) Rotate the body by an angle θ about the z axis.

(ii) Perform the second rotation by an angle λ about the x axis.

(iii) Finally rotate by ψ about the z axis again.

All the rotations are taken to be anticlockwise.

The matrices for stages (i) and (ii) are $\mathbf{M}_{11}$ and $\mathbf{M}_{13}$ respectively, and their product is $\mathbf{M}_{14}$ of eqn (15.81). For the final step, the matrix is like

$\mathbf{M}_{11}$ again, but with θ replaced by ψ. On premultiplying $\mathbf{M}_{14}$ by this, we obtain

$$\begin{pmatrix} \cos\psi\cos\theta - \sin\psi\cos\lambda\sin\theta & -\cos\psi\sin\theta - \sin\psi\cos\lambda\cos\theta & \sin\psi\sin\lambda \\ \sin\psi\cos\theta + \cos\psi\cos\lambda\sin\theta & -\sin\psi\sin\theta + \cos\psi\cos\lambda\cos\theta & -\cos\psi\sin\lambda \\ \sin\lambda\sin\theta & \sin\lambda\cos\theta & \cos\lambda \end{pmatrix}$$

$$(15.91)$$

The angles θ, λ and ψ can be determined from the orientation of a pair of vectors in the rotated body that were known before the rotation. These three parameters are known as Euler angles.

15.8 Physical applications of matrices

In this section, a few examples of matrices applied to physical problems are described. This will give an appreciation of the types of situations in which matrices can be used with advantage.

15.8.1 *Optics*

In a typical problem in geometrical optics, a light ray passes through simple optical elements at given positions; its angle of emergence from the system is required. For simplicity we initially restrict the problem to two dimensions.

An incident ray is specified by its height y_i and its direction $(dy/dx)_i$ at, say, $x = 0$. The answer involves $(dy/dx)_f$ and perhaps y_f at the far end of the system $x = x_f$. At least for simple optical elements and for small gradients dy/dx, the final parameters are given in terms of the initial ones by linear relations. We can thus write†

$$\begin{pmatrix} y_f \\ \left(\dfrac{dy}{dx}\right)_f \end{pmatrix} = \begin{pmatrix} M_{11} & M_{12} \\ M_{21} & M_{22} \end{pmatrix} \begin{pmatrix} y_i \\ \left(\dfrac{dy}{dx}\right)_i \end{pmatrix} \qquad (15.92)$$

where the position and gradient have been combined as a two-dimensional vector, and M_{ij} are elements of the matrix $\mathbf{M}$, which depend on the components of the optical system and their location.

In order to obtain $\mathbf{M}$ for a given system, we first consider the individual optical components separately, deduce a matrix for each of them, and

---

† There is no particular reason why we should write the first element of our two component vector as y, rather than dy/dx. If we use the alternative convention, then the matrix $\mathbf{M}$ for each optical component would be replaced by $\mathbf{M}'$, such that $M'_{ij} = M_{3-i,3-j}$.

then calculate **M** by multiplying the various individual matrices. We adopt the convention that the ray is travelling in the positive x direction.

The effect of a thin lens (see fig. 15.8(a)) is, to a good approximation, to change the direction of a ray but not its position. The standard formula for thin lenses is

$$\frac{1}{u} + \frac{1}{v} = \frac{1}{f} \tag{15.93}$$

where f is the focal length, and u and v are respectively the distances from the lens of an object and its image. On multiplying the equation by y and being careful about sign conventions, we deduce that the effect of the lens is to change the gradient of an incident ray by a fixed amount y/f, independent of the initial gradient. Thus the transformation equations are

$$\left. \begin{aligned} y_f &= y_i \\ \text{and} \quad \left(\frac{dy}{dx}\right)_f &= \left(\frac{dy}{dx}\right)_i - \frac{y_i}{f} \end{aligned} \right\} \tag{15.94}$$

and hence the matrix for a thin lens is

$$\mathbf{M}_l = \begin{pmatrix} 1 & 0 \\ -1/f & 1 \end{pmatrix} \tag{15.95}$$

A somewhat similar situation occurs if the light ray passes from one medium of refractive index μ_1 to another with μ_2. Then

$$\left. \begin{aligned} y_f &= y_i \\ \text{and} \quad \left(\frac{dy}{dx}\right)_f \mu_2 &= \left(\frac{dy}{dx}\right)_i \mu_1 \end{aligned} \right\} \tag{15.96}$$

where the second equation is Snell's Law in the small angle approximation. Thus the matrix for a boundary is

$$\mathbf{M}_b = \begin{pmatrix} 1 & 0 \\ 0 & \mu_1/\mu_2 \end{pmatrix} \tag{15.97}$$

The refractive indices (and the focal length of the lens in $\mathbf{M}_l$) could depend on the colour of the light.

As a final example, a light ray travelling a distance d in a uniform medium (or in empty space) results in

$$\left. \begin{aligned} y_f &= y_i + d\left(\frac{dy}{dx}\right)_i \\ \text{and} \quad \left(\frac{dy}{dx}\right)_f &= \left(\frac{dy}{dx}\right)_i \end{aligned} \right\} \tag{15.98}$$

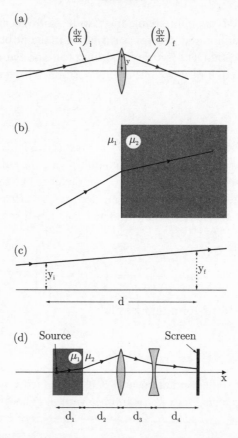

Fig. 15.8. Matrices and optical systems: (a)–(c) show the way simple optical elements affect the displacement y and direction dy/dx of a ray travelling from left to right: (a) A thin lens: y is unaltered but the direction is changed according to eqn (15.94). For a convex lens, the focal length f is positive and the gradient of a ray with positive y is reduced. (b) The boundary between two media with different indices of refraction μ: again y is unaltered, and dy/dx is changed according to eqn (15.96). (c) A region of empty space (or constant index of refraction): the direction of the ray is unaltered, but y changes (see eqns (15.98)). (d) An optical system consisting of a boundary between two regions, two lenses, and regions of empty space: the overall effect of the system is found by multiplying the matrices describing the separate components, as in eqn (15.100).

so that the required matrix is

$$\mathbf{M}_s = \begin{pmatrix} 1 & d \\ 0 & 1 \end{pmatrix} \qquad (15.99)$$

We now consider the optical system of fig. 15.8(d), consisting of:

(i) a light source at $x = 0$;
(ii) the region up to $x = d_1$ in water, the remainder being in air;
(iii) two thin lenses, one convex and the other concave, of focal lengths f_1 and $-f_2$ at $x = d_1 + d_2$ and $d_1 + d_2 + d_3$ respectively; and
(iv) a screen for recording the light at a distance d_4 beyond the second lens.

The net effect of this composite system is given by eqn (15.92), with

$$\mathbf{M} = \begin{pmatrix} 1 & d_4 \\ 0 & 1 \end{pmatrix} \begin{pmatrix} 1 & 0 \\ 1/f_2 & 1 \end{pmatrix} \begin{pmatrix} 1 & d_3 \\ 0 & 1 \end{pmatrix} \begin{pmatrix} 1 & 0 \\ -1/f_1 & 1 \end{pmatrix}$$
$$\times \begin{pmatrix} 1 & d_2 \\ 0 & 1 \end{pmatrix} \begin{pmatrix} 1 & 0 \\ 0 & \mu \end{pmatrix} \begin{pmatrix} 1 & d_1 \\ 0 & 1 \end{pmatrix} \qquad (15.100)$$

It is important to remember to write the matrix corresponding to the first optical component at the right-hand end of the matrix product; for our convention of the ray's direction, the matrices thus appear in the reverse order from the corresponding physical components. Another easily forgotten point is that it is necessary to include a matrix for each region of empty space.

The matrix multiplication is a little tedious, but at least it provides a defined procedure for determining the effect of the system. In more realistic situations, the matrices would be more complicated, and the multiplications would be performed by computer.

Once the matrix $\mathbf{M}$ has been determined for the system, it can be used to obtain the positions at which a bunch of rays diverging from a point source at $x = 0$ will hit the screen; it also gives their divergence after they pass through a hole in the screen. This approach is also capable of dealing with a source with a small extension in y.

If it is found that our optical system is in some way unsatisfactory (e.g. the image on the screen is too much out of focus; or it is too demagnified; or the outgoing beam has too large a divergence; etc.), it is very easy to investigate the effect of modifying it by changing the appropriate matrices. Thus if the second lens is moved to the right, all that it is necessary is to decrease d_4 and increase d_3 appropriately, reperform the matrix multiplications, and see whether the new matrix product corresponds to better behaviour. Alternatively an extra component could be included in the optical system, which would involve enlarging the matrix product (15.100) appropriately. In any case, especially with computer multiplications, this procedure greatly reduces the work of investigating the performance of an optical system. Of course it does not obviate the need eventually to test the actual arrangement, but if the matrices have provided a good approximation to the properties of the components, the complete system should behave very closely to what is predicted.

Real optical systems are, of course, three-dimensional. A ray then also needs z and dz/dx to describe it, so that its vector has four components, and the corresponding matrices are of size 4×4.

A very useful variant of this application is to high energy charged particles emerging from an accelerator. The aim is to make a beam of particles of a given momentum (equivalent to colour in the optical analogy), perhaps to bend them round some obstacles in the way, and to focus them into our apparatus. Magnets, quadrupoles and collimators play the rôle of prisms, lenses and slits. Again it is far preferable to perform initial studies on whether the system will do what is needed by multiplying a few matrices in a computer, rather than moving around massive components and using lots of beam time at the accelerator.

15.8.2 *Electrical networks*

Fig. 15.9 shows a simple electrical network, containing three impedances Z_A, Z_B and Z_C. The input current and voltage on the left are I_1 and V_1. We can determine I_2 and V_2, the current and voltage on the output side, since by Kirchhoff's laws

$$\left. \begin{array}{l} V_2 = V_1 - I_1 Z_A - I_2 Z_B \\ \text{and} \quad V_1 = I_1 Z_A + (I_1 - I_2) Z_C \end{array} \right\} \tag{15.101}$$

These are readily solved for V_2 and I_2 in terms of V_1 and I_1 yielding

$$\left. \begin{array}{l} V_2 = V_1(1 + Z_B/Z_C) + I_1(-Z_A - Z_B(Z_A + Z_C)/Z_C) \\ \text{and} \quad I_2 = -V_1/Z_C + I_1(Z_A + Z_C)/Z_C \end{array} \right\} \tag{15.102}$$

These can be written in matrix notation as†

$$\begin{pmatrix} V_2 \\ I_2 \end{pmatrix} = \begin{pmatrix} 1 + Z_B/Z_C & -Z_A - Z_B(Z_A + Z_C)/Z_C \\ -1/Z_C & (Z_A + Z_C)/Z_C \end{pmatrix} \begin{pmatrix} V_1 \\ I_1 \end{pmatrix} \tag{15.103}$$

The 2×2 matrix containing the impedances thus describes the effect of the circuit on an input voltage and current.

This in itself is not a great advantage, but it can be useful if many similar networks are linked together. The overall effect of the composite network is then obtained simply by multiplying the relevant set of individual matrices.

† As in the optical example of the previous section, it is a matter of free choice whether we write I or V as the first element of our vector.

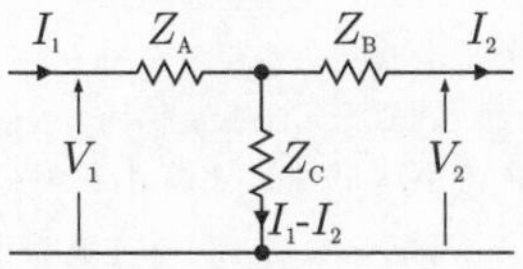

Fig. 15.9. Matrices and electric networks. The relations between the output voltage and current $(V_2$ and $I_2)$ and the input ones $(V_1$ and $I_1)$ are given by the matrix in eqn (15.103).

15.8.3 Anisotropic media

Many media are isotropic, which means that their properties are independent of direction within the medium. Thus, for example, if we consider the electric polarisation **p** produced by an electric field **e** in such a medium,

$$p_x = ke_x \tag{15.104}$$

for a field applied along the x direction, and similarly

$$\left. \begin{array}{l} p_y = ke_y \\ \text{and} \quad p_z = ke_z \end{array} \right\} \tag{15.104'}$$

for fields parallel to the other axes. The relevant point is that k is the *same* constant for any direction. As a result of this, if we apply a field in an arbitrary direction **e**, then

$$\mathbf{p} = k\mathbf{e} \tag{15.105}$$

where k is the same constant of proportionality as before, and **p** is parallel to **e**.

In contrast, for an anisotropic medium, the properties are direction-dependent. For some such materials, there are three independent directions known as the principal axes for which the polarisation is parallel to the electric field, but with different constants of proportionality for the three directions. If these axes are mutually perpendicular and aligned with the coordinate axes, then

$$\left. \begin{array}{l} p_x = k_1 e_x \\ p_y = k_2 e_y \\ \text{and} \quad p_x = k_3 e_z \end{array} \right\} \tag{15.106}$$

Because the ks are not all equal, however, for a general **e** given by

$$\mathbf{e} = e_x\mathbf{i} + e_y\mathbf{j} + e_z\mathbf{k}$$

the resulting polarisation will be

$$\mathbf{p} = k_1 e_x \mathbf{i} + k_2 e_y \mathbf{j} + k_3 e_z \mathbf{k} \qquad (15.107)$$

Thus, in general, $\mathbf{p}$ will not be parallel to $\mathbf{e}$. Instead it is related to it by

$$\begin{pmatrix} p_x \\ p_y \\ p_z \end{pmatrix} = \begin{pmatrix} k_1 & 0 & 0 \\ 0 & k_2 & 0 \\ 0 & 0 & k_3 \end{pmatrix} \begin{pmatrix} e_x \\ e_y \\ e_z \end{pmatrix} \qquad (15.108)$$

or

$$\mathbf{p} = \mathbf{K}\mathbf{e} \qquad (15.108')$$

where $\mathbf{K}$ is a 3×3 matrix rather than simply a constant as before. This matrix thus describes the anisotropy of the medium, and it is its presence in eqn (15.108') which results in $\mathbf{p}$ not being parallel to $\mathbf{e}$.

The matrix $\mathbf{K}$ in eqn (15.108) is fairly trivial, with all the off-diagonal elements being zero. This arose because our material had its principal axes parallel to the coordinate axes. For a more general orientation, the matrix would have looked more interesting with non-zero off-diagonal elements too. The general form of eqn (15.108') would still have applied, again implying non-parallel $\mathbf{p}$ and $\mathbf{e}$. (Indeed we would not expect this property to be affected simply by our arbitrary choice of the coordinate system.)

Section 15.9.5.1.2 describes the effect on a matrix like $\mathbf{K}$ in eqn (15.108) when the axes used for defining the problem are rotated; this will change it from its diagonal form and make the off-diagonal elements non-zero. In Section 15.9.5.1.1, we consider a very similar situation, but with the special directions not being orthogonal to each other. When the polarisation is described with respect to the standard x, y and z axes, the matrix $\mathbf{K}$ is not diagonal, and not even symmetric. We will also discover how, from its non-diagonal matrix, to find the special directions in which the properties of the anisotropic material are particularly simple (see Chapter 16).

15.8.4 Error matrices

In experimental sciences, it is very important to quote the accuracy of any experimental measurement, e.g. $x \pm \sigma_x$. If a particular experiment measures, say, three quantities x_1, x_2 and x_3, not only the errors on each of them are relevant, but also the three possible pairwise correlations among their uncertainties. These can be combined into a symmetric 3×3

matrix

$$E = \begin{pmatrix} \sigma_1^2 & \text{cov}(1,2) & \text{cov}(1,3) \\ \text{cov}(1,2) & \sigma_2^2 & \text{cov}(2,3) \\ \text{cov}(1,3) & \text{cov}(2,3) & \sigma_3^2 \end{pmatrix} \tag{15.109}$$

where the elements of E are the individual errors squared (along the diagonal), and the covariances $\text{cov}(i,j)$ which express the correlations; in the absence of correlations, these off-diagonal terms are zero.

If we change variables from x_1, x_2, x_3 to a new set u_1, u_2, u_3, the errors and correlations on the new variables can also be combined into an error matrix F. Provided that the errors are small, E and F are related by the matrix equation

$$F = \tilde{T}ET \tag{15.110}$$

Here T is a matrix connected with the transformation between the two sets of variables, with elements

$$T_{ij} = \frac{\partial u_j}{\partial x_i} \tag{15.111}$$

and $\tilde{T}$ is the transpose of the matrix T.

Error matrices can also be used for calculating the accuracy of a new variable y which is defined as a function of the previous set x_1, x_2, x_3 (or u_1, u_2, u_3) e.g.

$$y^2 = x_1^2 + x_2^2 + x_3^2$$

Manipulations with error matrices can be performed by standard matrix multiplications, without worrying about the details of what happens to individual errors and covariances at each stage. They thus greatly simplify error calculations in situations where more than one variable is involved.

15.9 Matrix properties

In the final section of this chapter, we state and/or derive various properties about matrices, which are useful for dealing with eigenvalues (see the next chapter) and for other matrix manipulations.

15.9.1 Simple definitions

The transpose of a matrix is obtained by interchanging its rows and columns. It is represented by putting a tilde over the matrix. From its definition, the elements of any matrix $\mathbf{M}$ and its transpose $\tilde{\mathbf{M}}$ are related by

$$\tilde{\mathbf{M}}_{ij} = \mathbf{M}_{ji} \tag{15.112}$$

For example

$$\widetilde{\begin{pmatrix} 1 & 0 \\ 3 & -2 \end{pmatrix}} = \begin{pmatrix} 1 & 3 \\ 0 & -2 \end{pmatrix}$$

It is readily shown that the transpose of a product $\mathbf{AB}$ is given by

$$\widetilde{\mathbf{AB}} = \tilde{\mathbf{B}}\tilde{\mathbf{A}} \tag{15.113}$$

This is similar to the rule for the inverse of a product

$$(\mathbf{AB})^{-1} = \mathbf{B}^{-1}\mathbf{A}^{-1} \tag{15.114}$$

Hints on the derivation of these are given in Problem 15.6.

A symmetric matrix is a square matrix whose elements occurring symmetrically about its leading diagonal (the one starting with the $(1,1)$ element) are equal, i.e.

$$\mathbf{S}_{ij} = \mathbf{S}_{ji} \tag{15.115}$$

or

$$\mathbf{S} = \tilde{\mathbf{S}} \tag{15.115'}$$

Such matrices are very common in physical applications.

A diagonal matrix is one whose only non-zero elements are along its leading diagonal. Examples are $\mathbf{M}_1$ to $\mathbf{M}_4$, $\mathbf{M}_8$ and the identity matrix $\mathbf{I}$. It is particularly simple to invert a diagonal matrix, or to determine its eigenvalues.

A matrix such as

$$\left(\begin{array}{cc|c} 1 & 3 & 0 \\ -4 & 2 & 0 \\ \hline 0 & 0 & 6 \end{array}\right)$$

is said to be block diagonal. This is because it can be partitioned by a vertical line and a horizontal one such that the off-diagonal sub-matrices contain only zeroes. If a block diagonal matrix is used to relate two vectors, the components corresponding to the separate parts of the matrix are completely decoupled from each other.† As far as matrix

---

† For example, the matrix $\mathbf{M}_{11}$ of eqn (15.79), corresponding to a rotation about the z axis, mixes the x and y coordinates, but leaves z unaffected — see eqns (15.78).

operations are concerned (e.g. finding the inverse), the matrix can almost be regarded as separate independent sub-matrices, in the above case 2×2 and 1×1.

15.9.2 Products of vectors

The scalar product of two vectors **a** and **b** is written in matrix notation as $\tilde{\mathbf{a}}\mathbf{b}$, where **a** and **b** are now single column matrices. Then $\tilde{\mathbf{a}}\mathbf{a}$ is simply the square of the length of the vector **a**.

The vector product $\mathbf{a} \wedge \mathbf{b}$ can be written in matrix notation as

$$\mathbf{Mb} = \begin{pmatrix} 0 & -a_z & a_y \\ a_z & 0 & -a_x \\ -a_y & a_x & 0 \end{pmatrix} \begin{pmatrix} b_x \\ b_y \\ b_z \end{pmatrix} \tag{15.116}$$

where the form of the matrix **M** in the above equation is chosen so as to ensure that the product gives the correct components for $\mathbf{a} \wedge \mathbf{b}$.

An extension of the scalar product arises with the result of multiplying two square 3×3 matrices, where we regard each as consisting of three vectors, i.e.

$$\tilde{\mathbf{P}}\mathbf{S} = \begin{pmatrix} p_1 & p_2 & p_3 \\ q_1 & q_2 & q_3 \\ r_1 & r_2 & r_3 \end{pmatrix} \begin{pmatrix} s_1 & t_1 & u_1 \\ s_2 & t_2 & u_2 \\ s_3 & t_3 & u_3 \end{pmatrix}$$

$$= \begin{pmatrix} \mathbf{p} \cdot \mathbf{s} & \mathbf{p} \cdot \mathbf{t} & \mathbf{p} \cdot \mathbf{u} \\ \mathbf{q} \cdot \mathbf{s} & \mathbf{q} \cdot \mathbf{t} & \mathbf{q} \cdot \mathbf{u} \\ \mathbf{r} \cdot \mathbf{s} & \mathbf{r} \cdot \mathbf{t} & \mathbf{r} \cdot \mathbf{u} \end{pmatrix} \tag{15.117}$$

Note that the matrices **P** and **S** are both written with each vector occurring as a column. However, eqn (15.117) involves $\tilde{\mathbf{P}}$, so the components of each vector there occupy a row.

For the special case of the matrix product $\tilde{\mathbf{S}}\mathbf{S}$, and where **s**, **t** and **u** happen to be orthogonal, the final matrix is diagonal. If **s**, **t** and **u** are also unit vectors, then

$$\tilde{\mathbf{S}}\mathbf{S} = \mathbf{I} \tag{15.118}$$

From the definition of an inverse, it then follows that

$$\tilde{\mathbf{S}} = \mathbf{S}^{-1}$$

These results will be of use in Section 15.9.4 in connection with rotation matrices.

15.9.3 *Matrix equations*

If we are confronted with the matrix equation

$$\mathbf{AB} = \mathbf{AC} \tag{15.119}$$

are we allowed to deduce that

$$\mathbf{B} = \mathbf{C}? \tag{15.120}$$

We might guess, in analogy with products of vectors (where, for example, $\mathbf{a} \wedge \mathbf{b} = \mathbf{a} \wedge \mathbf{c}$ does not in general imply that $\mathbf{b} = \mathbf{c}$ — see Section 3.3.5), that this conclusion was unjustified.

However, it would appear that in order to show that $\mathbf{B} = \mathbf{C}$, all that is necessary is to premultiply eqn (15.119) by $\mathbf{A}^{-1}$. The only situations in which this cannot be done is when $\mathbf{A}$ either is not square, or is singular; in either case, $\mathbf{A}^{-1}$ does not exist.

An example of eqn (15.120) not being valid is

$$\underset{\mathbf{A}}{(1-1)} \quad \underset{\mathbf{B}}{\begin{pmatrix} 6 \\ 3 \end{pmatrix}} = (3) = \underset{\mathbf{A}}{(1-1)} \quad \underset{\mathbf{C}}{\begin{pmatrix} 4 \\ 1 \end{pmatrix}}$$

Here $\mathbf{A}$ does not have an inverse because it is not square. Another failure is

$$\underset{\mathbf{A}}{\begin{pmatrix} 1 & 2 \\ 2 & 4 \end{pmatrix}} \quad \underset{\mathbf{B}}{\begin{pmatrix} 1 \\ 1 \end{pmatrix}} = \begin{pmatrix} 3 \\ 6 \end{pmatrix} = \underset{\mathbf{A}}{\begin{pmatrix} 1 & 2 \\ 2 & 4 \end{pmatrix}} \quad \underset{\mathbf{C}}{\begin{pmatrix} 3 \\ 0 \end{pmatrix}}$$

In this case, $\mathbf{B} \neq \mathbf{C}$ because $\det(\mathbf{A}) = 0$ and $\mathbf{A}$ is thus singular.

Given that scalar and vector products can be represented by matrices, we might wonder why the result (15.120) does not apply to these products. The reason is that in scalar products, the matrices representing the vectors are not square; while the matrix describing a vector product (see eqn (15.116)) is singular.

15.9.4 *Rotation matrix*

A rotation, represented by a matrix $\mathbf{R}$, has the property that it leaves the length of any vector $\mathbf{v}$ unchanged, i.e.

$$\tilde{\mathbf{v}}'\mathbf{v}' = \tilde{\mathbf{v}}\mathbf{v} \tag{15.121}$$

But

$$\left. \begin{array}{l} \mathbf{v}' = \mathbf{Rv} \\ \text{and} \quad \tilde{\mathbf{v}}' = \tilde{\mathbf{v}}\tilde{\mathbf{R}} \end{array} \right\}$$

so that

$$\tilde{\mathbf{v}}\tilde{\mathbf{R}}\mathbf{R}\mathbf{v} = \tilde{\mathbf{v}}\mathbf{v}$$

$$= \tilde{\mathbf{v}}\mathbf{I}\mathbf{v} \tag{15.122}$$

where in the last step we can introduce the identity matrix $\mathbf{I}$ without affecting the equation.

From this, we deduce that

$$\tilde{\mathbf{R}}\mathbf{R} = \mathbf{I} \tag{15.123}$$

This is not because we multiply eqn (15.122) by $\tilde{\mathbf{v}}^{-1}$ on the left and by $\mathbf{v}^{-1}$ on the right; this is impossible because $\mathbf{v}$ is a column vector, and only square matrices have inverses. However, eqn (15.122) is true for all possible vectors $\mathbf{v}$, which is sufficient to deduce (15.123).

Since $\mathbf{R}^{-1}$, the inverse of $\mathbf{R}$, is defined by

$$\mathbf{R}^{-1}\mathbf{R} = \mathbf{I}$$

comparison with eqn (15.123) yields

$$\mathbf{R}^{-1} = \tilde{\mathbf{R}} \tag{15.124}$$

That is, to find the inverse of a rotation matrix, we merely have to take its transpose.

Eqn (15.123) can be written in component form as

$$\sum_k \tilde{R}_{ik} R_{kj} = \sum_k R_{ki} R_{kj} = \delta_{ij} \tag{15.123'}$$

This means that if we take two different columns of a rotation matrix and multiply the corresponding elements together in pairs, their sum will be zero; while if we sum the squares of the elements of a given column (i.e. $i = j$ in the above equation), the answer is unity. This property remains true if the word "column" is everywhere replaced by "row".

If we compare these results with those in Section 15.9.2 above, we see that this corresponds to the columns of $\mathbf{R}$ representing orthogonal vectors of unit length. Again this is not surprising. The operations

$$\mathbf{R} \begin{pmatrix} 1 \\ 0 \\ 0 \end{pmatrix}, \quad \mathbf{R} \begin{pmatrix} 0 \\ 1 \\ 0 \end{pmatrix} \quad \text{and} \quad \mathbf{R} \begin{pmatrix} 0 \\ 0 \\ 1 \end{pmatrix} \tag{15.125}$$

give the effect of the rotation on mutually orthogonal unit vectors along the x, y and z axes, and they remain so after the rotation. But the products in (15.125) respectively pick out the first, second and third columns of **R**. These columns thus give the new positions of these unit vectors with respect to the old x, y and z directions; that is why these columns represent mutually orthogonal normalised vectors.

A matrix which satisfies (15.123) is said to be orthogonal.

15.9.5 Transformations of matrices

Almost the whole of this chapter has been devoted to the influence of a transformation **T** on a vector **v**. We now derive the effect of such a transformation on a matrix **N**. We might guess that since a vector is changed according to

$$\mathbf{v}' = \mathbf{Tv} \qquad (15.126)$$

the result would be

$$\mathbf{N}' = \mathbf{TN}$$

This is wrong. The correct answer depends on the way the matrix is used.

15.9.5.1 Matrix relation between vectors

We consider first the case where the matrix **N** describes what happens to, say, a vector **y**, i.e.

$$\mathbf{z} = \mathbf{Ny} \qquad (15.127)$$

After the transformation **T**, this equation becomes

$$\mathbf{z}' = \mathbf{N}'\mathbf{y}' \qquad (15.127')$$

Then using (15.126) for both **y** and **z** yields

$$\mathbf{z}' = \mathbf{Tz} = \mathbf{TNy} = \mathbf{TNT}^{-1}\mathbf{Ty} = \mathbf{TNT}^{-1}\mathbf{y}'$$
$$= \mathbf{N}'\mathbf{y}' \qquad (15.128)$$

The last equality gives

$$\mathbf{N}' = \mathbf{TNT}^{-1} \qquad (15.129)$$

and this shows the way **N** alters as the result of such a transformation.

15.9.5.1.1 Electric polarisability

As an example to illustrate this, we return to the material with electric polarisability as discussed in Section 15.8.3, and assume that it has "special directions"

$$\left.\begin{array}{c} \mathbf{v}_1 = (1,0,0) \\ \mathbf{v}_2 = (1,1,0)/\sqrt{2} \\ \text{and} \quad \mathbf{v}_3 = (0,0,1) \end{array}\right\} \tag{15.130}$$

along which the polarisation $\mathbf{p}$ is parallel to the electric field $\mathbf{e}$, and with constants of proportionality μ_1, μ_2 and μ_3 respectively, i.e.

$$\mathbf{p} = \begin{pmatrix} \mu_1 & 0 & 0 \\ 0 & \mu_2 & 0 \\ 0 & 0 & \mu_3 \end{pmatrix} \mathbf{e}$$

with respect to axes along the directions $\mathbf{v}_1$, $\mathbf{v}_2$ and $\mathbf{v}_3$. This matrix is equivalent to $\mathbf{N}$ in eqn (15.127). In this example, the "special directions" are not mutually perpendicular.

From eqns (15.130), the $\mathbf{v}_3$ axis is perpendicular to the other two, which are in the x–y plane and separated by 45°. The relations between coordinates (or between components of $\mathbf{e}$ (or $\mathbf{p}$)) in these two systems are then

$$\left.\begin{array}{c} x = v_1 + v_2/\sqrt{2} \\ y = v_2/\sqrt{2} \\ \text{and} \quad z = v_3 \end{array}\right\} \tag{15.131}$$

These can be derived by examining fig. 15.10, or by using the standard method for expressing a vector in terms of three non-orthogonal ones (see Section 3.6). The inverse relations are

$$\left.\begin{array}{c} v_1 = x - y \\ v_2 = \sqrt{2}y \\ \text{and} \quad v_3 = z \end{array}\right\} \tag{15.132}$$

The matrix describing the transformation (15.131)

$$\mathbf{r} = \mathbf{M}\mathbf{v}$$

is

$$\mathbf{M} = \begin{pmatrix} 1 & 1/\sqrt{2} & 0 \\ 0 & 1/\sqrt{2} & 0 \\ 0 & 0 & 1 \end{pmatrix} \tag{15.133}$$

Thus $\mathbf{M}$ is not a rotation matrix (in that it does not satisfy eqn (15.124)); indeed we do not expect it to be (see fig. 15.10).

If we now use eqn (15.129) to express the polarisability matrix with

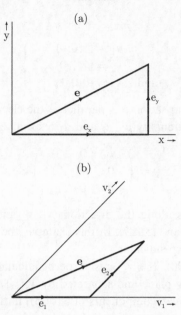

Fig. 15.10. An anisotropic medium has special directions given by $(1,0,0)$ and $(1,1,0)/\sqrt{2}$; the third direction along the z axis decouples from the other two in the x–y plane and is ignored. The electric field **e** can be thought of as being composed of either e_1 and e_2 along the special directions; or e_x and e_y along the axes. They are related by eqns (15.135). The matrix corresponding to this transformation allows us to convert the diagonal polarisability matrix **N** in the v_1, v_2 system into the non-symmetric matrix **N'** in the x, y system. The eigenvectors of **N'** then give the special directions v_1 and v_2 (see Section 16.4.4).

respect to the standard xyz coordinate axes, we obtain

$$\mathbf{N'} = \mathbf{MNM}^{-1}$$
$$= \begin{pmatrix} \mu_1 & -\mu_1 + \mu_2 & 0 \\ 0 & \mu_2 & 0 \\ 0 & 0 & \mu_3 \end{pmatrix} \qquad (15.134)$$

i.e. it is no longer diagonal, and is not even symmetric. In this particular case, the matrix is block diagonal, with the v_3 or z component being decoupled from the other two.

With a bit of effort, we can see that the matrix **N'** in (15.134) gives the correct description of the system in the xyz axes. Since the third components decouple, we can ignore them below and concentrate on the first two.

We can apply eqns (15.131) to the electric field to write

$$e_x = e_1 + e_2/\sqrt{2} \atop e_y = e_2/\sqrt{2} \Bigg\}$$ (15.135)

The corresponding components of $\mathbf{p}$ are given by multiplying each of the e_1, e_2 (and e_3) contributions by its appropriate μ (since for these directions $\mathbf{p}$ is proportional to $\mathbf{e}$). Then

$$\text{and} \quad {p_x = \mu_1 e_1 + \mu_2 e_2/\sqrt{2} \atop p_y = \mu_2 e_2/\sqrt{2}} \Bigg\}$$

We finally need eqns (15.132) in order to write the $\mathbf{e}$ components in terms of e_x and e_y, rather than e_1 and e_2:

$$p_x = \mu_1(e_x - e_y) + \mu_2 e_y = \mu_1 e_x + (\mu_2 - \mu_1)e_y$$
$$p_y = \mu_2 e_y$$

Thus these relations between $\mathbf{p}'$ and $\mathbf{e}'$ in the $xy(z)$ system are exactly as given by the matrix relation†

$$\mathbf{p}' = \mathbf{N}'\mathbf{e}'$$

15.9.5.1.2 *Rotating a diagonal matrix*

As a second example, we consider the effect of a rotation $\mathbf{R}$ on a diagonal matrix $\mathbf{D}$. This becomes

$$\mathbf{D}' = \mathbf{R}\mathbf{D}\mathbf{R}^{-1}$$

which is no longer diagonal. It is, however, symmetric, because

$$\tilde{\mathbf{D}}' = \tilde{\mathbf{R}}^{-1}\tilde{\mathbf{D}}\tilde{\mathbf{R}}$$
$$= \mathbf{R}\mathbf{D}\mathbf{R}^{-1}$$
$$= \mathbf{D}'$$ (15.136)

To obtain the second line, we have twice made use of the fact that $\tilde{\mathbf{R}} = \mathbf{R}^{-1}$ for a rotation matrix; and $\tilde{\mathbf{D}} = \mathbf{D}$ for a diagonal matrix. The fact that $\mathbf{D}'$ is symmetric is one of the reasons that symmetric matrices play an important rôle physically (see also Chapter 16).

The question then arises as to why the matrix $\mathbf{N}'$ (15.134) is not symmetric, given that $\mathbf{N}$ was diagonal. The reason is that the transformation matrix $\mathbf{T}$ in that case did not correspond to a rotation, in that the original

---

† Perhaps we should not be too surprised by this agreement because what we have just been doing is essentially to go through the derivation of the relation $\mathbf{N}' = \mathbf{M}\mathbf{N}\mathbf{M}^{-1}$ for our particular case. It does, however, give slightly more insight into the derivation of $\mathbf{N}'$ from $\mathbf{N}$, as compared with simply performing the necessary matrix operations.

non-orthogonal axes a, b, c were converted to orthogonal ones x, y, z. It is only for rotations that diagonal matrices always remain symmetric.

15.9.5.2 *Quadratic forms*

We now consider a second type of usage of a matrix, to examine its transformation properties. This is illustrated by the quadratic form considered in Section 8.3.3, e.g.

$$Ax^2 + 2Bxy + Cy^2 = 1 \qquad (8.22)$$

This can be written as

$$\tilde{\mathbf{r}}\mathbf{Q}\mathbf{r} = 1$$

where $\mathbf{Q}$ is a symmetric matrix, describing the quadratic form.

On transforming the axes such that

$$\mathbf{r}' = \mathbf{T}\mathbf{r} \qquad (15.137)$$

the equation becomes

$$
\begin{aligned}
1 &= (\widetilde{\mathbf{T}^{-1}\mathbf{r}'})\mathbf{Q}(\mathbf{T}^{-1}\mathbf{r}') \\
&= \tilde{\mathbf{r}}'(\tilde{\mathbf{T}}^{-1}\mathbf{Q}\mathbf{T}^{-1})\mathbf{r}' \\
&= \tilde{\mathbf{r}}'\mathbf{Q}'\mathbf{r}'
\end{aligned}
$$

$$\text{with} \quad \mathbf{Q}' = \tilde{\mathbf{T}}^{-1}\mathbf{Q}\mathbf{T}^{-1} \qquad (15.138)$$

In the special case where the transformation is a rotation, $\mathbf{T}$ is a rotation matrix and hence $\mathbf{T}^{-1} = \tilde{\mathbf{T}}$. Then

$$\mathbf{Q}' = \mathbf{T}\mathbf{Q}\mathbf{T}^{-1} \quad \text{(rotation only)} \qquad (15.139)$$

which happens to be the same as the earlier result (15.129) for the way a matrix relating two vectors transforms. However, the more general result (15.138) applies for any linear transformation on a quadratic form, e.g. for a stretching of the axes.

It is worth remembering that for a transformation that is not a rotation, the way a matrix transforms depends on the way in which the matrix is used.

Finally we note that eqns (15.129) and (15.139) can be used for finding out what happens to a matrix under a rotation that is either active or passive (see Section 2.2.3.1, for example). The only difference between these cases is that the transformation matrix $\mathbf{T}$ must be correctly chosen for the appropriate situation.

15.9.6 *Trace of a matrix*

The trace of a square matrix $\mathbf{M}$, denoted by $\mathrm{Tr}(\mathbf{M})$, is defined as the sum of the components along the leading diagonal. It has the property that under a rotation, the value of the trace does not change. This follows from the result that

$$\mathrm{Tr}(\mathbf{AB}) = \mathrm{Tr}(\mathbf{BA}) \tag{15.140}$$

since

$$\mathrm{Tr}(\mathbf{AB}) = \sum_i (\mathbf{AB})_{ii} = \sum_i \sum_j A_{ij} B_{ji}$$

while

$$\mathrm{Tr}(\mathbf{BA}) = \sum_j (\mathbf{BA})_{jj} = \sum_j \sum_i B_{ji} A_{ij}$$

After the rotation, the matrix $\mathbf{M}$ becomes $\mathbf{RMR}^{-1}$, whose trace is

$$\mathrm{Tr}(\mathbf{RMR}^{-1}) = \mathrm{Tr}(\mathbf{MR}^{-1}\mathbf{R}) \quad \text{(from (15.140))}$$
$$= \mathrm{Tr}(\mathbf{M}) \tag{15.141}$$

This proves the required result. It is useful for calculating the angle of rotation for a rotation matrix (see Section 16.4.2.2); and also as a simple check on a calculation of the effect of a rotation on a matrix $\mathbf{M}$.

Problems

15.1 Evaluate the following matrix products (where possible)

(i) $\begin{pmatrix} 1 & 2 \\ 3 & 4 \end{pmatrix}\begin{pmatrix} 4 & 3 \\ 2 & 1 \end{pmatrix}$ (ii) $\begin{pmatrix} 4 & 3 \\ 2 & 1 \end{pmatrix}\begin{pmatrix} 1 & 2 \\ 3 & 4 \end{pmatrix}$

(iii) $(1 \ 2)\begin{pmatrix} 3 \\ 4 \end{pmatrix}$ (iv) $\begin{pmatrix} 3 \\ 4 \end{pmatrix}(1 \ 2)$

(v) $\begin{pmatrix} 1 \\ 2 \end{pmatrix}(3 \ 4)$ (vi) $\begin{pmatrix} 1 & 2 \\ 3 & 4 \end{pmatrix}\begin{pmatrix} 5 \\ 6 \end{pmatrix}$

(vii) $\begin{pmatrix} 5 \\ 6 \end{pmatrix}\begin{pmatrix} 1 & 2 \\ 3 & 4 \end{pmatrix}$ (viii) $(5 \ 6)\begin{pmatrix} 1 & 2 \\ 3 & 4 \end{pmatrix}$

Find examples where $\mathbf{AB} \neq \mathbf{BA}$.

15.2 **A** and **B** are general 2×2 matrices

$$\begin{pmatrix} a_{11} & a_{12} \\ a_{21} & a_{22} \end{pmatrix} \text{ and } \begin{pmatrix} b_{11} & b_{12} \\ b_{21} & b_{22} \end{pmatrix}$$

By explicit multiplication of the matrices and then solving the four simultaneous equations for the elements b_{ij}, determine the matrix **B** such that $\mathbf{AB} = \mathbf{A}$. Is your solution as expected? Similarly determine **B** such that $\mathbf{AB} = \mathbf{I}$. Compare your answer with that expected from eqn (15.69)

15.3 Use conventional algebra to solve the simultaneous equations

$$a_1 x + b_1 y = d_1$$
$$a_2 x + b_2 y = d_2$$

Now convert them into a matrix equation, and write down the solution for the vector $\begin{pmatrix} x \\ y \end{pmatrix}$ as the product of a matrix and a vector. By explicit multiplication, show that this gives the same result as conventional algebra.

15.4 Obtain the inverse of the matrix **X** of fig. 15.5. Find the determinants of **X** and of its inverse, and check that they are reciprocals of each other. Calculate the matrix product $\mathbf{X}^{-1}\mathbf{X}$.

15.5 Find the matrix **R** for a rotation of $180°$ about the y axis (in three-dimensional space). Determine the inverse of **R**. Do your matrices **R** and $\mathbf{R}^{-1}$ have the right relationship to each other?

15.6 Show that for general matrices **A** and **B**,

$$(\mathbf{AB})^{-1} = \mathbf{B}^{-1}\mathbf{A}^{-1} \tag{a}$$

and

$$\widetilde{\mathbf{AB}} = \tilde{\mathbf{B}}\tilde{\mathbf{A}} \tag{b}$$

(Hints: For (a), consider the matrix equation that defines what $(\mathbf{AB})^{-1}$ is. For (b), write out the element $(\widetilde{\mathbf{AB}})_{ij}$ in terms of a sum of products involving elements of **A** and **B**, and then of $\tilde{\mathbf{A}}$ and $\tilde{\mathbf{B}}$.)

15.7 A vector **v** is rotated with angular velocity ω, where the direction of ω gives the axis of rotation. The rate of change of the vector (with respect to fixed axes) is given by the vector equation

$$\frac{d\mathbf{v}}{dt} = \omega \wedge \mathbf{v}$$

If this is written as a matrix equation

$$\frac{d\mathbf{v}}{dt} = \mathbf{M}\mathbf{v}$$

determine the components of the matrix $\mathbf{M}$. Evaluate the determinant of $\mathbf{M}$.

15.8 Convince yourself that the matrix $\mathbf{M}_{15}$ of eqn (15.82), with $\theta = \lambda = 90°$, agrees with the diagrams of the lower part of fig. 15.7.

15.9

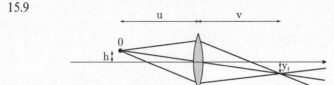

An optical system consists of a light source O at a distance u from a thin lens. The source is a height h above the axis of the lens. Rays leave O with variable directions dy/dx. Derive the matrix $\mathbf{M}$ which relates the height y and the direction dy/dx of a ray at a distance v beyond the lens to the corresponding quantities at O. Use this matrix to show that the position y_f of the rays at v is independent of the direction of the rays at O provided that

$$\frac{1}{u} + \frac{1}{v} = \frac{1}{f}$$

and that the vertical position of the rays at v is given by

$$y_f/h = 1 - v/f = -v/u$$

15.10 A matrix

$$\mathbf{N} = \begin{pmatrix} 1 & 0 & 2 \\ 0 & 0 & 0 \\ 0 & 0 & 1 \end{pmatrix}$$

relates two vectors $\mathbf{w}$ and $\mathbf{v}$ according to $\mathbf{w} = \mathbf{N}\mathbf{v}$. Another matrix

$$\mathbf{Q} = \begin{pmatrix} 1 & 0 & 0 \\ 0 & 1 & 0 \\ 0 & 0 & 1 \end{pmatrix}$$

is used to define a quadratic form $\tilde{\mathbf{r}}\mathbf{Q}\mathbf{r} = 1$.

The coordinates are now transformed according to $\mathbf{r}' = \mathbf{T}\mathbf{r}$,

Matrices

where

$$\mathbf{T} = \begin{pmatrix} \frac{1}{2} & 0 & 0 \\ 0 & 1 & 0 \\ 0 & 0 & 1 \end{pmatrix}$$

Determine the transformed matrices $\mathbf{N}'$ and $\mathbf{Q}'$ from eqns (15.129) and (15.138) respectively. Compare with simple calculations not using matrices for the relationship between $\mathbf{w}'$ and $\mathbf{v}'$, and for the quadratic form in the transformed axes.

16

Eigenvectors and eigenvalues

This chapter continues our discussion of matrices, and considers what are known as eigenvalues and eigenvectors. We start by discussing eigenvalue problems in general, and then turn to matrices in particular. By investigating both their mathematical properties and some of their physical applications, we should be able to achieve a good understanding of what eigenvalues and eigenvectors are.

16.1 The general eigenvalue problem

The eigenvalue idea is not restricted to matrices, but is of wider application. The most general formulation is to write

$$Fy = \lambda y \qquad (16.1)$$

where F is some sort of operator which is applied to y, and λ is a constant. The crucial point is that after operating on y with F, the result is proportional to y. The number λ is called the eigenvalue, and y is the eigenvector or eigenfunction, depending on the type of problem.

16.1.1 Differential equations

Before considering the matrix situation, we look at the case where F is a differential operator. A very simple example is

$$F = \frac{d^2}{dx^2}$$

so that the eigenvalue equation is

$$\frac{d^2y}{dx^2} = \lambda y \qquad (16.2)$$

For a general function $y(x)$, this equation will usually not be satisfied. For example, if $y = x^3$,

$$\frac{d^2y}{dx^2} = 6x \neq \lambda y$$

Thus $y = x^3$ is not an eigenfunction of this differential operator. On the other hand, $e^{\pm px}$, $\sin kx$ and $\cos lx$ are all eigenfunctions, with eigenvalues λ of p^2, $-k^2$ and $-l^2$ respectively.

Such eigenfunction problems are usually combined with boundary conditions for y, e.g.

$$y = 0 \text{ at } x = 0 \text{ and at } x = \pi \tag{16.3}$$

This could, for example, represent the displacement on a stretched string which is held at both ends. The eigenvalue equation then provides eigenfunctions $y(x)$ for possible displacements of the string, which vibrate at a single frequency ω; the eigenvalue is related to the value of ω.

In this case, suitable choices for $y(x)$ are restricted to $\sin kx$, where k is any positive integer. There are thus an infinite number of eigenfunctions, each of them being a sine wave with a different wave number and hence having a different eigenvalue. Since our eigenvalue equation (16.2) is linear in y, a solution $\sin kx$ can be multiplied by any constant, and it still remains an eigenfunction, with an unchanged eigenvalue. (This corresponds to the fact that the frequency of a wave on the string is independent of its amplitude.) Thus the eigenfunctions are

$$y_k(x) = A_k \sin kx \quad k = 1, 2, \ldots \tag{16.4}$$

Provided $k \neq k'$, these different eigenfunctions satisfy the equation

$$\int_0^\pi y_k(x) y_{k'}(x) dx = 0 \tag{16.5}$$

where the end-points of the region of integration correspond to where the boundary conditions (16.3) are imposed. The functions $y_k(x)$ and $y_{k'}(x)$ are thus called "orthogonal", as a somewhat extrapolated use of the term from the fact that the scalar product of two orthogonal vectors **a** and **b** is zero, i.e.

$$\sum_{i=1}^3 a_i b_i = 0$$

where the sum extends over the three components of each of the vectors.

For $k = k'$, eqn (16.5) clearly does not hold, since the integrand is now

$y_k^2(x)$ and hence is always positive. The magnitude of such an integral depends on the arbitrary choice of A_k. If they are chosen as $\sqrt{2/\pi}$, then

$$\int_0^\pi y_k^2(x)dx = 1 \tag{16.6}$$

and the $y_k(x)$ are described as being normalised. Eqns (16.5) and (16.6) can be combined as

$$\int_0^\pi y_k(x)y_{k'}(x)dx = \delta_{kk'} \tag{16.7}$$

where the Kronecker symbol is defined as

$$\delta_{kk'} = \begin{cases} 0 \text{ for } k \neq k' \\ 1 \text{ for } k = k' \end{cases}$$

Eqn (16.7) then applies, whether or not k and k' are equal.†

Another property of the eigenfunctions is that

$$\int_0^\pi y_k(x)Fy_{k'}(x)dx = \delta_{kk'}\lambda_k \tag{16.8}$$

where F is the differential operator, and λ_k is the eigenvalue corresponding to the eigenfunction $y_k(x)$. This follows directly from (16.1) and (16.7).

The set of functions $y_k(x)$ is complete in the sense that any general function $\phi(x)$ that is not an eigenfunction (but which satisfies the boundary conditions) can be expanded in terms of the $y_k(x)$, i.e.

$$\phi(x) = \Sigma b_k y_k(x) \tag{16.9}$$

where the constant coefficients b_k are obtained by

$$b_k = \int_0^\pi \phi(x)y_k(x)dx$$

For our particular example where the $y_k(x)$ are normalised sine waves, this procedure is simply that of Fourier series.

The above properties of the eigenfunctions have direct analogies with those of the eigenvectors in the corresponding matrix problem.

---

† For more complicated differential operators, a degenerate situation can exist where different eigenfunctions can have the same eigenvalue, i.e. $\lambda_k = \lambda_{k'}$, for $k \neq k'$. Then any linear combination of the degenerate eigenfunctions is also an eigenfunction. By exploiting this arbitrariness, we can choose suitable linear combinations such that eqn (16.7) remains true, even in the degenerate case. (Compare the discussion at the end of Section 16.3.2.1, for the corresponding situation with matrices.)

16.1.2 An example: Schrödinger's Equation

In quantum mechanics, an electron is described by a wave function $\psi(x, y, z)$, such that the probability of observing the electron within a small volume δV at some specific location is $|\psi|^2 \delta V$. The wave function of an electron with energy E is given by the solution $\psi(x, y, z)$ of Schrödinger's Equation, which is

$$\left[-\frac{h^2}{8\pi^2 m} \left(\frac{\partial^2}{\partial x^2} + \frac{\partial^2}{\partial y^2} + \frac{\partial^2}{\partial z^2} \right) + V \right] \psi = E\psi \qquad (16.10)$$

Here h is Planck's constant; m is the mass of the electron; and $V(x, y, z)$ is the potential in which an electron finds itself (e.g. $-e^2/4\pi\varepsilon_0 r$ for a hydrogen atom, where e is the electron's charge and $r = \sqrt{x^2 + y^2 + z^2}$ is its distance from the proton at the centre of the atom).

Eqn (16.10) is thus an eigenvalue equation, with the solutions being the eigenfunctions ψ_k and the eigenvalues the corresponding energies E_k. According to the philosophy behind this equation, the only allowed states of the electron in a hydrogen atom are those which are eigenfunctions of Schrödinger's Equation, and the eigenvalues give the energies of these states. This then explains the discrete energy spectra observed in atomic physics, and gives good agreement with the actual values of the energies in a simple case like hydrogen. (For larger atoms, it does not give wrong answers. It is just that multibody problems are much more difficult to solve accurately.)

Not only are the energies given, but so are the wave functions and hence the probability distributions. This enables other atomic properties to be calculated (e.g. atomic sizes; which decay processes are suppressed; etc.), thereby making the theory more powerful.

For further discussion of Schrödinger's Equation, see also Sections 11.3.5 and 11.5.4; quantum mechanics is also mentioned in Sections 12.12.4.3 and 14.10.4.

16.2 Matrices: finding eigenvalues and eigenfunctions

We now turn to the case of matrices. An eigenvector **r** of a matrix **M** is given by

$$\mathbf{Mr} = \lambda\mathbf{r} \qquad (16.11)$$

where λ is the corresponding eigenvalue.

In general, operating on a vector **q** with **M** results in a new vector **q'** which is not parallel to **q**. It is only for eigenvectors that the result is parallel to the original vector. λ gives the factor by which the matrix operation increases the length of the eigenvector.

As eqn (16.11) is linear in **r**, once we have a solution we could multiply it by any constant and it would still be a solution of (16.11). A useful convention is to define the eigenvectors to have unit length, i.e.

$$\tilde{\mathbf{r}}\mathbf{r} = 1 \qquad (16.12)$$

16.2.1 General approach

In order to find the eigenvalues, eqn (16.11) is rewritten as

$$(\mathbf{M} - \lambda\mathbf{I})\mathbf{r} = 0 \qquad (16.13)$$

For this to have a solution other than $\mathbf{r} = 0$, the determinant of the coefficients in eqns (16.13) for **r** must be zero (see Section 1.3), i.e.

$$|\mathbf{M} - \lambda\mathbf{I}| = 0 \qquad (16.14)$$

Thus if eqn (16.11) refers to three-dimensional space, and in terms of coordinates is

$$\left.\begin{array}{l} m_{11}x + m_{12}y + m_{13}z = \lambda x \\ m_{21}x + m_{22}y + m_{23}z = \lambda y \\ m_{31}x + m_{32}y + m_{33}z = \lambda z \end{array}\right\} \qquad (16.11')$$

then eqn (16.14) is equivalent to

$$\begin{vmatrix} m_{11} - \lambda & m_{12} & m_{13} \\ m_{21} & m_{22} - \lambda & m_{23} \\ m_{31} & m_{32} & m_{33} - \lambda \end{vmatrix} = 0 \qquad (16.14')$$

When multiplied out, this provides a cubic equation for λ, which has three solutions (although they do not all have to be different or real).

For other values of λ, eqns (16.11) have only the uninteresting solution

$$x = y = z = 0$$

The three eigenvalues are the special ones which make one of the equations a linear combination of the others. In effect we then have two equations, which enable us to find the ratios y/x and z/x. Together with the normalisation convention (16.12), this defines the eigenvectors. Since there are three values of λ, there are also three eigenvectors.†

---

† The reason there are three of each is that this is a three-dimensional problem. In contrast there were an infinite number of eigenfunctions in the differential operator problem. That was because the functions $y(x)$ were defined at an infinite number of points x.

To make the above clearer, we will now perform a specific worked example.

16.2.2 Worked example

If our matrix is diagonal, the diagonal elements are already the eigenvalues; and the eigenvectors are the axes (i.e. $(1,0,0)$, $(0,1,0)$ and $(0,0,1)$). This problem is too simple, so we need one with non-zero off-diagonal elements. Since it leads to several simple properties, we consider the case of a symmetric matrix, and take the example

$$\mathbf{M} = \begin{pmatrix} 4 & 2 & -1 \\ 2 & 4 & 1 \\ -1 & 1 & 3 \end{pmatrix} \tag{16.15}$$

The determinant equation for the eigenvalues is

$$\begin{vmatrix} 4-\lambda & 2 & -1 \\ 2 & 4-\lambda & 1 \\ -1 & 1 & 3-\lambda \end{vmatrix} = 0 \tag{16.16}$$

Explicit evaluation of the determinant leads to the equation

$$24 - 34\lambda + 11\lambda^2 - \lambda^3 = 0$$

Since this is a cubic equation, we need some insight to solve it. The coefficients are such that it is apparent that the equation is satisfied when $\lambda = 1$. This implies that $\lambda - 1$ is a factor of our cubic, and hence the remaining factors can be found as

$$(1 - \lambda)(24 - 10\lambda + \lambda^2) = (1 - \lambda)(4 - \lambda)(6 - \lambda) = 0 \tag{16.16'}$$

whence

$$\lambda = 1 \text{ or } 4 \text{ or } 6 \tag{16.17}$$

These are the three different eigenvalues. We now need to find the eigenvector for each of them.

Substitution of $\lambda = 1$ into the equivalent of eqn (16.11') for our matrix $\mathbf{M}$ yields

$$\left.\begin{array}{c} 3x + 2y - z = 0 \\ 2x + 3y + z = 0 \\ (-x + y + 2z = 0) \end{array}\right\} \tag{16.18}$$

(We write the third equation in brackets, as the choice $\lambda = 1$ ensures that it contains no new information. It is worth checking that this is indeed so, since otherwise we have made some mistake, probably in the evaluation of λ. In this case, the third equation is simply the second minus the first.)

Adding the first two equations of (16.18) eliminates z, and yields

$$5x + 5y = 0$$

whence

$$x = -y \qquad (16.19)$$

Finally substituting this in the top equation of (16.18) gives

$$x = z \qquad (16.20)$$

Thus eqns (16.19) and (16.20) express the information about the eigenvector corresponding to the eigenvalue $\lambda = 1$. We do not expect to be able to find the three components of the vector uniquely. This is because (i) any multiple of an eigenvector is also satisfactory as an eigenvector, and (ii) we have only two independent equations in (16.18) for our three variables. Thus the eigenvector is in the direction

$$x : y : z = 1 : -1 : 1$$

The normalisation convention would then fix this as $(1, -1, 1)/\sqrt{3}$.

We now have to repeat the procedure for determining the eigenvectors for each of the other two eigenvalues. For $\lambda = 4$ and $\lambda = 6$, the normalised eigenvectors turn out to be $(-1, 1, 2)/\sqrt{6}$ and $(1, 1, 0)/\sqrt{2}$ respectively.

The determination of eigenvalues and eigenvectors is thus a slow but well-defined procedure. If we are worried that we may have made a numerical error, we can always check that our solutions do satisfy the eigenvalue equation (16.11). For example

$$\begin{pmatrix} 4 & 2 & -1 \\ 2 & 4 & 1 \\ -1 & 1 & 3 \end{pmatrix} \begin{pmatrix} 1 \\ 1 \\ 0 \end{pmatrix} = \begin{pmatrix} 6 \\ 6 \\ 0 \end{pmatrix} = 6 \begin{pmatrix} 1 \\ 1 \\ 0 \end{pmatrix} \qquad (16.21)$$

thereby verifying that the vector $\begin{pmatrix} 1 \\ 1 \\ 0 \end{pmatrix}$ is indeed an eigenvector, with eigenvalue 6.

If we take the scalar product of any pair of eigenvectors, the answer is zero. This means that the three eigenvectors are mutually orthogonal. This is a general property for a symmetric matrix (see Section 16.3.2.1). This provides another useful check for the eigenvectors of a symmetric matrix.

16.3 Mathematical properties of eigenvectors and eigenvalues

16.3.1 Symmetric, non-symmetric and rotation matrices

Let us assume that we start with a 3×3 matrix $\mathbf{D}$ in diagonal form. (This is often what is obtained when we use the most obvious axes for describing a system, e.g. electric polarisation in an inhomogeneous medium — see Section 15.8.3.) Then we can simply read off its eigenvalues from the diagonal elements, and the eigenvectors are given by $\begin{pmatrix} 1 \\ 0 \\ 0 \end{pmatrix}, \begin{pmatrix} 0 \\ 1 \\ 0 \end{pmatrix}$ and $\begin{pmatrix} 0 \\ 0 \\ 1 \end{pmatrix}$, i.e. by the axes with respect to which the system is defined.

We now investigate what happens when we use some transformed system, such that a vector $\mathbf{v}$ becomes

$$\mathbf{v}' = \mathbf{Tv} \tag{16.22}$$

The effect on the diagonal matrix $\mathbf{D}$ is to convert it to some new matrix $\mathbf{M}$, where

$$\mathbf{M} = \mathbf{TDT}^{-1} \tag{16.23}$$

(see eqn (15.129)).

If the transformation is just a rotation, the new matrix $\mathbf{M}$ will be symmetric (see eqn (15.136)). For a general transformation, the resulting new matrix $\mathbf{M}$ need not be symmetric (see, for example, Section 15.9.5.1.1). We thus associate symmetric matrices with diagonal matrices that are described in a rotated system; and non-symmetric matrices with general transformations of a diagonal matrix. Because rotations are simpler and more intuitive, we consider the properties of symmetric matrices first in Section 16.3.2, and non-symmetric ones afterwards (see Section 16.3.3).

For rotation matrices, two of the eigenvalues and the corresponding eigenvectors are complex. (Do not worry about this; it turns out that this is reasonable — see Section 16.4.2.1.) It is thus not particularly useful to think of them as being reducible in their simplest form to diagonal matrices. We thus consider rotation matrices separately in Section 16.4.2.

16.3.1.1 Effect of transformation on eigenvalues

After a transformation $\mathbf{T}$, a matrix $\mathbf{M}$ which relates two vectors becomes

$$\mathbf{M}' = \mathbf{TMT}^{-1}$$

(Compare Section 15.9.5.1. See also eqn (15.139) for quadratic forms under rotation.) How are the eigenvalues of $\mathbf{M}$ and $\mathbf{M}'$ related?

Intuitively we can often regard eigenvalues and eigenvectors as being associated with particular physical properties of the situation that the matrix describes. (An example is provided in the paragraph after next.) Then we would expect the properties of $\mathbf{M}$ and $\mathbf{M}'$ to be simply related. This we find to be so.

If $\mathbf{v}$ is one of the eigenvectors of $\mathbf{M}$ with eigenvalue λ, then

$$\mathbf{M}\mathbf{v} = \lambda\mathbf{v}$$

Then when we apply the transformation described by $\mathbf{T}$, we might expect $\mathbf{T}\mathbf{v}$ to be an eigenvector of $\mathbf{M}'$. To check if this is so, we calculate

$$\mathbf{M}'(\mathbf{T}\mathbf{v}) = \mathbf{T}\mathbf{M}\mathbf{T}^{-1}\mathbf{T}\mathbf{v}$$
$$= \mathbf{T}\mathbf{M}\mathbf{v}$$
$$= \lambda(\mathbf{T}\mathbf{v}) \tag{16.24}$$

This demonstrates that our guess about the eigenvector was correct, and that the eigenvalue is λ as before.

For example, if a matrix is used to describe an ellipsoid, the eigenvectors give the directions of the axes, and the eigenvalues are related to the lengths of the axes (see Section 16.4.3). If a rotation is applied, the lengths of the axes of the ellipsoid are unaffected; and its axes, of course, remain in the same orientations with respect to the body of the ellipsoid, and are thus just rotated by the transformation. This corresponds to the statements above that eigenvalues are unchanged, and eigenvectors are simply transformed.

The above is not true for matrices which transform according to eqn (15.138), with $\mathbf{T}^{-1} \neq \tilde{\mathbf{T}}$. Thus if a quadratic form such as an ellipse is transformed to new axes corresponding to a stretch along a direction θ which is not along one of the ellipse's axes, the new eigenvectors and eigenvalues do not bear such a simple relation to the corresponding quantities before the transformation.

16.3.1.2 Eigenvalues of the inverse matrix

The inverse $\mathbf{M}^{-1}$ has identical eigenvectors to those of $\mathbf{M}$, and its eigenvalues are the reciprocals of those of $\mathbf{M}$. This can be seen with the help of the results of the previous section.

Thus we can imagine changing $\mathbf{M}$ by a transformation $\mathbf{T}$ such that it becomes diagonal, i.e.

$$\mathbf{M}' = \mathbf{TMT}^{-1} = \begin{pmatrix} \lambda_1 & 0 & 0 \\ 0 & \lambda_2 & 0 \\ 0 & 0 & \lambda_3 \end{pmatrix} \tag{16.25}$$

This is particularly easy to invert, and gives

$$(\mathbf{M}')^{-1} = \begin{pmatrix} 1/\lambda_1 & 0 & 0 \\ 0 & 1/\lambda_2 & 0 \\ 0 & 0 & 1/\lambda_3 \end{pmatrix} \tag{16.26}$$

Now since the eigenvalues of a matrix are invariant with respect to a transformation, these diagonal elements are the eigenvalues of $\mathbf{M}$ and $\mathbf{M}^{-1}$, respectively.

The eigenvectors of both $\mathbf{M}'$ and $(\mathbf{M}')^{-1}$ are along the axes in the transformed system. We can go back to the original system via the transformation $\mathbf{T}^{-1}$. This affects both sets of eigenvectors in the same way, and so they remain identical.

An alternative and more direct (but less intuitive) way of demonstrating these facts is as follows. If $\mathbf{v}$ is an eigenvector of $\mathbf{M}$, then

$$\mathbf{Mv} = \lambda\mathbf{v}$$

On premultiplying by $\mathbf{M}^{-1}$, we thus find

$$\mathbf{v} = \lambda\mathbf{M}^{-1}\mathbf{v}$$

or

$$\mathbf{M}^{-1}\mathbf{v} = \frac{1}{\lambda}\mathbf{v} \tag{16.27}$$

which shows that $\mathbf{M}^{-1}$ has an eigenvector $\mathbf{v}$ with eigenvalue $1/\lambda$.

16.3.2 Symmetric matrices

16.3.2.1 Orthogonal eigenvectors

For a symmetric matrix $\mathbf{S}$, the eigenvectors† $\mathbf{v}_m$ and $\mathbf{v}_n$ corresponding to different eigenvalues are orthogonal, i.e.

$$\tilde{\mathbf{v}}_m\mathbf{v}_n = 0 \quad (\lambda_m \neq \lambda_n) \tag{16.28}$$

Together with the normalisation convention (16.12) for eigenvectors, this

† The subscripts m and n here label different eigenvectors.

enables us to write

$$\tilde{\mathbf{v}}_m \mathbf{v}_n = \delta_{mn} \quad \text{(including } m = n) \tag{16.29}$$

in analogy with eqn (16.7) for eigenfunctions of the differential operator.

The proof of the orthogonality condition (16.28) starts from the eigenvalue equation

$$\mathbf{S}\mathbf{v}_m = \lambda_m \mathbf{v}_m$$

Then

$$\tilde{\mathbf{v}}_n \mathbf{S}\mathbf{v}_m = \lambda_m \tilde{\mathbf{v}}_n \mathbf{v}_m \tag{16.30}$$

where $\mathbf{v}_n$ is a different eigenvector. We now take the transpose of this equation, and obtain

$$\tilde{\mathbf{v}}_m \mathbf{S}\mathbf{v}_n = \lambda_m \tilde{\mathbf{v}}_m \mathbf{v}_n \tag{16.31}$$

where we have explicitly taken account of the fact that $\mathbf{S}$ is symmetric, so that $\tilde{\mathbf{S}} = \mathbf{S}$; and made use of the result in Problem 15.6, for the transpose of a matrix product.

However, had we interchanged the rôles of m and n in the above discussion, instead of eqn (16.30) we would have obtained

$$\tilde{\mathbf{v}}_m \mathbf{S}\mathbf{v}_n = \lambda_n \tilde{\mathbf{v}}_m \mathbf{v}_n \tag{16.30'}$$

On comparing eqns (16.31) and (16.30′), we find that

$$(\lambda_m - \lambda_n)\tilde{\mathbf{v}}_m \mathbf{v}_n = 0 \tag{16.32}$$

so that

$$\left. \begin{array}{l} \text{if} \quad \lambda_m \neq \lambda_n \\ \text{then} \quad \tilde{\mathbf{v}}_m \mathbf{v}_n = 0 \end{array} \right\} \tag{16.33}$$

That is, if $\mathbf{v}_m$ and $\mathbf{v}_n$ correspond to different eigenvalues, they must be orthogonal. It is straightforward to check that the eigenvectors of the matrix (16.15) satisfy this property.

An interesting situation arises when two different eigenvectors happen to have the same eigenvalue, e.g.

$$\left. \begin{array}{l} \mathbf{M}\mathbf{v}_m = \lambda \mathbf{v}_m \\ \text{and} \quad \mathbf{M}\mathbf{v}_n = \lambda \mathbf{v}_n \end{array} \right\} \tag{16.34}$$

for the same value of λ. Eqn (16.32) is still true but is satisfied trivially because $\lambda_m = \lambda_n$. We can then no longer necessarily deduce that $\mathbf{v}_m$ and $\mathbf{v}_n$ are orthogonal.

However, in this degenerate case, any linear combination of $\mathbf{v}_m$ and $\mathbf{v}_n$

is also an eigenvector, corresponding to the same eigenvalue λ. This is readily derived from eqns (16.34) since

$$\mathbf{M}(\alpha\mathbf{v}_m + \beta\mathbf{v}_n) = \lambda(\alpha\mathbf{v}_m + \beta\mathbf{v}_n) \tag{16.35}$$

Thus while we always have the freedom to choose the length of eigenvectors, we here also have the extra arbitrariness of which linear combinations to accept. If we want to, we can thus maintain the orthogonality condition, even in this degenerate situation, by suitable values for (α_1, β_1) and (α_2, β_2); in that case, eqn (16.29) is always true. This problem is discussed physically in Section 16.4.3.

Because the different eigenvectors of a symmetric matrix are orthogonal, eqn (16.30) can be rewritten as

$$\tilde{\mathbf{v}}_n\mathbf{S}\mathbf{v}_m = \lambda_m\delta_{mn}$$

This is equivalent to the differential operator relationship (16.8).

16.3.2.2 Matrix $\mathbf{V}$ of eigenvectors

We can construct a new 3×3 matrix $\mathbf{V}$ which consists of the three normalised eigenvectors $\mathbf{v}$ of $\mathbf{M}$, each written as a column, i.e.

$$V_{im} = (\mathbf{v}_m)_i$$

where V_{im} is the relevant element of the matrix $\mathbf{V}$, $\mathbf{v}_m$ is the mth eigenvector and the suffix i for the rows of $\mathbf{V}$ refers to the x, y or z component of $\mathbf{v}_m$. Thus for the matrix (16.15),

$$\mathbf{V} = \begin{pmatrix} 1/\sqrt{3} & -1/\sqrt{6} & 1/\sqrt{2} \\ -1/\sqrt{3} & 1/\sqrt{6} & 1/\sqrt{2} \\ 1/\sqrt{3} & 2/\sqrt{6} & 0 \end{pmatrix} \tag{16.36}$$

For any symmetric matrix $\mathbf{M}$, the corresponding $\mathbf{V}$ has the property that

$$\tilde{\mathbf{V}}\mathbf{V} = \mathbf{I} \tag{16.37}$$

This is because, from eqn (15.117),

$$\tilde{\mathbf{V}}\mathbf{V} = \begin{pmatrix} \mathbf{v}_1 \cdot \mathbf{v}_1 & \mathbf{v}_1 \cdot \mathbf{v}_2 & \mathbf{v}_1 \cdot \mathbf{v}_3 \\ \mathbf{v}_2 \cdot \mathbf{v}_1 & \mathbf{v}_2 \cdot \mathbf{v}_2 & \mathbf{v}_2 \cdot \mathbf{v}_3 \\ \mathbf{v}_3 \cdot \mathbf{v}_1 & \mathbf{v}_3 \cdot \mathbf{v}_2 & \mathbf{v}_3 \cdot \mathbf{v}_3 \end{pmatrix} \tag{16.38}$$

Because the eigenvectors $\mathbf{v}_1$, $\mathbf{v}_2$ and $\mathbf{v}_3$ are normalised and mutually orthogonal, the scalar products in (16.38) are all either 1 or 0, and the matrix becomes simply $\mathbf{I}$.

Eqn (16.37) implies that matrix $\mathbf{V}$ is a rotation matrix. From the definition of $\mathbf{V}$ and by direct evaluation, we find that

$$\mathbf{V} \begin{pmatrix} 1 \\ 0 \\ 0 \end{pmatrix} = \mathbf{v}_1 \tag{16.39}$$

so that the old x axis is transformed by $\mathbf{V}$ to the direction of the first eigenvector (and correspondingly for y and z). Then $\tilde{\mathbf{V}}$ is the inverse rotation which reorientates the three mutually perpendicular eigenvectors along the coordinate axes.

We now consider the matrix product $\tilde{\mathbf{V}}\mathbf{M}\mathbf{V}$. This turns out to be a diagonal matrix with the eigenvalues along the diagonal, i.e.

$$\tilde{\mathbf{V}}\mathbf{M}\mathbf{V} = \begin{pmatrix} \lambda_1 & 0 & 0 \\ 0 & \lambda_2 & 0 \\ 0 & 0 & \lambda_3 \end{pmatrix} \tag{16.40}$$

We can obtain some insight on eqn (16.40) as follows. The matrix $\tilde{\mathbf{V}}$ corresponds to a rotation, with $\mathbf{M}$ being transformed as

$$\mathbf{M}' = \tilde{\mathbf{V}}\mathbf{M}(\tilde{\mathbf{V}})^{-1}$$
$$= \tilde{\mathbf{V}}\mathbf{M}\mathbf{V} \tag{16.41}$$

where the first equation follows from eqn (15.129), and the second relies on the fact that $\mathbf{V}$ is a rotation matrix, so $\tilde{\mathbf{V}} = \mathbf{V}^{-1}$. Thus $\mathbf{M}'$ is the transformed matrix, described with respect to axes aligned along the eigenvectors. In this case the matrix achieves its simplest form, and is diagonal; and the diagonal elements are the eigenvalues. (Which eigenvalue appears first is determined by the arbitrary choice of the ordering of the eigenvectors $\mathbf{v}_m$ in the definition of the matrix $\mathbf{V}$.)

A more calculational way of seeing how (16.40) arises is obtained by first considering $\mathbf{M}\mathbf{V}$. If we write

$$\mathbf{V} = (\mathbf{v}_1, \mathbf{v}_2, \mathbf{v}_3) \tag{16.42}$$

where each $\mathbf{v}_m$ on the right-hand side of (16.42) is to be considered as a column vector, then

$$\mathbf{M}\mathbf{v} = (\lambda_1\mathbf{v}_1, \lambda_2\mathbf{v}_2, \lambda_3\mathbf{v}_3) \tag{16.43}$$

since the $\mathbf{v}_m$ are eigenvectors. Then

$$\tilde{\mathbf{V}}\mathbf{M}\mathbf{V} = \begin{pmatrix} \lambda_1\mathbf{v}_1 \cdot \mathbf{v}_1 & \lambda_2\mathbf{v}_1 \cdot \mathbf{v}_2 & \lambda_3\mathbf{v}_1 \cdot \mathbf{v}_3 \\ \lambda_1\mathbf{v}_2 \cdot \mathbf{v}_1 & \lambda_2\mathbf{v}_2 \cdot \mathbf{v}_2 & \lambda_3\mathbf{v}_2 \cdot \mathbf{v}_3 \\ \lambda_1\mathbf{v}_3 \cdot \mathbf{v}_1 & \lambda_2\mathbf{v}_3 \cdot \mathbf{v}_2 & \lambda_3\mathbf{v}_3 \cdot \mathbf{v}_3 \end{pmatrix}$$

$$= \begin{pmatrix} \lambda_1 & 0 & 0 \\ 0 & \lambda_2 & 0 \\ 0 & 0 & \lambda_3 \end{pmatrix} \tag{16.40}$$

Our final result for the symmetric matrix case is that, because its eigenvectors are mutually orthogonal, any general vector can readily be expressed as a linear combination of these eigenvectors. This parallels the property (16.9) for expressing functions in terms of the eigenfunctions of the differential operator.

16.3.3 Non-symmetric matrices

We now turn to non-symmetric matrices, and compare and contrast their properties with those of symmetric ones.

The proof in Section 16.3.2.1 about mutually orthogonal eigenvectors relied on the fact that the matrix $\mathbf{S}$ was symmetric (see the way in which eqn (16.31) follows from eqn (16.30)). Thus for a non-symmetric matrix, the eigenvectors need not all be mutually perpendicular (although some of them may be).

The matrix $\mathbf{V}$ of normalised eigenvectors thus does not satisfy eqn (16.37), since the scalar products in (16.38) that previously vanished now do not do so. This means that $\mathbf{V}$ is not a rotation matrix. This is reasonable, since there is no rotation which would transform the orthogonal x, y and z axes to the non-orthogonal eigenvectors.

Although $\tilde{\mathbf{V}}\mathbf{V} \neq \mathbf{I}$, $\mathbf{V}^{-1}\mathbf{V} = \mathbf{I}$ but this follows trivially from the definition of an inverse.

If instead of $\tilde{\mathbf{V}}\mathbf{M}\mathbf{V}$ we construct† $\mathbf{V}^{-1}\mathbf{M}\mathbf{V}$, we find that it is diagonal, with the eigenvalues along the diagonal. This is because we have again transformed the matrix, and are describing it using the simplest set of axes along the eigenvectors. Of course, the transformation here is not a rotation.

Although the three eigenvectors are not mutually orthogonal, we can

---

† This of course makes no difference for the symmetric matrix case, since there $\mathbf{V}^{-1} = \tilde{\mathbf{V}}$.

still express any general vector **q** in terms of them. For example

$$\mathbf{q} = \alpha \mathbf{v}_1 \wedge \mathbf{v}_2 + \beta \mathbf{v}_2 \wedge \mathbf{v}_3 + \gamma \mathbf{v}_3 \wedge \mathbf{v}_1 \qquad (16.44)$$

(compare Section 3.6).

16.4 Physical examples of eigenvectors

By considering eigenvectors of matrices corresponding to specific physical situations, we shall obtain a better understanding of their significance, and of their properties.

16.4.1 Simple two-dimensional matrices

We start by looking at some of the simple two-dimensional matrices that were discussed in Section 15.5. Their effect on eigenvectors and on general vectors is shown in fig 16.1.

The stretching matrix $\mathbf{M}_4$ of eqn (15.48) is already diagonal, and its diagonal elements k and l are its eigenvalues. The corresponding eigenvectors are $\begin{pmatrix} 1 \\ 0 \end{pmatrix}$ and $\begin{pmatrix} 0 \\ 1 \end{pmatrix}$ respectively. This is because it is these vectors along the x and y axes respectively which are stretched by the given factors, but left unchanged in direction. A general vector $\begin{pmatrix} x \\ y \end{pmatrix}$ after stretching becomes $\begin{pmatrix} kx \\ ly \end{pmatrix}$, and hence is rotated; it is thus not an eigenvector.

More interesting is the matrix $\mathbf{M}_5$ of eqn (15.53), corresponding to a stretch by a factor r along a direction with a gradient $\tan\theta$. Its eigenvalues are r and 1, with eigenvectors $\begin{pmatrix} \cos\theta \\ \sin\theta \end{pmatrix}$ and $\begin{pmatrix} -\sin\theta \\ \cos\theta \end{pmatrix}$ respectively. This is a statement of the fact that the stretch r is in the expected direction, while a vector perpendicular to this is left unaffected. This is an example of the general theorem that a symmetric matrix has orthogonal eigenvectors.

Similarly the two eigenvectors of a two-dimensional reflection matrix correspond to a vector parallel to the line of reflection (which is left unaffected, and hence has eigenvalue 1) and a vector perpendicular to this, which is reversed and so its eigenvalue is -1.

The identity matrix **I** is already diagonal, and its eigenvalues are both 1. This is an example of degenerate eigenvalues, and hence although the

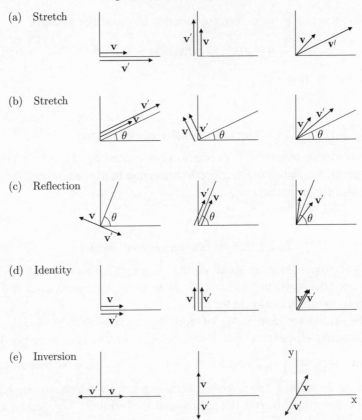

(a) Stretch

(b) Stretch

(c) Reflection

(d) Identity

(e) Inversion

Fig. 16.1. The effect of applying specific matrices to various vectors. The first two columns correspond to eigenvectors; the third is for a general vector. The original vectors are denoted by **v** and the transformed ones by **v**′. Where **v** and **v**′ are parallel, they are shown slightly separated for clarity. The matrices correspond to: (a) a stretch of k along x, and l along y; (b) a stretch of r along a direction θ, and no change in the perpendicular direction; (c) a reflection about a line at an angle θ; (d) the identity operation of leaving vectors unchanged; and (e) inversion through the origin. In (d), the eigenvalues are identical, and also in (e). In degenerate cases, any linear combination of the two eigenvectors is also an eigenvector. Thus **v**′ is respectively parallel or antiparallel to **v** in these cases, even for a general vector.

vectors $\begin{pmatrix} 1 \\ 0 \end{pmatrix}$ and $\begin{pmatrix} 0 \\ 1 \end{pmatrix}$ are eigenvectors, so is any linear combination of them. This is sensible, since any vector is unaltered by the identity matrix, and hence is an eigenvector with eigenvalue 1. If we want to choose two that are perpendicular, we can do so (e.g. the x and y axes) so as to maintain the property that a symmetric matrix has orthogonal eigenvectors.

In a similar way the matrix

$$\begin{pmatrix} -1 & 0 \\ 0 & -1 \end{pmatrix}$$

corresponding to an inversion though the origin also is such that any vector is an eigenvector, this time with degenerate eigenvalue -1.

16.4.2 Three-dimensional rotation matrices

16.4.2.1 How many eigenvectors?

We start with the simplest form of the three-dimensional rotation matrix (see M_{11} of eqn (15.79)). It must have three eigenvectors, but the only one we could guess would be the one corresponding to the rotation axis; a vector in this direction is unaffected by the rotation and hence has eigenvalue unity. What about the others?

If we go through the standard procedure for finding the eigenvalues λ, we obtain the equation

$$(\lambda - 1)(\lambda^2 - 2\lambda \cos\theta + 1) = 0$$

which on solving for λ yields

$$\lambda = 1 \text{ or } e^{\pm i\theta} \tag{16.45}$$

Thus one of them is unity, as expected, but the other two are complex. The corresponding eigenvectors are

$$\left. \begin{array}{lll} \text{For } \lambda = 1 & : & \begin{pmatrix} 0 \\ 0 \\ 1 \end{pmatrix} \\ \text{For } \lambda = e^{\pm i\theta} & : & x = \pm iy, \ z = 0 \end{array} \right\} \tag{16.46}$$

The last two eigenvectors are complex, and need not concern us further. This agrees with our physical insight that there should be only one (real) eigenvector. It also explains why, when a rotation matrix is expressed like M_{11} in its simplest form, it is not diagonal (because we do not use the last two complex eigenvectors as a basis for defining M).

16.4.2.2 Axis and angle of rotation

Now we consider rotation matrices expressed with respect to general axes, rather than having the rotation axis along the z axis (see also Section 15.7.2).

The first problem is to recognise such a rotation matrix. This is easily done because the inverse $\mathbf{R}^{-1}$ of a rotation matrix is just its transpose $\tilde{\mathbf{R}}$ (see eqn (15.124)). Thus we need to check that

$$\left.\begin{array}{c} \tilde{\mathbf{R}}\mathbf{R} = \mathbf{I} \\ \text{and} \quad \det(\mathbf{R}) = +1 \end{array}\right\} \tag{16.47}$$

The last condition is necessary because if the determinant is -1, three orthogonal vectors $\mathbf{u}_m$ would remain orthogonal after the transformation, but one set would be left-handed and the other right. There is then no way that a rotation could take $\mathbf{u}_m$ into $\mathbf{R}\mathbf{u}_m$. A matrix satisfying $\tilde{\mathbf{M}}\mathbf{M} = \mathbf{I}$, but having $\det(\mathbf{M}) = -1$, would correspond to the combined operations of rotation and inversion about the origin.

Having discovered that a matrix corresponds to a rotation, the next step is to determine the angle of rotation. This can be achieved by finding the eigenvalues λ. Then apart from $\lambda = 1$, the others are $e^{\pm i\theta}$, and so we can read off the angle of rotation. (There is, however, an ambiguity of the sign of the rotation angle, which we describe below how to resolve.)

However, it is easier to make use of the fact that the trace of a matrix is unaffected by a rotation (see Section 15.9.6). When the rotation matrix is expressed in its simplest form, the trace is $1 + 2\cos\theta$, as can be seen from eqn (15.79), and this remains true whatever axes are used to describe the rotation. Thus we just have to add the diagonal elements of a rotation matrix, equate the sum to $1 + 2\cos\theta$, and hence find θ. Again we have a two-fold ambiguity.

The axis of rotation $\mathbf{a}$ is the eigenvector corresponding to $\lambda = 1$ and can be found by the standard method of determining eigenvectors. As usual it can point in either direction. We simply choose one arbitrarily; the other would result in the opposite sign for θ. (Similarly the rotation of the earth can be regarded as being in a left-handed sense about an axis emerging from the North Pole, or right-handedly about one coming out of the South Pole.)

To resolve the sign ambiguity, it is simplest to construct any vector $\mathbf{p}$ perpendicular to $\mathbf{a}$. Then $\mathbf{R}\mathbf{p}$ should make an angle of θ with $\mathbf{p}$ — if not, we have made a mistake somewhere. The sense of rotation is determined by whether **Rp** is in the direction of $\mathbf{a} \wedge \mathbf{p}$ or not, i.e. whether

$$(\mathbf{Rp}) \cdot (\mathbf{a} \wedge \mathbf{p}) > 0$$

To illustrate this with a trivial example, the rotation matrix $\mathbf{M}_{11}$ corresponds to a rotation of θ about the z axis. The x axis is an example of a vector $\mathbf{p}$ perpendicular to the rotation axis. For $\theta < \pi$, after the rotation, **Rp** will have a positive component along $\mathbf{z} \wedge \mathbf{x} = \mathbf{y}$.

16.4.3 Quadratic forms

16.4.3.1 Connection with eigenvalues and eigenvectors

Very useful insights into eigenvectors and eigenvalues are provided by quadratic forms. These can be considered in any number of dimensions, but we will confine ourselves to three. These quadratic forms are then generalisations of the two-dimensional figures such as ellipses, hyperbolae and parabolae. These new shapes are most easily thought of as being generated from their corresponding two-dimensional cousins. For example, one of the conic sections in the x–y plane is centred at the origin and arranged in its "most simple orientation" (so that its equation contains no xy term), and is then rotated about the x axis. Finally the resulting surface is stretched in the z direction.

Thus whereas an ellipse is given by

$$\frac{x^2}{c^2} + \frac{y^2}{d^2} = 1 \tag{16.48}$$

the corresponding ellipsoid is

$$\frac{x^2}{c^2} + \frac{y^2}{d^2} + \frac{z^2}{e^2} = 1 \tag{16.49}$$

The lengths of the semi-axes of the ellipse are c and d; those for the ellipsoid are c, d and e (see fig. 16.2). For both, the axes of the figure coincide with the coordinate axes. In order for eqn (16.49) to represent an ellipsoid (rather than a hyperboloid or the like), c^2, d^2 and e^2 must all be positive, which is one reason why they have been written as squared quantities.

All this is trivial. The problems of:

(i) recognising whether or not our equation represents an ellipsoid,
(ii) finding the directions of the axes, and
(iii) determining the lengths of the axes

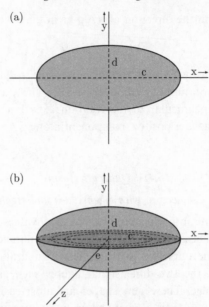

Fig. 16.2. (a) The ellipse of eqn (16.48), with semi-axes of length c and d. (b) The ellipsoid of eqn (16.49), with semi-axes of length c, d and e.

are not so simple if the figure is rotated with respect to the coordinate axes.† Thus an ellipse would be of the form

$$ax^2 + by^2 + 2cxy = 1 \tag{16.50}$$

while a general quadratic form in three dimensions is

$$ax^2 + by^2 + cz^2 + 2dxy + 2exz + 2fyz = 1 \tag{16.51}$$

We shall use eigenvectors and eigenvalues to help us solve these problems.

The reason for the convention of including factors of 2 in the above

---

† For ease of the algebra, we consider surfaces and curves that are centred on the origin. If this is not so, the equation needs to be transposed by a shift of the origin so that terms that are linear in x, y (and z) disappear.

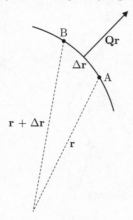

Fig. 16.3. The points A and B are on the surface of the quadratic form (16.52), with position vectors $\mathbf{r}$ and $\mathbf{r} + \Delta\mathbf{r}$ respectively. Thus the vector $\Delta\mathbf{r}$ lies in the surface. According to eqn (16.55), $\mathbf{Qr}$ is normal to $\Delta\mathbf{r}$, and hence is normal to the surface.

equations becomes clear when we write it in matrix form. Then†

$$1 = (x \ y \ z)\begin{pmatrix} a & d & e \\ d & b & f \\ e & f & c \end{pmatrix}\begin{pmatrix} x \\ y \\ z \end{pmatrix}$$

$$= \tilde{\mathbf{r}}\mathbf{Qr} \tag{16.52}$$

A useful property of $\mathbf{Q}$ is that the vector $\mathbf{Qr}$ is in the direction of the normal to the surface at $\mathbf{r}$. To see this, we consider two neighbouring points $\mathbf{r}$ and $\mathbf{r} + \Delta\mathbf{r}$ on the surface (see fig. 16.3). Then in addition to eqn (16.52),

$$(\tilde{\mathbf{r}} + \widetilde{\Delta\mathbf{r}})\mathbf{Q}(\mathbf{r} + \Delta\mathbf{r}) = 1 \tag{16.53}$$

On subtracting eqn (16.52),

$$\widetilde{\Delta\mathbf{r}}\mathbf{Qr} + \tilde{\mathbf{r}}\mathbf{Q}\Delta\mathbf{r} = 0 \tag{16.54}$$

where we have ignored the term $\widetilde{\Delta\mathbf{r}}\mathbf{Q}\Delta\mathbf{r}$, which is very much smaller. In fact, eqn (16.54) could have been obtained more quickly simply by differentiating eqn (16.52).

---

† Another convention has slipped in here. The matrix $\mathbf{Q}$ in eqn (16.52) is chosen to be symmetric, because symmetric matrices have particularly simple properties. In fact, in order for (16.52) to reproduce (16.51), all that is required for the off-diagonal elements is that, for example, $Q_{12} + Q_{21} = 2d$. In order to have $\mathbf{Q}$ symmetric, the choice $Q_{12} = Q_{21} = d$ is made.

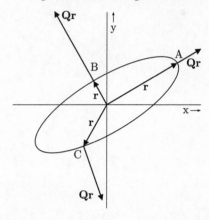

Fig. 16.4. At the ends of the axes of the ellipsoid (e.g. points A and B), the normal to the surface $\mathbf{Qr}$ is parallel to the radial vector $\mathbf{r}$ at that point. For a general point (e.g. C), this is not so.

Now the two terms in eqn (16.54) are equal as:

(i) they are the transpose of each other; and
(ii) each is simply a scalar number.

Thus

$$\widetilde{\Delta \mathbf{r}}\mathbf{Qr} = 0 \qquad (16.55)$$

This implies that the two vectors $\Delta\mathbf{r}$ and $\mathbf{Qr}$ are perpendicular. Since $\Delta\mathbf{r}$ is a vector in the surface, it follows that $\mathbf{Qr}$ is normal to it.

We now make use of this result to find the directions of the axes. They can be defined as the points on the surface where the normal is parallel to the radial vector at that point (see fig. 16.4), i.e.

$$\mathbf{Qr} = \lambda\mathbf{r} \qquad (16.56)$$

This is, of course, the standard eigenvalue equation. It shows that the directions of the axes of the ellipsoid are eigenvectors of $\mathbf{Q}$.

In this situation, rather than the three eigenvectors $\mathbf{v}_m$ arbitrarily having unit length, it is more sensible to normalise them so that their end-points lie on the surface of the ellipsoid. Then

$$1 = \tilde{\mathbf{v}}_m\mathbf{Q}\mathbf{v}_m = \lambda_m\tilde{\mathbf{v}}_m\mathbf{v}_m \qquad (16.57)$$

since $\mathbf{v}_m$ is an eigenvector. Eqn (16.57) then implies that the square of the length of the eigenvector is $1/\lambda_m$, i.e. the lengths of the semi-axes of the ellipsoid are $1/\sqrt{\lambda_m}$.

Table 16.1. *Ellipsoids, eigenvalues and eigenvectors.*

Is the quadratic form an ellipsoid?	$\lambda_1, \lambda_2, \lambda_3 > 0$
Where are the axes?	Along the eigenvectors
How long are the semi-axes?	$1/\sqrt{\lambda_m}$

For an ellipsoid (rather than an alternative quadratic form), the lengths of the three semi-axes must, of course, be real. Thus all three eigenvalues of $\mathbf{Q}$ are required to be positive.

The answers to the three problems posed earlier about the ellipsoid are summarised in Table 16.1

An alternative method of solving this problem is by Lagrangian multipliers (see Section 8.3.3). The end-points of the axes are then the answers to the question "As we wander around over the surface of the ellipsoid, which points are such that their distance from the origin is a stationary value?" The largest axis of the ellipsoid then gives the maximum distance, the smallest one is the minimum, while the middle length axis corresponds to a saddle point. It is interesting that the solution of the problem by Langrangian multipliers involves identical algebra to that required for finding the eigenvectors of the matrix.

It should also be possible to find the axes and their lengths by performing the appropriate three-dimensional rotations, in order to reorientate the ellipsoid along the axes.

16.4.3.2 Specific examples

We now illustrate these ideas with two examples, the first being a trivial one.

For the ellipsoid of eqn (16.49), the matrix is

$$\mathbf{Q} = \begin{pmatrix} 1/c^2 & 0 & 0 \\ 0 & 1/d^2 & 0 \\ 0 & 0 & 1/e^2 \end{pmatrix} \tag{16.58}$$

The eigenvectors lie along the coordinate axes. The eigenvalues are $1/c^2$, $1/d^2$ and $1/e^2$, so the lengths of the semi-axes are c, d and e.

For the more realistic matrix $\mathbf{M}$ of eqn (16.15), the corresponding quadratic form is

$$4x^2 + 4y^2 + 3z^2 + 4xy - 2xz + 2yz = 1 \tag{16.59}$$

The eigenvalues λ_m of this matrix $\mathbf{M}$ are 1, 4 and 6 (see Section 16.2.2)

Since these are all positive, the quadratic form (16.59) does correspond to an ellipsoid.

The axes are in the directions of the eigenvectors, i.e. $(1, -1, 1)$, $(-1, 1, 2)$ and $(1, 1, 0)$. These are, of course, mutually perpendicular. The length of the first semi-axis can be found by taking an arbitrary point $(v, -v, v)$ along the first eigenvector, and choosing v so that it lies on the ellipsoid, i.e. the coordinates of the point satisfy eqn (16.59). This determines v as $\pm 1/\sqrt{3}$. Then since the end-points of this axis are at $\pm(1, -1, 1)/\sqrt{3}$, the length of the semi-axis is 1. Alternatively, and more quickly, this could have been obtained from $1/\sqrt{\lambda_1}$. Since the eigenvalues are 1, 4 and 6, the lengths of the three semi-axes are 1, $1/2$ and $1/\sqrt{6}$.

If eqn (16.59) is rewritten with respect to new rotated axes along the ellipsoid's axes, it becomes

$$x'^2 + 4y'^2 + 6z'^2 = 1 \qquad (16.59')$$

There is, of course, an intrinsic ambiguity as to which coefficient is associated with which of the new coordinates. This corresponds to six possible choices of which of the ellipsoid's axes are aligned with x', etc.

16.4.3.3 Degenerate eigenvalues

We now consider the case where two of the eigenvalues are equal, but the third is different and smaller. Then the third axis is larger than the other two which are equal to each other. The ellipsoid has thus more or less the shape of a rugby ball. The axis corresponding to λ_3 is uniquely determined, but in the plane perpendicular to it, the cross-section of the ball is circular. In this case, any direction in this plane is suitable as an eigenvector or axis of the ellipsoid. This agrees with the properties of the matrix, in that any linear combination of the pair of degenerate eigenvectors (i.e. with the same eigenvalue) is also an eigenvector. We can still make the conventional choice of selecting two perpendicular directions in this plane as eigenvectors, and then all three are mutually orthogonal.

The earth has a shape where the distance between the Poles is smaller than a diameter in the equatorial plane; but all equatorial diameters are equal (to the extent that we can ignore the odd mountain in Equador). This corresponds to another degenerate ellipsoid, but in contrast with the rugby ball, now $\lambda_1 = \lambda_2 < \lambda_3$. If we are required to find the axes for the earth, we would of course choose the polar direction, but any two independent vectors from the centre of the earth to the equator would

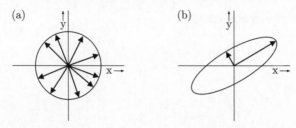

Fig. 16.5. (a) For the case of equal eigenvalues, the ellipse becomes a circle, with any radial vector being suitable as an axis. (b) For an ellipse, the directions of the axes are uniquely defined (up to a sign ambiguity).

be adequate. Again a conventional choice would be perpendicular axes, such as those to points on the surface with longitudes 20° E and 110° E.

For the completely degenerate situation of $\lambda_1 = \lambda_2 = \lambda_3$, our ellipsoid becomes a sphere, for which any radial vector would be an axis. This time we would probably choose any three mutually orthogonal directions for the axes.

These degenerate cases contrast with the situation where all three λ_n are different. The ellipsoid's shape is then rather like that of a rugby ball that has been placed on the ground and then deformed by someone heavy sitting on it. In the plane perpendicular to the long axis, the cross section is now elliptical, and the axes are uniquely defined with no arbitrariness involved.

The corresponding situation in two dimensions is illustrated in fig. 16.5.

16.4.4 Anisotropic media and non-symmetric matrices

We consider the case of an anisotropic medium where the "special directions" along which the electric polarisation **p** is parallel to the electric field **e** are non-orthogonal, and given by the unit vectors of (15.130). Then with respect to the coordinate axes, the relation between a general **e** and the resulting polarisation **p** is

$$\mathbf{p} = \mathbf{N}'\mathbf{e}$$

where $\mathbf{N}'$ is the matrix of eqn (15.134).

The eigenvalues of $\mathbf{N}'$ turn out to be μ_1, μ_2 and μ_3, and the corresponding eigenvectors are:

$$\text{for } \mu_1, \quad \begin{pmatrix} 1 \\ 0 \\ 0 \end{pmatrix}$$

$$\text{for } \mu_2, \quad \begin{pmatrix} 1 \\ 1 \\ 0 \end{pmatrix} \qquad (16.60)$$

$$\text{and for } \mu_3, \quad \begin{pmatrix} 0 \\ 0 \\ 1 \end{pmatrix}$$

This is a very satisfactory result: the eigenvalues give the directions in *xyz* space which are special for the material and the eigenvalues are the polarisabilities along these directions.

16.4.5 Normal modes

The subject of normal modes is considered in detail in Chapter 13. They are relevant to situations involving, for example, the free oscillations of a set of N coupled pendula, where the restoring force on one pendulum depends not only on its own displacement, but also on those of some (or all) of the others. Then the N simultaneous differential equations describing the displacements x_i are

$$\left. \begin{array}{l} m_1 \ddot{x}_1 = k_{11} x_1 + k_{12} x_2 + \cdots \\ m_2 \ddot{x}_2 = k_{21} x_1 + k_{22} x_2 + \cdots \\ \vdots \end{array} \right\} \qquad (16.61)$$

Here m_i are the masses of each pendulum's bob, and the k_{ij} are coefficients relating the restoring force to the x_i. In the absence of coupling between the pendula, only the diagonal coefficients k_{ii} are non-zero, the equations separate into uncoupled differential equations, and the pendula oscillate independently. It is the existence of the coupling which turns it into an interesting normal modes situation, where the pendula affect each other. Newton's Third Law (that action and reaction are equal and opposite) generally ensures that

$$k_{ij} = k_{ji} \qquad (16.62)$$

The normal mode approach is to look for solutions where all the different x_i oscillate harmonically at the same angular frequence ω. Then

Table 16.2. *The significance of eigenvectors and eigenvalues.*

Matrix describes:	Eigenvectors	Eigenvalues
Rotation	axis of rotation (+2 complex directions)	1 (and $e^{\pm i\theta}$)
Quadratic form	axes of figure	(lengths of semi-axes)$^{-2}$
Electric field in non-isotropic medium	special directions	polarisability along these directions
Normal modes	vector of relative displacements	$-\omega_m^2$

each $\ddot{x}_i$ can be replaced by $-\omega^2 x_i$, with ω^2 being the same for all x_i. Eqns (16.61) become

$$-\omega^2 \mathbf{x} = \mathbf{Kx} \qquad (16.63)$$

where $\mathbf{x}$ is an N-dimensional vector of the x_i, and the $N \times N$ matrix

$$\mathbf{K} = \begin{pmatrix} k_{11}/m_1 & k_{12}/m_1 & \dots \\ k_{21}/m_2 & k_{22}/m_2 & \dots \\ \vdots & & \end{pmatrix} \qquad (16.64)$$

Given eqn (16.62), the matrix $\mathbf{K}$ is symmetric if all the m_i are equal.

Eqn (16.63) is an eigenvalue equation, with N different eigenvalues $-\omega_m^2$. For each value of ω_m, there is a corresponding eigenvector $\mathbf{x}_m$; its N components give the relative amplitudes of the oscillations of each of the N individual pendula. Thus each eigenvalue $-\omega_m^2$ is characterised by its own pattern of the vibrations of the pendula. The eigenvectors simply correspond to the normal modes of the system, in which all components oscillate at the same frequency.

It is possible to choose N new vectors $\mathbf{y}_m$, each with N components and being defined by one of the eigenvectors as a suitable linear combination of the individual displacements $x_1, x_2, \dots, x_N$, such that with respect to the $\mathbf{y}_m$ the matrix $\mathbf{K}$ becomes diagonal. Then the differential equations for the $\mathbf{y}_m$ decouple from each other, and can be trivially solved. Again we see that a system in a normal mode oscillates at its eigenfrequency ω_m.

We thus conclude that, in applications where non-diagonal matrices are involved, eigenvectors and eigenvalues often have direct physical significance. The examples of the last four sections are summarised in Table 16.2.

Problems

16.1 The matrix M_5 of eqn (15.53) corresponds to a stretch by a factor k along a direction at an angle θ with respect to the x axis. Determine the eigenvalues and eigenvectors of M_5, and interpret your result.

16.2 (i) Evaluate the eigenvalues and eigenvectors of the matrix

$$M = \begin{pmatrix} 4 & 2 & -1 \\ 2 & 4 & 1 \\ -1 & 1 & 3 \end{pmatrix}$$

What are the three angles between the pairs of eigenvectors? Check your answers against those given in Section 16.2.2.

(ii) Construct a 3×3 matrix V, where each column corresponds to a normalised eigenvector of M. Evaluate $\tilde{V}V$ and $\tilde{V}MV$. Interpret your results.

16.3 Evaluate M^{-1}, the inverse of the matrix M of Problem 16.2(i). Calculate the eigenvalues and eigenvectors of M^{-1}, and compare them with those of M.

16.4 What tests is it necessary to perform in order to determine whether a given matrix is a three-dimensional rotation matrix?

Confirm that the matrix V of eqn (16.36) (and Problem 16.2(ii)) is a rotation matrix.

16.5 For the rotation matrix

$$R = \begin{pmatrix} 0 & -1 & 0 \\ 0 & 0 & 1 \\ -1 & 0 & 0 \end{pmatrix}$$

find the axis and the angle of rotation. Construct a vector p perpendicular to the axis of rotation, and confirm that Rp makes the expected angle with p.

16.6 (i) The equation

$$5x^2 + 4xy + 2y^2 = \frac{11}{16}$$

describes an ellipse in two dimensions. Write this equation in the matrix form

$$\tilde{r}Qr = 1$$

For the point on the ellipse with coordinates $(\frac{1}{4}, \frac{1}{4})$ evaluate:
(a) the vector Qr, and
(b) the direction of the tangent to the ellipse.

Confirm that these are perpendicular to each other.

Explain why **Qr** is useful for finding the axes of the ellipse. Find the directions of the axes and their lengths.

(ii) Now use the method of Lagrangian multipliers to find the directions of the ellipse's axes. Compare the algebra with that required for method (i).

16.7 Determine the eigenvalues of the matrix

$$\begin{pmatrix} 4 & 0 & 3 \\ 0 & 1 & 0 \\ 1 & 0 & 2 \end{pmatrix}$$

Find/choose three eigenvectors that are mutually perpendicular.

16.8 In a particular anisotropic medium, the relation between the polarisation **p** and the electric field **e** that produces it is **p** = **Ne**, where

$$\mathbf{N} = \begin{pmatrix} 3 & 2 & 0 \\ 0 & 1 & 0 \\ 0 & 0 & 2 \end{pmatrix}$$

Find the eigenvalues and eigenvectors of the non-symmetric matrix **N**. What is their physical significance? Are the eigenvectors orthogonal?

Write down the 3×3 matrix **V** consisting of the three normalised eigenvectors each arranged as a column. What is the appropriate matrix product involving **V** and **N**, such that the result is diagonal? By performing the required matrix multiplications, verify that it is so.

16.9 Rewrite the equations of motion for the two pendulum bobs of Problem 13.2 in the matrix form

$$\ddot{\mathbf{x}} = \mathbf{M}\mathbf{x}$$

where **x** is the two component vector $\begin{pmatrix} x \\ y \end{pmatrix}$, and **M** is a 2 × 2 matrix chosen to reproduce the equations of motion.

Find the eigenvalues and eigenvectors of **M**, and interpret them in terms of your previous solution for Problem 13.2.

Are the eigenvectors orthogonal to each other? To what feature of the matrix **M** is this related?

Appendix C

This appendix contains a brief summary of the crucial equations for each of the chapters.

Chapter 9 Integrals

$$\int f(x,y)dl = \int f(x,y)\sqrt{1+\left(\frac{dy}{dx}\right)^2}\,dx \tag{9.3}$$

$$\text{Area enclosed by curve} = \oint xdy$$
$$= \frac{1}{2}\oint(xdy - ydx) \tag{9.11}$$

$$\int \rho(x,y,z)da = \int\int \rho(x,y,z)\sqrt{1+\left(\frac{\partial z}{\partial x}\right)^2+\left(\frac{\partial z}{\partial y}\right)^2}\,dxdy \tag{9.25}$$

with $z = z(x,y)$

Care with limits of integration.

$$\int\int f(x,y)dxdy = \int\int f(u,v)|J|dudv \tag{6.68}$$

where

$$J = \frac{\partial(x,y)}{\partial(u,v)} = \begin{vmatrix} \dfrac{\partial x}{\partial u} & \dfrac{\partial x}{\partial v} \\ \dfrac{\partial y}{\partial u} & \dfrac{\partial y}{\partial v} \end{vmatrix} \tag{6.67}$$

Two-dimensional polars:

$$J = r \qquad (6.69)$$

Cylindrical polars:

$$J = r \qquad (9.50)$$

Spherical polars:

$$J = r^2 \sin \theta \qquad (9.51)$$

Chapter 10 Vector operators

Gradient:

$$\text{grad}\phi = \nabla \phi \qquad (10.24)$$

Divergence:

$$\text{div}\mathbf{v} = \nabla \cdot \mathbf{v} \qquad (10.25)$$

Curl:

$$\text{curl}\mathbf{v} = \nabla \wedge \mathbf{v} \qquad (10.26)$$

where for Cartesian coordinates

$$\nabla = \mathbf{i}\frac{\partial}{\partial x} + \mathbf{j}\frac{\partial}{\partial y} + \mathbf{k}\frac{\partial}{\partial z} \qquad (10.23)$$

$\text{grad}\phi$ is in direction of steepest ascent

$$\delta\phi = \text{grad}\phi \cdot d\mathbf{l} \qquad (10.7)$$

$$\text{div}\mathbf{v} = \lim_{\delta V \to 0}\left(\frac{1}{\delta V}\int \mathbf{v} \cdot d\mathbf{a}\right) \qquad (10.8)$$

$$(\text{curl}\mathbf{v})_\perp = \lim_{\delta a \to 0}\left(\frac{1}{\delta a}\int_c \mathbf{v} \cdot d\mathbf{l}\right) \qquad (10.13)$$

$$\text{grad}(r^n) = nr^{n-1}\hat{\mathbf{r}} \qquad (10.6)$$

$$\text{div}\mathbf{r} = 3 \qquad (10.11)$$

$$\text{curl}\mathbf{r} = 0$$

$$\text{curl}(r^n\hat{\boldsymbol{\tau}}) = (n+1)r^{(n-1)}\mathbf{k} \qquad (10.21)$$

$$\nabla^2 r^n = n(n+1)r^{(n-2)}$$

Double operators: see Table 10.2, e.g.

$$\text{curl}(\text{curl}\mathbf{v}) = \nabla(\nabla \cdot \mathbf{v}) - (\nabla \cdot \nabla)\mathbf{v}$$

Operators on product fields:

$$\text{grad}(\phi\psi) = \phi\text{grad}\psi + \psi\text{grad}\phi$$

$$\text{grad}(\mathbf{u} \cdot \mathbf{v}) = (\mathbf{v} \cdot \nabla)\mathbf{u} + (\mathbf{u} \cdot \nabla)\mathbf{v} + \mathbf{v} \wedge \text{curl}\mathbf{u} + \mathbf{u} \wedge \text{curl}\mathbf{v}$$

$$\text{div}(\phi\mathbf{v}) = (\text{grad}\phi) \cdot \mathbf{v} + \phi\text{div}\mathbf{v}$$

$$\text{div}(\mathbf{u} \wedge \mathbf{v}) = \mathbf{v} \cdot \text{curl}\mathbf{u} - \mathbf{u} \cdot \text{curl}\mathbf{v}$$

$$\text{curl}(\phi\mathbf{v}) = \phi\text{curl}\mathbf{v} + (\text{grad}\phi) \wedge \mathbf{v}$$

$$\text{curl}(\mathbf{u} \wedge \mathbf{v}) = (\mathbf{v} \cdot \nabla)\mathbf{u} + (\text{div}\mathbf{v})\mathbf{u} - (\mathbf{u} \cdot \nabla)\mathbf{v} - (\text{div}\mathbf{u})\mathbf{v}$$

Vector operators in various coordinate systems: see Table 10.4

Divergence Theorem:

$$\int \text{div}\mathbf{v}\,dV = \int \mathbf{v} \cdot \mathbf{da} \tag{10.51}$$

Stokes' Theorem:

$$\int \text{curl}\mathbf{v} \cdot \mathbf{da} = \int \mathbf{v} \cdot \mathbf{dl} \tag{10.55}$$

Green's Theorems:

$$\int (P\,dx + Q\,dy) = \int\int \left(\frac{\partial Q}{\partial x} - \frac{\partial P}{\partial y}\right) dx\,dy \tag{10.61}$$

$$\int\int (P\,dydz + Q\,dxdz + R\,dxdy) = \int\int\int \left(\frac{\partial P}{\partial x} + \frac{\partial Q}{\partial y} + \frac{\partial R}{\partial z}\right) dx\,dy\,dz \tag{10.64}$$

$$\int s\frac{\partial t}{\partial n}\,da = \int (s\nabla^2 t + \nabla s \cdot \nabla t)\,dV \tag{10.65}$$

Unit radial vector:

$$(x\mathbf{i} + y\mathbf{j} + z\mathbf{k})/R \tag{10.69}$$

Unit tangential vector (in two-dimensions):

$$(-y\mathbf{i} + x\mathbf{j})/\sqrt{x^2 + y^2} \tag{10.75}$$

Continuity equation:

$$\frac{\partial \rho}{\partial t} = -\text{div}\mathbf{j} \tag{10.87}$$

Heat conduction equation:

$$\kappa \nabla^2 T = s\rho \frac{\partial T}{\partial t} \tag{10.91}$$

Laplace's equation:

$$\nabla^2 T = 0 \tag{10.92}$$

Maxwell's equations:

$$\text{div}\mathbf{D} = \rho \tag{M1}$$

$$\text{div}\mathbf{B} = 0 \tag{M2}$$

$$\text{curl}\mathbf{E} = -\frac{\partial \mathbf{B}}{\partial t} \tag{M3}$$

$$\text{curl}\mathbf{H} = \mathbf{j} + \frac{\partial \mathbf{D}}{\partial t} \tag{M4}$$

Wave equation:

$$\nabla^2 \mathbf{E} = \mu_0 \varepsilon_0 \frac{\partial^2 \mathbf{E}}{\partial t^2} \tag{10.104}$$

Chapter 11 Partial differential equations

d'Alembert's change of variables

$$\left. \begin{array}{l} u = x + ct \\ \text{and} \quad v = x - ct \end{array} \right\} \tag{11.16}$$

turns the wave equation

$$\frac{\partial^2 z}{\partial x^2} = \frac{\rho}{T} \frac{\partial^2 z}{\partial t^2} = \frac{1}{c^2} \frac{\partial^2 z}{\partial t^2} \tag{11.10}$$

into

$$4\frac{\partial^2 z}{\partial u \partial v} = 0 \tag{11.21}$$

$\nabla^2 V$ in different coordinate systems:
Cartesian:

$$\frac{\partial^2 V}{\partial x^2} + \frac{\partial^2 V}{\partial y^2} + \frac{\partial^2 V}{\partial z^2} \tag{11.7}$$

cylindrical polars:

$$\frac{1}{r}\frac{\partial}{\partial r}\left(r\frac{\partial V}{\partial r}\right) + \frac{1}{r^2}\frac{\partial^2 V}{\partial \theta^2} + \frac{\partial^2 V}{\partial z^2} \tag{11.26}$$

spherical polars:

$$\frac{1}{r^2}\frac{\partial}{\partial r}\left(r^2\frac{\partial V}{\partial r}\right) + \frac{1}{r^2\sin\theta}\frac{\partial}{\partial\theta}\left(\sin\theta\frac{\partial V}{\partial\theta}\right) + \frac{1}{r^2\sin^2\theta}\frac{\partial^2 V}{\partial\phi^2} \quad (11.28)$$

Separation of variables:

e.g. look for solution

$$z(x,t) = F(x)G(t) \quad (11.29)$$

substitute into partial differential equation;
divide through by $F(x)G(t)$;
remember the crucial sentence;
choose constant of separation sensibly, to satisfy boundary conditions.

Heat conduction equation:

$$\frac{\partial^2 T}{\partial x^2} = \frac{s\rho}{\kappa}\frac{\partial T}{\partial t} \quad (11.5)$$

with possible solutions

$$T(x,t) = a\sin(mx+\beta)e^{-(\kappa m^2/s\rho)t} \quad (11.41)$$

$$T(x,t) = ax + b$$

$$T(x,t) = ce^{-\sqrt{m/2}x}\cos\left(\frac{\kappa m}{s\rho}t - \sqrt{\frac{m}{2}}x\right) \quad (11.50')$$

$$= ce^{-\beta x}\cos(\beta x - \omega t)$$

Laplace's equation:

$$\nabla^2 V = 0 \quad (11.15)$$

with possible solutions

$$V = V_0[a\sin mx + (1-a)\cos mx][b\sin ny + (1-b)\cos ny]$$
$$\times [ce^{-\sqrt{m^2+n^2}z} + (1-c)e^{+\sqrt{m^2+n^2}z}] \quad (11.59)$$

$$V(r,\theta) = (f + g\ln r) + \Sigma\{d_m r^m[e_m\cos m\theta + (1-e_m)\sin m\theta]$$
$$+ f_m r^{-m}[g_m\cos m\theta + (1-g_m)\sin m\theta]\} \quad (11.75)$$

$$V(r, \theta, \phi) = \sum_l \sum_m \{a_{l,m} r^l [c_{l,m} \cos m\phi + (1 - c_{l,m}) \sin m\phi]$$
$$+ b_{l,m} r^{-(l+1)} [c'_{l,m} \cos m\phi + (1 - c'_{l,m}) \sin m\phi]\} Y_l^m (\cos \theta) \tag{11.80}$$

Wave equation: separable solutions = stationary waves, e.g.

$$z(x, t) = a \sin kx \cos kct \tag{11.81}$$

d'Alembert's approach = travelling waves:

$$z(x, t) = g(x - ct) + h(x + ct) \tag{11.93'}$$

Schrödinger's equation:

$$-\frac{\hbar^2}{2m} \nabla^2 \psi + V\psi = i\hbar \frac{\partial \psi}{\partial t} \tag{11.83}$$

Chapter 12 Fourier series

$\sin mt$ and $\cos nt$ are orthogonal over the range 0–2π $\hspace{1cm}$ (12.10)

$$\left.\begin{array}{ll} \int_0^{2\pi} \sin mt \sin lt \, dt = 0 & m \neq l \\ \int_0^{2\pi} \cos nt \cos kt \, dt = 0 & n \neq k \\ \text{and} \quad \int_0^{2\pi} \sin mt \cos nt \, dt = 0 & \text{any } m, n \end{array}\right\} \tag{12.11}$$

$$\int_0^{2\pi} \sin^2 mt \, dt = \int_0^{2\pi} \cos^2 nt \, dt = \pi$$

For a function that repeats with period 2π,

$$f(t) = a_0/2 + \sum_{n=1}^{\infty} a_n \cos nt + \sum_{n=1}^{\infty} b_n \sin nt \tag{12.1}$$

$$a_n = \frac{1}{\pi} \int_0^{2\pi} f(t) \cos nt \, dt \tag{12.12'}$$

and

$$b_n = \frac{1}{\pi} \int_0^{2\pi} f(t) \sin nt \, dt \tag{12.12}$$

For period τ,

$$f(t) = a_0/2 + \sum_{n=1}^{\infty} a_n \cos(2\pi nt/\tau) + \sum_{n=1}^{\infty} b_n \sin(2\pi nt/\tau) \tag{12.3}$$

$$a_n = \frac{2}{\tau} \int_0^{\tau} f(t) \cos(2\pi nt/\tau) dt$$

Least squares approach to fitting $f(t)$ gives identical coefficients.
Draw function and estimate first couple of coefficients, etc. before doing any integration.
Coefficients decrease like $n^{-(m+1)}$, where $d^m f/dt^m$ is lowest order discontinuous derivative.
Perform simple checks on calculated coefficients.

Fourier transform for non-repeating functions:

$$y(x) = \frac{1}{\sqrt{2\pi}} \int_{-\infty}^{+\infty} a(k)e^{ikx} dk \tag{12.55}$$

$$a(k) = \frac{1}{\sqrt{2\pi}} \int_{-\infty}^{+\infty} y(x)e^{-ikx} dx \tag{12.56}$$

Product of widths of $y(x)$ and $a(k) \approx$ unity.
Uncertainty Principle:

$$\Delta E \Delta t \approx h/2\pi$$

$$\Delta p \Delta x \approx h/2\pi$$

Chapter 13 Normal modes

The crucial points to remember are not the formulae, but the basic steps involved in solving normal mode problems, as set out in Section 13.6:

  (i) set up equations of motion;
  (ii) replace d^2/dt^2 by $-\omega^2$;
  (iii) rewrite in matrix form;
  (iv) find eigenvalues;
  (v) find eigenvectors;
  (vi) express initial conditions in terms of normal modes;
  (vii) subsequent motion is sum of normal modes, each at own frequency, with coefficient as found in (vi).

Chapter 14 Waves

Waves travelling to the right:

$$y = A \sin[k(x - vt)] \tag{14.1}$$

$$y = A \sin(kx - \omega t) \tag{14.10}$$

$$\lambda = 2\pi/k$$

$$\tau = 2\pi/kv = \lambda/v = 2\pi/\omega = 1/\nu$$

Stationary wave:

$$y = A \sin kx \cos \omega t \tag{14.13}$$

Wave equation:

$$\frac{\partial^2 y}{\partial x^2} = \frac{1}{v^2} \frac{\partial^2 y}{\partial t^2} \tag{14.20}$$

Transverse waves on string:

$$v = \sqrt{T/\rho} \tag{14.25}$$

Longitudinal waves:

$$v = \sqrt{\lambda/\rho} \tag{14.80}$$

Adiabatic sound waves:

$$v = \sqrt{\gamma p/\rho} \tag{14.83'}$$

Group velocity in dispersive medium:

$$g = \frac{d\omega}{dk} \tag{14.36}$$

$$= v + k \frac{dv}{dk} \tag{14.36'}$$

where k is defined in the medium.

$$KE/l = \frac{1}{4} \rho A^2 \omega^2 = PE/l \tag{14.40) and (14.42}$$

Energy flux:

$$U = \frac{1}{2} \rho \omega^2 A^2 v \tag{14.51}$$

Reflection at boundary:

$$\left.\begin{array}{l} R = \dfrac{k_1 - k_2}{k_1 + k_2} A \\[2mm] T = \dfrac{2k_1}{k_1 + k_2} A \end{array}\right\} \qquad (14.63)$$

Right-handed circular polarisation:

$$y = A\sin(kx - \omega t + \phi_1) \qquad (14.69)$$

and

$$z = A\sin(kx - \omega t + \phi_1 - \pi/2) \qquad (14.74)$$

Beat frequency:

$$\Delta\omega = \omega_1 - \omega_2$$

Young's slits:

$$\Delta x = \lambda L/2s \qquad (14.94)$$

where slit separation is $2s$.

Chapter 15 Matrices

Matrix multiplication:

$$A_{ij} = \sum_k B_{ik} C_{kj} \qquad (15.24)$$

Two-dimensional rotation matrix:

$$\mathbf{M}_6 = \begin{pmatrix} \cos\theta & -\sin\theta \\ \sin\theta & \cos\theta \end{pmatrix} \qquad (15.55)$$

Inverse:

$$\mathbf{A}\mathbf{A}^{-1} = \mathbf{A}^{-1}\mathbf{A} = \mathbf{I} \qquad (15.66)$$

$$A_{kl} = (-1)^{k+l} c_{lk}/d \qquad (15.69)$$

where c_{lk} is the minor and d the determinant of the matrix

Singular matrix:

$$det(\mathbf{M}) = \mathbf{0}$$

Three-dimensional rotation about the z axis:

$$\mathbf{M}_{11} = \begin{pmatrix} \cos\theta & -\sin\theta & 0 \\ \sin\theta & \cos\theta & 0 \\ 0 & 0 & 1 \end{pmatrix} \qquad (15.79)$$

$$\widetilde{AB} = \tilde{B}\tilde{A} \tag{15.113}$$

$$(AB)^{-1} = B^{-1}A^{-1} \tag{15.114}$$

Rotation matrix is orthogonal:

$$R^{-1} = \tilde{R} \tag{15.124}$$

Transformations of matrices:
For $z = Ny$:

$$N' = TNT^{-1} \tag{15.129}$$

(if N is diagonal, and T is rotation matrix, N' is symmetric).
For quadratic form:

$$Q' = \tilde{T}^{-1}QT^{-1} \tag{15.138}$$

If T is rotation matrix:

$$Q' = TQT^{-1} \tag{15.139}$$

Trace of matrix:

$$Tr(AB) = Tr(BA) \tag{15.140}$$

Trace of a matrix is unaffected by rotation.

Chapter 16 Eigenvectors and eigenvalues

$$Mr = \lambda r \tag{16.11}$$

$$|M - \lambda I| = 0 \tag{16.14}$$

Determine eigenvectors for each value of λ, from $N - 1$ simultaneous equations of (16.11); normalisation arbitrary.

For linear transformation T on vectors:

$$\left. \begin{array}{l} \text{eigenvectors } v \to Tv \\ \text{eigenvalues } \lambda \text{ unaffected} \end{array} \right\} \tag{16.24}$$

For inverse M^{-1}:

$$\left. \begin{array}{l} \text{eigenvectors unaffected} \\ \text{eigenvalues } \lambda \to 1/\lambda \end{array} \right\} \tag{16.27}$$

Symmetric matrix S has orthogonal eigenvectors $\tag{16.29}$

Rotation matrix **R**:

$$\text{eigenvalues: } \lambda = 1 \text{ or } e^{\pm i\theta} \tag{16.45}$$

$$\text{Real eigenvector} = \text{axis of rotation}$$

$$\text{Tr}(\mathbf{R}) = 1 + 2\cos\theta$$

Quadratic form:

$$\tilde{\mathbf{r}}\mathbf{Q}\mathbf{r} = 1$$

$$\mathbf{Q}\mathbf{r} = \text{ normal to surface} \tag{16.55}$$

$$\text{Eigenvectors in directions of axes} \tag{16.56}$$

$$\text{Length of axes } = 2/\sqrt{\lambda} \tag{16.57}$$

Degenerate eigenvalues:

$$\text{Linear combination of eigenvectors} = \text{eigenvector.}$$

Normal modes:

$$-\omega^2\mathbf{x} = \mathbf{K}\mathbf{x} \tag{16.63}$$

Eigenvalues give normal mode frequencies.
Eigenvectors are normal mode amplitudes.

Index